Spatial Information Systems

Qualitative Spatial Change

ANTONY GALTON

OXFORD
UNIVERSITY PRESS

This book has been printed digitally and produced in a standard specification in order to ensure its continuing availability

OXFORD
UNIVERSITY PRESS

Great Clarendon Street, Oxford OX2 6DP

Oxford University Press is a department of the University of Oxford.
It furthers the University's objective of excellence in research, scholarship, and education by publishing worldwide in

Oxford New York

Auckland Bangkok Buenos Aires Cape Town Chennai
Dar es Salaam Delhi Hong Kong Istanbul Karachi Kolkata
Kuala Lumpur Madrid Melbourne Mexico City Mumbai Nairobi
São Paulo Shanghai Taipei Tokyo Toronto

Oxford is a registered trade mark of Oxford University Press in the UK and in certain other countries

Published in the United States
by Oxford University Press Inc., New York

Reprinted 2003

ISBN 0-19-823397-3

Printed in Great Britain by
Antony Rowe Ltd., Eastbourne

To March and Dips

Acknowledgements

Writing this book has taken me several years, but the bulk of the work was accomplished during the period October–December 1996, when I was fortunate enough to be granted a term's paid study leave by my employer, the University of Exeter. During this period I wrote first drafts of all the chapters, which I subsequently sent to various colleagues in universities both in Britain and abroad. In this way I received valuable feedback that has enabled me to vastly improve the original drafts (in some cases to the extent of complete rewriting). Other colleagues provided valuable input by allowing me to engage in discussion on many topics covered by this book, both by e-mail and face to face at conferences and workshops. Yet others have made an important contribution by drawing my attention to published work which I should have been aware of but which had somehow escaped my notice. The following list of people who have helped in one or more of these ways is doubtless incomplete; my apologies to anyone who feels they should have been included but was not. Thanks, then, alphabetically, to Juan Carlos Augusto, Brandon Bennett, Tony Cohn, Christoph Dornheim, Ivo Düntsch, Carola Eschenbach, Christian Freksa, Nicola Guarino, Jean-François Hangouët, Kathleen Hornsby, Colin Johnson, Zoltán Juhász, Stephen Kuhn, Oliver Lemon, Carlos Martínez-Mascarúa, Richard Meathrel, Patrick Olivier, Ian Pratt, Barry Smith, Mike Smyth, John Stell, Achille Varzi, Laure Vieu, Julian Webster, and Mike Worboys.

Because of the wide range of topics covered in this book, I am acutely conscious of the possibility of errors and oversights; naturally I retain full responsibility for these. It has been difficult to strike a balance between diligently pursuing every last error—a process that might take up many more years—and aiming to have the book published before too much of it becomes out of date. This book is the fruit of my investigations during the 1990s; the passing of that decade seems an appropriate time to make it available to the wider research community.

A.P.G.
Exeter, May 2000

Contents

1

Introduction

1.1. Who cares about spatial change?

Spatial change occurs whenever objects possess different spatial attributes at different times. Involving as it thus does some of the most fundamental elements of our experience—space, time, and objects—it is inevitable that in various ways we all care about spatial change. All of us experience spatial change in ourselves and in others every day of our lives. Throughout our environment, things are constantly changing their positions, shapes, sizes, orientations, and since these are spatial attributes the changes we see in them are examples of spatial change. Spatial change is clearly ubiquitous; but this in itself does not explain why anyone should want to read or write a whole book on the subject!

There is, none the less, a growing community of scholars with an interest in developing rigorous and systematic accounts of space, time, and spatial change. This is not to say that we do not already have such accounts. The Greeks brought the abstract study of geometry to an astonishing level of depth and sophistication; precise mathematical accounts of space, time, and motion lay at the heart of the intellectual revolution associated with the names of Galileo and Newton, not to mention the later revolution associated with that of Einstein. Our mathematical grasp of the way things move is wonderfully precise: precise enough for us to have sent spacecraft to almost all the major bodies of the solar system (and some of the minor ones) to send back breath-taking photographs of these distant worlds which in ancient times were—to the extent that they were not gods ruling over our destinies—nothing more than the merest pin-pricks of light.

Despite these impressive achievements, it has become apparent in recent years that there are many aspects of space, time, and spatial change which these high-precision mathematical tools are not well adapted to handling. This realization has been engendered by the burgeoning of a new set of intellectual paradigms embodied in the techniques and aspirations of information technology. In seeking to use computers to further our spatio-temporal information-handling capacities, we have found it necessary to try to spell out explicitly all kinds of things which were hitherto implicit in our common-sense view of the world, and so banal as to go almost unnoticed. Of particular importance here are the *qualitative* aspects of the spatio-temporal organization of the world, requiring a range of descriptive tools very different from the precise quantitative tools employed, for example, by mathematical physics.

The following disciplines have all contributed, and continue to contribute, to the emerging discipline of *qualitative spatio-temporal information theory*:

- *Artificial Intelligence* (AI), which looks for ways of endowing computer systems with at least some of the capacities which are traditionally regarded as the preserve of animate intelligence. In particular, one important strand in AI research concerns automated reasoning about spatial and temporal aspects of the world.
- *Robotics*, sometimes viewed as a branch of AI, in which the goal is to construct automata with the capacity to manipulate objects and perform manœuvres in a physical environment in real time. A prime desideratum for such a capacity is the mastery of a sophisticated spatio-temporal model of the environment in which the robot is to work.
- *Computer vision*, which is related to both AI and robotics, but is specifically concerned with the complex task, achieved apparently effortlessly by the human visual system, of building an internal model of the world from the ever-changing patterns of retinal activation, a model in which space and time play essential roles.
- *Cognitive science*, which is related to both AI and psychology, and is concerned with the information-processing capacities of the human brain, amongst other things with understanding how people come to conceptualize the spatio-temporal aspects of the world and thereby acquire those capacities which researchers in AI, robotics, and computer vision are seeking to emulate.
- *Philosophy*, which can act as a handmaiden to the other disciplines listed here but which can also be regarded as an important area of study in its own right. Philosophy addresses such questions as what space and time really are, whether an external framework within which we have our being and experience the world, or a framework imposed by our own modes of experience on a fundamentally unknowable external reality.

- *Linguistics*, which is concerned with the systematic study of the structure and organization of human language, and in particular with the way in which languages encode—or help to form—aspects of our experience such as the spatial and temporal attributes of, and relations among, the objects of our discourse.
- *Logic* is concerned with how to systematize within a deductive framework the most fundamental relations amongst our concepts, including, of course, our concepts of space and time; temporal logic, in particular, has played an important role in several of the disciplines listed above, and continues to be accorded considerable importance by many researchers.
- *Geography*, the spatial science *par excellence*, is intimately concerned with the problem of how best to collect, record, and make sense of the complex ever-changing mass of geographical data confronting us in the modern world.
- *Geographical Information Systems* (GIS) address the problems of using the computer as an essential tool in the information-processing tasks facing geographers, requiring the development of sophisticated and highly general abstract models of spatial data.

In all these disciplines there arise problems which come within the general area of qualitative spatial reasoning, and which have provided inspiration for the material in this book. Inevitably the treatment accorded to this material is somewhat biased in the directions suggested by my own interests and competences: having originally been trained as a mathematician and having then turned via logic and philosophy to AI, I bring to the topics handled in this book a particular perspective which is doubtless very different from what one might expect from, say, a linguist, or a geographer. None the less, I hope that what I have to say will prove to be of some interest to at least some people from each of the disciplines listed above.

1.2. The plan of this book

The remainder of this chapter is devoted to a discussion of a number of general themes which recur in various contexts throughout the rest of the book: idealization, granularity, and the distinction between qualitative and quantitative representations. The remaining chapters then explore in turn the various ingredients required for a theory of spatial change. As noted above, spatial change occurs whenever objects possess different spatial attributes at different times. I shall systematically unpick this definition into its simplest ingredients. For a theory of spatial change we require theories of Space, of Objects, of Spatial Attributes, and of Time and Change. Each of these elements has a chapter to itself in what follows. Another chapter surveys the many varieties of Spatial Change which different combinations of characters taken from the above categories give rise to. The last two chapters concentrate on an aspect of particular importance in the discussion of change, namely Continuity.

1.3. The nature of qualitative representations

> Is the distinction between a difference in degree and a difference in kind a difference in degree or a difference in kind?
>
> Culled unattributed from the Internet, *c.*1990

1.3.1. Qualitative versus quantitative

It is harder than one might think to provide watertight definitions of the terms 'qualitative' and 'quantitative'. Here are some dictionary definitions:

> **qualitative**. Concerned with, depending on, quality.
> **quality**. Degree of excellence, relative nature or kind or character.
> **quantitative**. Measured or measurable by, concerned with, quantity.
> **quantity**. The property of things that is estimable by some sort of measure, the having of size, extension, weight, amount, or number.
>
> *The Concise Oxford Dictionary*, Fifth Edition (1964)

What kinds of things are described as qualitative or quantitative? Not physical objects: it makes no sense to describe a table as either qualitative or quantitative, for example. On the other hand, the table has many different properties, some of which can be described as qualitative and others as quantitative. The table has a height of 72 centimetres, a weight of 20 kilograms, and the surface area of its top is one square metre; these things—height, weight, and area — are all 'measured by quantity' and therefore quantitative. It is made of wood, dark brown in colour, and rather ugly: these things are 'concerned with quality' and therefore qualitative. Some properties are not so easy to characterize as clearly either qualitative or quantitative. The table has four legs and a rectangular top. That it has four legs, as opposed to three or five, is measured by quantity ('the having of ... number'); but on the other hand the difference between a three-legged table, a four-legged table, and a five-legged table would seem to be a difference in 'relative nature or kind or character'. Again, that its top is rectangular certainly involves measurement: its perimeter must be divided into four straight edges, each equal in length to the one opposite it, and meeting its neighbours at right angles. But again the difference between a rectangular table and a triangular one or a circular one would appear to be qualitative, a difference in relative nature or kind or character.

This last example, concerning the shape of the table-top, suggests an important further aspect of the distinction between the qualitative and the quantitative. If I want to distinguish a rectangular table-top from a circular one, I shall not do so by means of any measurement, but by drawing attention to gross qualitative distinctions such as the fact that the rectangular top, but not the circular, has corners, where the edge sharply changes direction; and again, that apart from at the corners the edge of the rectangular top is everywhere straight, whereas that of the circular top is nowhere straight. If on the other hand I want to distinguish the rectangular table-top from one which has the form of an irregular quadrilateral, then I shall draw attention to the fact that whereas in the rectangle the opposite sides are equal in length, in the irregular figure they differ by measurable amounts; and that whereas in the rectangle the angles are all 90°, in the irregular figure at least two of the angles are measurably different from this amount. Thus the distinction between 'rectangular' and 'circular' seems to be entirely qualitative, whereas that between 'rectangular' and 'irregular quadrilateral' seems to be at least partly quantitative. An even clearer case of a quantitative distinction would be between our rectangular table and another table which is exactly similar apart from being 5 per cent larger in all its linear dimensions: this would seem to be a purely quantitative difference. This leads us to the thought that it is not so much *properties* that must be characterized as either qualitative or quantitative, but *differences*. We often contrast 'differences in degree' with 'differences in kind', which is just another way of adverting to quantitative as against qualitative differences.

So far we have established that an object may differ qualitatively from some objects and quantitatively from others. But the issue is subtler than this, for a given pair of objects can differ qualitatively in some respects and quantitatively in others. Two geometrically similar tables, one 5 per cent larger than the other, the larger made of pine and the other of beech, differ both quantitatively (with respect to their size) and qualitatively (with respect to their material constitution). More subtly still, the very same feature with respect to which two objects differ can be described both quantitatively and qualitatively. Thus, the rectangular table and the irregular quadrilateral table differ with respect to the feature 'shape of top'; this difference can be described quantitatively by giving the various measurements, or it can be described qualitatively by referring to the fact that the rectangular table has both reflectional and rotational symmetry, whereas the irregular table has neither. And the supposedly qualitative difference between the rectangular table and the circular table can be partly characterized quantitatively by contrasting the ratio of the square of the perimeter to the area of the top, which comes to about 12.5 in the case of the circular table but must be at least 16 for the rectangular one.

Having given some indication of the difficulties in pinning down what we mean by 'qualitative', I shall now lay down a number of stipulations concerning the way in which I shall use the term in this book. The next few sections will take these stipulations in turn and discuss their ramifications.

1. *Precise qualitative descriptions are more important than vague ones.*
2. *Quantitative properties form a continuous space, qualitative properties form a discrete space*
3. *Qualitative differences are supervenient upon quantitative ones.*
4. *The divisions of qualitative space correspond to salient discontinuities in our apprehension of quantitative space.*

1.3.2. Vagueness and precision

Precise qualitative descriptions are more important than vague ones. An example of the latter type is 'John is tall'. This is indeed vague, but note that it is the *description* 'tall' that is vague, and not any property of John himself, who is precisely what he is and not anything else. There is no such thing as a vague property, only vague descriptions of properties. Moreover 'tall' is a vague description of a *quantitative* property, not of a qualitative one. It only counts as qualitative because it is not explicitly numerical (unlike, say, 'John is about six feet tall', which is a vague quantitative statement).

A more truly qualitative description of John's height is 'John is taller than Mary'. The truth or falsity of this statement can be determined without the use of numerical measurement; we need only have John and Mary stand up alongside one another and make a visual comparison—or if the difference in height is too small for this, to lay a spirit level across the tops of their heads. And unlike 'John is tall', the statement 'John is taller than Mary' is in this sense not at all vague.

Compare the three vague descriptions

John is tall
John is of middling height
John is short

with the three precise ones

John is taller than Mary
John is the same height as Mary
John is shorter than Mary.

The former divide up the space of possible heights for John into three ill-defined bands, as shown in the left half of Figure 1.1. The latter divide it up into two clearly defined bands separated by a sharp boundary, as in the right half of the figure.

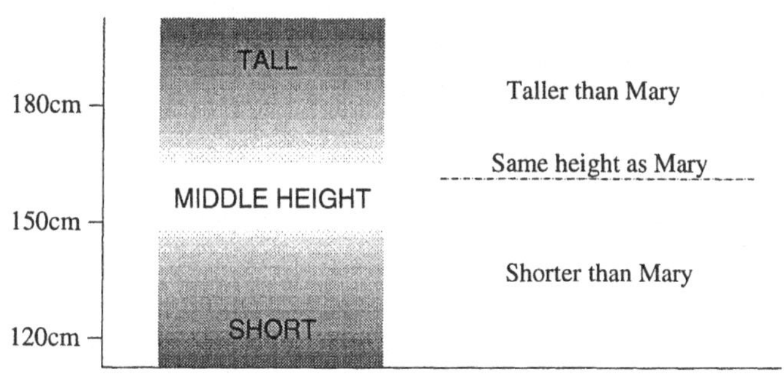

Figure 1.1. Vague and precise qualitative descriptions

The essence of a qualitative description is that it defines some kind of boundary between the things it applies to and things it does not. The boundary may be vague or precise, but this has nothing to do with the way things are in themselves: nature does not provide us with precise boundaries between types of things, but then it does not provide vague ones either. We define the boundaries in accordance with how we see the world; the vagueness comes from *our* inability to decide where best to draw a line. The difference between a vague description and a precise one (whether qualitative or quantitative—for as we have seen the distinction between vague and precise cuts across that between qualitative and quantitative) is that in the latter we have clear and unambiguous criteria for its application.

Many qualitative descriptions occupy a middle ground between the quite precise and the hopelessly vague. An example is provided by colour terms. To say that something is 'blue' might be considered vague on the grounds that there does not seem to be a sharp cut-off point between green and blue. One reason we say this is that people sometimes have disputes as to whether or not a particular object should be called green or blue. But note that this dispute does not in any way support the idea that there is no sharp cut-off here. What *would* support it would be if there were a wide band of hues intermediate between green and blue which everyone agreed could not be clearly assigned to either colour. But in the disputes we actually hear, one party insists that the thing is green and the other that it is blue ('How *can* you say it's blue? It's so obviously green!'). Here we are dealing with individual differences in colour vision; these are well documented in the psycho-physical literature. So we must set such disputes aside and consider a single individual; for that individual we must look for the range of hues in the neighbourhood of the green/blue borderline which they are unable to name as definitely green or blue. The range of such hues is much narrower than you might imagine, but none the less cannot be reduced to zero.

1.3.3. Continuity and discreteness

Quantitative properties form a continuous space, qualitative properties form a discrete space. This follows from the observation that qualitative descriptions provide boundaries in conceptual space. There is a discrete transition between one side of the boundary and the other—or, if the boundary itself is significant, between the boundary and the spaces on each side of it.

It also means that there is a tension between 'qualitative' and 'quantitative' in the case of descriptions using integers rather than real numbers. Integers form a discrete space, real numbers a continuous one. This might seem to imply that properties expressed using integers are qualitative. I have already noted this in the case of triangles, quadrilaterals, and pentagons. It seems natural to regard these classes as qualitatively distinct, and not merely quantitatively so. But this feeling of naturalness tends to disappear as the numbers get larger: the difference between a polygon with eighty sides and one with eighty-one sides seems more quantitative than qualitative. Similarly, 'two's company, three's a crowd'—an important qualitative difference—but the difference between a crowd of a hundred people and a crowd of a hundred and one is merely quantitative.

Many, perhaps most, qualitative spaces can be thought of as *partitions* of quantitative spaces. The idea is that some continuous range of variation is divided up into a discrete set of zones regarded as qualitatively distinct. A simple example is the partition of the continuous range of possible heights into the zones 'taller than Mary' and 'shorter than Mary', with 'same height as Mary' marking the boundary between them.

A more complex example is shown in Figure 1.2, in which the space of possible triangles (defined up to similarity) is partitioned into zones corresponding to the broad qualitative divisions of triangles into acute-angled, right-angled, obtuse-angled, and degenerate, and scalene, isosceles, and equilateral.[1] The quantitative space is portrayed using two different 'map projections'. In (*a*), the triangle is assumed to have its smallest angle at the origin and its longest side lying along the x-axis from the origin to the point (1,0): any triangle is similar to a unique triangle in this configuration. The triangle itself is completely determined by the co-ordinates X and Y of the third vertex, and these co-ordinates are plotted along the axes in (*a*). Essentially the same mapping, but with considerable redundancy, is portrayed in figure 4.5 of Shepard (1988). In (*b*), we see an alternative presentation of the same space, this time organized in terms of angles: the smallest angle (which must lie in the range $0 \leq \alpha \leq 60°$) is plotted along the vertical axis, the second smallest (which must lie in the range $0 \leq \beta \leq 90°$) along the horizontal. The space is partitioned in the same way as in (*a*).[2]

This way of partitioning triangle space brings to the fore the existence of a qualitative distinction which is not reflected in our terminology. The subspace corresponding to isosceles triangles is seen to consist of two disjoint line segments, meeting at the point corresponding to the equilateral triangles. One branch of this subspace contains those isosceles triangles whose equal angles are smaller than the third angle—this will be the case when the equal angles are less than 60°. The other branch contains those isosceles triangles whose equal angles are between 60° and 90°. Let us call the former class the 'fat' isosceles triangles and the latter class the 'thin' isosceles triangles. So

[1] A triangle is here understood to be generated by three distinct points; it is degenerate when the points are collinear. A more general possibility, not considered here, allows two or more of the points to coincide.

[2] For yet other ways of mapping triangle space, see Dryden and Mardia 1998: ch. 2.

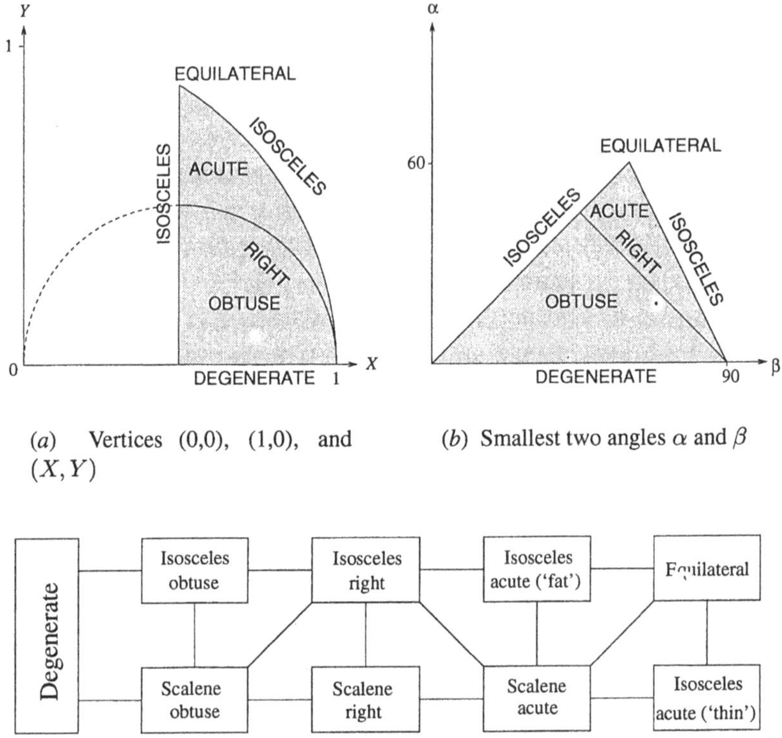

(*a*) Vertices (0,0), (1,0), and (X,Y)

(*b*) Smallest two angles α and β

(*c*) Abstraction of the qualitative information only

Figure 1.2. The space of all possible triangles up to similarity

every obtuse-angled isosceles triangle is fat (since its equal angles must be less than 45°), but amongst the acute-angled isosceles triangles, some are fat and some are thin.

In Figure 1.2*c* is shown a more abstract representation of the qualitative partition of triangle space, in which all the quantitative information has been filtered out, leaving only the adjacency relation between the various qualitative subspaces. No distinction is made here between qualitative subspaces which correspond to areas, lines, or points in the original partition. We shall meet many more examples of this way of representing a qualitative space by abstracting the quantitative information from what is originally a partition of a quantitative space. The regular pattern observed throughout most of the diagram seems to break down at the left-hand end: this can be remedied by dividing the class of degenerate triangles into isosceles and scalene varieties, the former corresponding to three points equally spaced along a line.

1.3.4. Supervenience

Qualitative differences are supervenient upon quantitative ones.[3] This means that whenever a qualitative difference can be found between two things, there must be a quantitative difference as well. You cannot have two things which are alike in all quantitative respects but different in some qualitative respect.

A blue ball and an otherwise identical green ball differ because there is at least one wavelength of light such that the percentage of that wavelength absorbed by the former ball differs numerically from the percentage of the same wavelength absorbed by the latter. This is a quantitative difference underlying what we perceive as a qualitative one.

Even properties such as the ugliness of the table obey this principle. Given two tables, one of which I find ugly and the other beautiful, there must surely be some measurable difference between them. This is not to claim that any measurable difference between the tables *explains* the beauty and the ugliness, although in some sense it seems we must say that the totality of measurable differences between the tables is *responsible* for the aesthetic difference between them, since if all these differences were to be removed, the aesthetic difference would disappear too.[4] And it does not mean that my aesthetic appreciation of the tables necessarily involves any awareness on my part of the quantitative difference between them.[5] (Of course, you might find ugly a table that I find beautiful, yet there is no quantitative difference between the table I see and the same table you see. This is beside the point, however, because the qualitative difference involved here is between you and me, between whom we can certainly find any number of measurable differences.)

The supervenience of the qualitative on the quantitative fits in well with the picture of qualitative spaces as partitions of quantitative ones. To travel

[3] 'Qualitative' resembles 'vague' in this respect; as Williamson (1994: §7.4) notes, vague facts supervene on precise ones.

[4] We must not forget the role of the observer, of course, but the important thing for our present point is that the qualitative aesthetic difference which the observer ascribes to the objects must reflect some measurable difference between the objects themselves.

[5] Compare Wittgenstein (1966): 'Architecture:– draw a door—"Slightly too large." You might say: "He has an excellent eye for measurement." No—he sees it hasn't the right expression—it doesn't make the right gesture.'

from one zone of qualitative space to a different zone, one has to move from one point of quantitative space to a different one.

It is only fair to add, however, that a case can be made for the opposite point of view, that the quantitative is supervenient on the qualitative. More exactly, we might say that quantitative *knowledge* is supervenient on qualitative knowledge. To see this, consider that quantitative knowledge depends on measurement, and measurement always ultimately involves perceptual discrimination. Thus, in order to measure the length of something with a ruler, I lay the ruler alongside it, with one end against the zero mark, and determine where the other end comes in relation to the calibrations on the ruler. I may observe that the end of the object comes between the marks corresponding to 12.4 centimetres and 12.5 centimetres, somewhat nearer to the former. I therefore conclude that the object's length is just over 12.4 centimetres. The judgements that I have to make are, primarily, whether the end of the object is to the left of, coincident with, or to the right of a given mark on the ruler. These are qualitative rather than quantitative judgements. Similar remarks could be made about all forms of measurement.

This does not mean that qualitative distinctions are not, after all, supervenient on quantitative ones. That would be to confuse the quantitative aspects of reality with our knowledge of those aspects. The latter depends on the former, and the nature of that dependence involves a qualitative element; but that in no way implies that the quantitative aspects of reality must themselves depend on something irreducibly qualitative.

1.3.5. Salience

Why did I choose to partition triangle space in just the way that I did? A key notion here is that of *salience*. The word 'salient' originally meant 'jumping', and salient features are ones that jump out at you, that leap to the eye. They are the things you notice straight away about a situation you are confronted with, prior to detailed examination. Within the class of triangles, features such as right angles or equal sides are perceptually salient, which is why we feel that categories such as 'isosceles triangle' and 'right-angled triangle' are qualitative in nature even though they can be categorized purely in terms of measurements. This does not mean that such properties are only of psychological interest: in the case of the triangles the reasons for the psychological salience involve mathematically significant properties such as symmetry—though a detractor might argue that 'mathematical significance' is also, in the final analysis, a form of psychological salience.

Since both counting and measurement almost always involve detailed examination, the numerical aspects of a situation tend not to be salient. Here we must make an exception in the case of small integers; it is well known that very small numbers are psychologically salient to human beings in the sense that we can distinguish them at a glance even in the absence of any pattern: this operation has been called 'subitizing' to distinguish it from counting proper (for a discussion, see Dehaene 1997: ch. 3). The fact of subitization perhaps explains why we tend to feel that differences between smaller integers are qualitative whereas differences between larger ones are quantitative.

One may take the supervenience of the qualitative on the quantitative in a severely reductionist spirit and say that therefore the world is essentially quantitative in nature, a doctrine first espoused, it seems, by Pythagoras and his circle. Another, quite different, expression of this idea is that the whole truth about the universe at any one time can be captured by specifying the exact position and momentum of every fundamental particle. This notion, deriving from the successes of Newtonian dynamics, was popularized by Laplace with the further twist that we only need the complete truth, in this sense, at *one* time to infer, using the laws of physics, the complete truth at any other time, past, present, and future.[6]

One reason why this 'numericist' approach to nature is profoundly unsatisfying to many people is that it entirely neglects salience. The world as we see it is strongly filtered through the many physiological and psychological mechanisms by which we perceive and conceptualize it. These mechanisms project on to the world a structure that is not intrinsic to it but originates from within ourselves. This has been widely recognized, but there has been much disagreement concerning the relationship between the structure we project on to the world and the world itself.

At the most pessimistic end of the spectrum is the idea that the world as it is in itself bears no relationship to the world as we conceive it; this is the world of Descartes's *genius aliquis malignus*, an 'evil genius' who systematically deludes us into thinking we inhabit a world utterly different from the world as it really is.[7] It is also the world of the 'brain in a vat', wired up to a computer that feeds it with all the stimuli needed to create the illusion of the world we think we live in (Davis 1974; Putnam 1981). Or again, it is Kant's world, in which—at least on one interpretation of Kant's difficult writing—the phenomena of our experience are sharply distinct from the

[6] Laplace, *A Philosophical Essay on Probabilities*, ch. 2.

[7] See Descartes's *First Meditation.*

things in themselves, or *noumena*, whose intrinsic nature is unknowable to us, because all our experience necessarily takes place within a framework, including such fundamental notions as space, time, and causality, that originates entirely within ourselves (though for Kant these latter notions are necessary prerequisites for us to have any experience at all, which is why they necessarily appear to us as synthetic *a priori* features of the world).

At the optimistic end of the spectrum is the idea, proposed by Descartes in preference to the 'evil genius' that he did not believe in, that an all-powerful and all-benevolent deity would not have given him a mind that was deluded about the nature of the world. This idea led directly to Descartes's famous dictum, around which his whole philosophy revolved, that 'whatever I see clearly and distinctly must be true'. Although in many ways Descartes was anything but naïve, his faith in the divinely assured non-delusory nature of his experiences is akin to that of the 'naïve realist' who holds that, quite simply, the world is just as it appears.

The problem with naïve realism is that the world appears differently to different people and even to the same person in different circumstances. For that reason naïve realism is untenable, and we have to find some way of moderating its optimism to produce a more realistic view of the relation between experience and reality. An attractive moderately optimistic viewpoint is provided by the theory of natural selection. Amongst creatures which construct internal representations of the world, those whose internal representations are more faithful to the reality out there will, under appropriate circumstances, have a competitive advantage over those with less faithful internal representations, and hence survive to pass on the genetic underpinning of their more faithful representations to succeeding generations. Since we are the products of billions of years of natural selection, there might seem to be good grounds for supposing that the perceptual apparatus which our ancestry has endowed us with is in large measure veridical. It may not tell us the whole truth, and it may not tell us nothing but the truth, but none the less it can surely be relied upon not to mislead us too drastically. This has been called the 'received view', within cognitive science, of the evolutionary basis for cognition (Varela *et al.* 1991).

The use of the word 'truth' in the preceding paragraph might give cause for some scruples. At issue is the nature of the fit between what our perceptual and cognitive faculties tell us about the world and what, if this means anything at all, the world is 'in itself'. Varela *et al.* (1991) are circumspect: since other equally successful species can categorize the world in ways that are radically incompatible with ours (for example, their colour spaces can and do differ from ours in dimensionality), it follows that human cognitive categories are at best *viable* or *effective*, and in no sense *optimal*. Only optimality could confer the uniqueness that would supply a minimal criterion for 'truth'. As Johnston (1999) puts it,

> If the attributes of mind, like sensations and feelings, are malleable emergent properties of brains, rather than rigid properties of the environment, then they are free to evolve and become increasingly refined over generations. But natural selection 'doesn't care' if such emergent properties are, or are not, an accurate reflection of the external world, or a product of any other criteria for that matter; it only answers to the single constraint of gene survival. As a consequence, our conscious minds have evolved into an array of highly refined active filters that amplify and distinguish between the biologically relevant aspects of the physical and social world. Our vision of the world may be an illusion, but it's an adaptive illusion.

The features of our perceptions of the world that we find salient are genuine features of the world itself only in the sense that, amongst creatures with our way of life, those for whom these features are salient have a competitive advantage over those for whom they are not. Of course, for creatures with a completely different way of life it is possible that the overlap between the features which they find salient and the features that we find salient is rather small. But this does not have to mean that one or other of us has got it wrong, only that we each have a partial view of the full complexity of the world we live in, a view that is, within its own terms of reference, pretty nearly correct.[8]

The distinctive characteristic of humans in this respect is that we have been able to overcome some of the limitations imposed on us by our inbuilt perceptual and conceptual frameworks. However veridical these may be within their own limits, they certainly do not tell the whole truth about the world. But they have been sufficient to enable us to construct instruments and theories that penetrate further into the nature of the physical world. We, and amongst earthly creatures only we, are able to view the night sky using radio waves, infra-red, ultra-violet, and X-rays; to observe the fine structure of

[8] The notion of correctness employed here is not intended to imply straightforward resemblance; the sense in which the reality 'out there' might be *like* our perceptions of it is more like the mathematical notion of a homomorphism—implying congruity of abstract structure—than any more down-to-earth idea of verisimilitude. As has often been pointed out, the idea that there can be a direct resemblance between any idea I may form of an object, and the object itself, is absurd, since the former is a mental entity, the latter—*pace* Berkeley—not.

crystals, to unravel the structure of DNA. To an extreme epistemological pessimist, there is no more reason to think we are not deluded with respect to all these discoveries than, according to the pessimist, we are deluded about what our ordinary senses tell us about the everyday world. While this is a logically consistent attitude to take, it is one that most people (including myself) find difficult to take seriously in practice. I shall therefore assume that the scientific world-picture, while undoubtedly still very incomplete, and very likely incorrect in many details, is not less true to the reality of the world (and in many respects more true to it) than our common-sense conceptions of it.

Coming back now to the problem of distinguishing qualitative from quantitative representations, we may observe that the scientific world-view is for the most part more quantitative than any common-sense view. None the less, it is very far from being entirely quantitative. Qualitative aspects loom as large in scientific knowledge as in everyday knowledge. The question is whether the increased quantitative component can lend support to the idea that the qualitative is supervenient on the quantitative. As an example, let us consider the case of *colour*.

We attribute colours to objects, but there is a long passage between a coloured object that I see out there in the world and the pattern of cortical activation that is the immediate physical correlate of my conscious awareness of a particular shade of blue. In the course of this passage the information undergoes many transformations. It is reasonable to suppose, though, that the end-product, my perception of colour, is telling the truth about some aspect of the object itself; for otherwise it is hard to see what the point of colour vision is. If we are to recognize such a thing as 'objective colour', it must encompass those physical properties of objects which can be systematically correlated with the colours we ascribe to them.

The best candidate for an objective correlate of the perceived colour in the object itself is the surface spectral reflectance, which is expressible as a mathematical function from wavelengths to percentages (Hilbert 1987). For each wavelength of visible light, the surface spectral reflectance determines what percentage of any light of that wavelength incident on the surface is reflected back. Under given conditions of illumination (which determines the composition by wavelength of the incident light), the surface spectral reflectance of an object determines the composition by wavelength of the light reflected off its surface, and hence of the light incident on the relevant portion of the retina of someone looking at the object. The human visual system is remarkably sophisticated, and is capable of making rather accurate estimates of the composition by wavelength of the ambient illumination; it can use this to work backwards from the composition of the light arriving at the retina to the surface spectral reflectance of the object off which that light has been reflected. The colour we finally perceive is our brain's best estimate of the surface spectral reflectance of the object itself.

In saying this, we are not implying that the correlation between surface spectral reflectances and perceived colours takes the form of a one-to-one mapping. In fact it falls very far short of doing so, as attested by the phenomenon of metamerism, by which objects with very different surface spectral reflectances can appear the same colour even when viewed in broad daylight. The surface spectral reflectance can be modelled by a function $f : [\lambda_0, \lambda_1] \rightarrow [0, 1]$, where λ_0 and λ_1 are respectively the shortest and longest wavelengths of visible light. This function assigns to each wavelength λ in the visible spectrum the fraction $f(\lambda)$ of the incident light of that wavelength that is reflected from the surface. A perfectly black surface would have $f(\lambda) = 0$ for every wavelength λ, whereas a perfectly white one would have $f(\lambda) = 1$. The space of all possible surface spectral reflectances is thus continuous, quantitative, and most naturally regarded as having infinitely many dimensions (one for each wavelength). After processing by the retina, this infinite-dimensional space has been collapsed into a three-dimensional space, each colour being represented by a triple (r, g, b) specifying the response of each of the three types of colour-sensitive receptor (cones) in the retina to the incoming combination of wavelengths.[9] This *component* space of representations is entirely a product of the human visual system and does not reflect any intrinsic properties of the space of spectral reflectances.[10] None the less, the mapping from the latter space into the former is quite systematic.

The (r, g, b) representation itself does not correspond well with our actual perceptions of colour: we do not, for example, perceive yellow as a mixture of red and green, although mixtures of red and green light do appear yellow to us. There is a third colour space which does correspond better to our perceptions: each colour is assigned a position along each of three axes, red/green, yellow/blue, and black/white, and the numerical values assigned to a colour in relation to these three axes are derived from the (r, g, b) values of the colour

[9] These do not correspond exactly with the red/green/blue values used in specifying colours for, e.g., computer monitor displays, but in what follows this distinction is ignored.

[10] Indeed, as Johnston (1999) points out, the colour differences which play such an important part in our experienced world represent an enormous amplification of what from any 'objective' point of view are essentially minuscule differences in the world itself.

in a systematic way—for details, see Hurvich 1981. In this *opponent* space, red and green are complementary, as are the other two pairs; it does not make sense to speak of a 'reddish green' or a 'bluish yellow'. On the other hand, we do perceive colours as reddish yellow (which we call orange), reddish blue (purple), bluish green, and greenish yellow, and we also recognize whitish or blackish (that is, light and dark) versions of any colour, exactly as predicted by the organization of the third colour space.

So far, all these colour spaces are essentially quantitative, in that positions in them are specified by precise numerical values. Even with regard to our subjective perceptions, quantitative measurement is by no means alien to us: for example, given three particular shades of blue, we can make good sense of the idea that one is exactly halfway between the other two, or that it is nearer to one of them than to the other by some more or less precisely specifiable amount. On the other hand, the names we give to colours—*red, green, yellow, blue, orange, purple, grey,* and so on—pick out broad divisions of colour space, and are felt to be entirely qualitative. Figure 1.3 shows a highly schematic representation of one cross-section, perpendicular to the black/white axis, of this three-dimensional colour space, divided into qualitative regions corresponding to these colour names. (Of the remaining 'basic' colour terms (Berlin and Kay 1969), *pink* and *white* would be found on cross-sections further towards the white end of the black/white axis, in the positions corresponding to red and grey respectively in the cross-section illustrated, while *brown* and *black* would be found on cross-sections towards the black end, in positions corresponding to yellow/orange/red and grey respectively.) Needless to say, the boundaries between these regions are not sharp, both varying between individuals, and, as suggested above, somewhat indeterminate even for one individual.

We may say, then, that the qualitative colour categories we assign names to in our language supervene on a complex of underlying quantitative distinctions. We have, indeed, an embarrassingly rich choice of quantitative colour spaces, depending on which stage in the passage between coloured object and human perception we select as the point at which the 'true' colour is located; different choices here feed different viewpoints in the continuing debate as to the correct application of the word 'colour'. None the less, the broad picture is compelling: our qualitative judgements concerning colour depend in a complex but systematic way on some underlying features of physical reality (which includes eyes and brains as well as photons and coloured objects) which can, in principle at least, be described quantitatively.

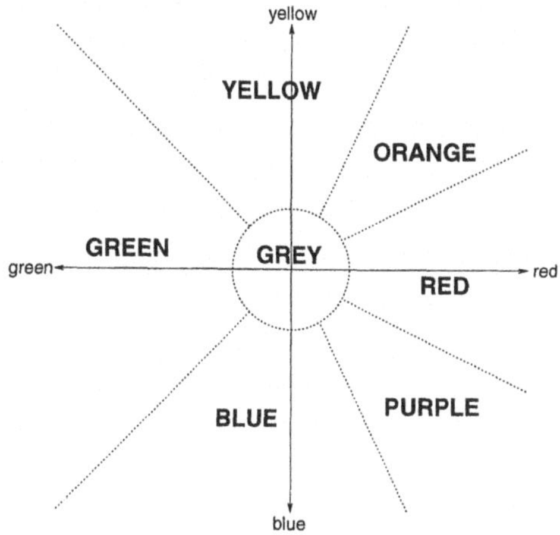

Figure 1.3. A cross-section of 3D colour space perpendicular to the black/white axis, with qualitative divisions corresponding to elements of English basic colour vocabulary

1.4. Phase spaces and mode spaces

The picture which we have obtained from both the triangles and the colour examples can be summarized as follows. We have an objectively determined space of variation that can be characterized in terms of some set of numerical parameters. I shall call this a quantitative state space or 'phase space'. We humans apprehend elements from this space by way of mechanisms which project upon this space a structure which may or may not correspond to the intrinsic structure of the space itself. The effect of this is to divide up the space into broad categories, such that the differences between elements belonging to different categories are more salient that the differences between elements belonging to the same category. I shall call this space a qualitative state space or 'mode space'. A mode space is thus a partition of a phase space. The elements of the former are supervenient on those of the latter. This is the broad picture that I shall adopt for the representation of qualitative and quantitative phenomena and their interrelationship.

The terms 'state', 'phase', and 'mode' all carry a suggestion of temporal variability. We tend not to speak of the state or phase of a system except in relation to the possibility that its state or phase might change. As employed here these terms are not to be understood as carrying this implication by necessity; however, our main interest in what follows will be precisely to describe changes in mode space, that is, qualitative change.

The concepts introduced here are closely related to a number of ideas that have appeared in the literature. A more specific notion of phase space was introduced into physics by Poincaré. A simple example concerns the dynamics of a moving particle. At any time, the particle has a position $\vec{s}$ and a velocity $\vec{v}$. If the particle is free to move in three dimensional space, then each of these quantities has three components, parallel to the coordinate axes. Thus the state of the particle is given by six real numbers, $(s_x, s_y, s_z, v_x, v_y, v_z)$, so the set of all possible states of the particle constitutes a six-dimensional phase space. Different force regimes will direct the state of the particle along different trajectories in this phase space. I shall consider an example similar to this in more detail in §1.4.1.

Our phase spaces are related to the 'quantity spaces' of Forbus (1984). A quantity space is a finite partially ordered set of numerical parameters used to describe the state of some dynamic system. Each assignment of actual values to the numerical parameters in a quantity space generates a phase of the system, and the set of all such assignments is therefore a phase space in exactly the sense described above. However, the phase spaces cited here cannot be represented directly as quantity spaces; they need to be transformed so that the constraints on the parameters take the form of partial orderings.

Davis (1990: ch. 4) makes use of a general notion of a 'measure space', which can consist of any partially or totally ordered set of entities; very often they will be real numbers or integers. An *$\mathcal{M}$-valued parameter* associates each state of the system under study with a value from the measure space $\mathcal{M}$. The range of each parameter of the system is divided up into intervals at salient 'landmark' values; a particular assignment of intervals to each parameter is called a 'mode' of the system. Thus the set of all possible modes for a given choice of landmark values defines a mode space in our sense. There is nothing about Davis's definition of a measure space which requires the parameter values to be numerical, but if they are then the phase space of the system consists of all the possible assignments of values to the parameters. Note that because the modes are determined by dividing up the ranges separately for each parameter, it is not possible directly to exhibit a mode space such as our triangle space within Davis's framework. This is because the equations of the inter-mode boundaries involve both of the parameters defining the system, whether these be X and Y or α and β. The parameters would need to be transformed in some way in order to bring this mode space into line with Davis's characterization.

I should add that this cannot be the last word concerning the nature of the qualitative. One form of qualitative distinction which does not seem to be supervenient on any quantitative distinction is that between an asymmetric three-dimensional object and its mirror image. It is not possible to distinguish between a left glove and its right-handed partner solely by means of numerical measurement.

1.4.1. Phase spaces: dynamics of a moving body

I shall now consider in more detail how this conception of phase and mode spaces can be related to the phase spaces introduced by Poincaré. I shall use the simple example of the dynamics of a body constrained to move in one dimension.[11] The position and state of motion of the body at any instant can be comprehensively described in terms of a numerical variable s giving its position on the line measured from some fixed zero point in an agreed positive direction, together with the series of derivatives

$$\frac{ds}{dt}, \frac{d^2s}{dt^2}, \frac{d^3s}{dt^3}, \ldots$$

of s with respect to time, giving the numerical values of the body's velocity, its acceleration, the rate of increase of its acceleration, and so on. Given the exact state of the system at time t in terms of this sequence of parameters, the subsequent evolution of the system is determined, so long as the system is closed.

In the case when the third and all higher derivatives are zero we have just three variables to consider, namely

$$s,\ v = \frac{ds}{dt},\ a = \frac{dv}{dt} = \frac{d^2s}{dt^2}.$$

Integrating the equations $v = ds/dt$ and $a = d^2s/dt^2$ leads to the well-known kinematical equations

$$v = v_0 + at,\ s = s_0 + v_0t + \tfrac{1}{2}at^2,$$

which are valid so long as the impressed force (and hence, by Newton's first law, the acceleration a) remains constant. This is the case, for example, of a body falling freely under gravity. The kinematical equations can be regarded

[11] For more details on the use of phase spaces in visualizing the qualitative behaviour of systems governed by differential equations, see Jordan and Smith 1987.

as parametric equations for curves in the s, v plane, with t as the parameter. Each s, v point is a phase, and the whole plane constitutes the phase space for the system under consideration. Depending on the values of s_0, v_0, and a, various qualitatively different behaviours can occur, as portrayed in Figure 1.4. If $a = 0$, then either the trajectory is along a straight line (above the s-axis moving to the right if $v_0 > 0$, below the s-axis moving left if $v_0 < 0$), or it consists of a single point on the s-axis itself (if $v = 0$), the position along the axis determined by the value of s_0. If $a \neq 0$, then eliminating t from the parametric equations gives

$$v^2 = (v_0^2 - 2as_0) + 2as,$$

the equation of a parabola, symmetrical about the s-axis and crossing it at the point $s = (v_0^2/2a) - s_0$, its orientation depending on the sign of a.

If instead of making a constant we postulate a restorative force always directed to the equilibrium position and proportional to the displacement from equilibrium, then we have the classic case of (undamped) simple harmonic motion, as exemplified (to a good degree of approximation, so long as the displacements remain small) by a frictionless pendulum or (more exactly) by a spring. In this case the system obeys the equation

$$\frac{d^2v}{dt^2} = -k^2 s,$$

which has solutions of the form

$$s = A \sin(kt + e).$$

The trajectories in phase space now consist of ellipses centred on the origin, crossing the axes at $s = \pm A, v = \pm kA$, as shown in Figure 1.5*a*.

From a qualitative point of view, the obvious landmark value for a real variable is zero. The whole of phase space can be partitioned into just nine modes, determined by setting each of s and v to any of negative, zero, or positive. The possible trajectories in our two examples are shown in Figures 1.5*b* (for simple harmonic motion) and 1.6 (for motion under constant force). In these diagrams, boxes with rounded corners represent modes which can only obtain for an instant before being replaced by the next mode in the trajectory; square boxes represent modes which can persist over an interval—equivalent to having an arrow from the mode looping back into the same mode.

The diagrams differ both with respect to which modes are connected and with respect to the direction of the arrows attached to the connections. They can all be regarded as subspaces of a more complex mode space that is obtained if acceleration is taken into account as a third parameter. We now have 27 modes in a cubic array, as shown in Figure 1.7. Here each mode is given a label in the form sva, where each of s, v, and a can be any of – (for 'negative'), 0, or + (for 'positive'). The trajectories shown in Figure 1.6*a–c* correspond to the three 'horizontal' planes $a > 0$, $a = 0$, and $a < 0$ of the full diagram. The modes for simple harmonic motion shown in Figure 1.5*b* correspond to those modes in the full diagram in which the a value is the negative of the s value, where $+$ and $-$ are negatives of each other and 0 is its own negative. These modes are situated on the diagonal plane whose four corners are $(-,+,+)$, $(+,+,-)$, $(+,-,-)$, and $(-,-,+)$. This example will be discussed further in Chapter 8, where it will provide a useful test bed for proposals concerning the qualitative modelling of continuous phenomena.

1.5. Idealization

Representation of real-world phenomena always involves an element of idealization, and a common criticism of such representations is that the idealizations involved are unrealistic or inappropriate. It is therefore important to consider briefly the nature of the idealizations that we shall be using.

The word 'idealization' comes from the Greek word *idea*, which was the word used by Plato for his 'Forms'. A Platonic Form is a perfect, unchanging, eternally existing entity of which the imperfect and changeable objects of the material world are crude copies. For Plato, a material object such as a tree, for example, owes its treelikeness to an imperfect resemblance to the ideal Tree in the domain of Forms. The more closely an object partakes of the properties of the ideal Tree, the more nearly perfect is it as an exemplar of the category of trees, but no mere material object can ever be entirely perfect in this respect. For Plato the Forms constitute the truest reality and the most worthy object of study for rational beings such as ourselves.

Now Plato had inscribed over the entrance to his Academy the words 'Let no one who is without geometry enter here', and it is perhaps to Plato's reverence for geometry that we must attribute the origin of the theory of Forms. For the subject-matter of geometry consists of perfect, unchangeable, timeless entities such as points and lines, triangles and squares, circles, cones, cubes, cylinders, and spheres. No perfect exemplars of any of these things can be found in the material world, but on the other hand material things can

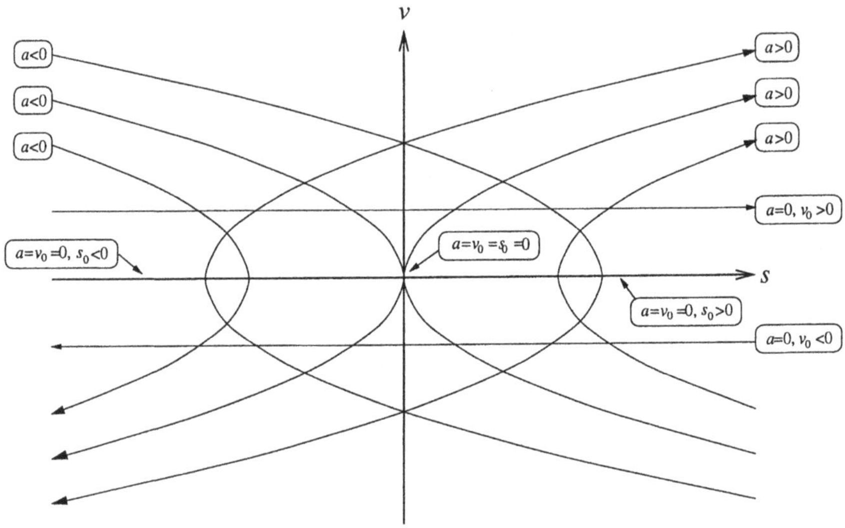

Figure 1.4. Trajectories in phase space for a body moving under constant force

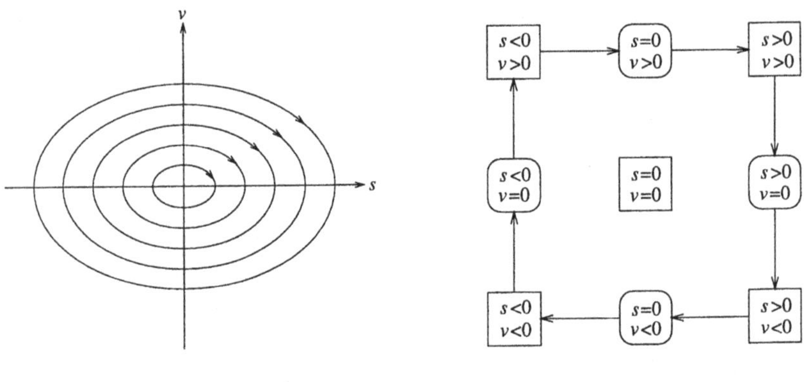

(*a*) Phase space

(*b*) Mode space

Figure 1.5. Trajectories for simple harmonic motion

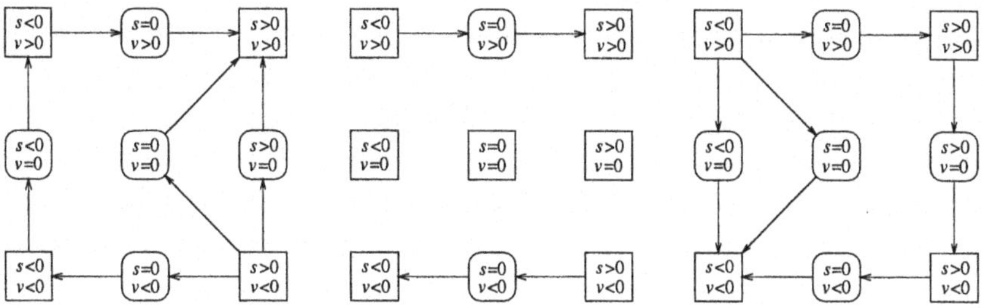

(*a*) Positive force (*b*) Zero force (*c*) Negative force

Figure 1.6. Trajectories of body moving under constant force in s, v-mode space

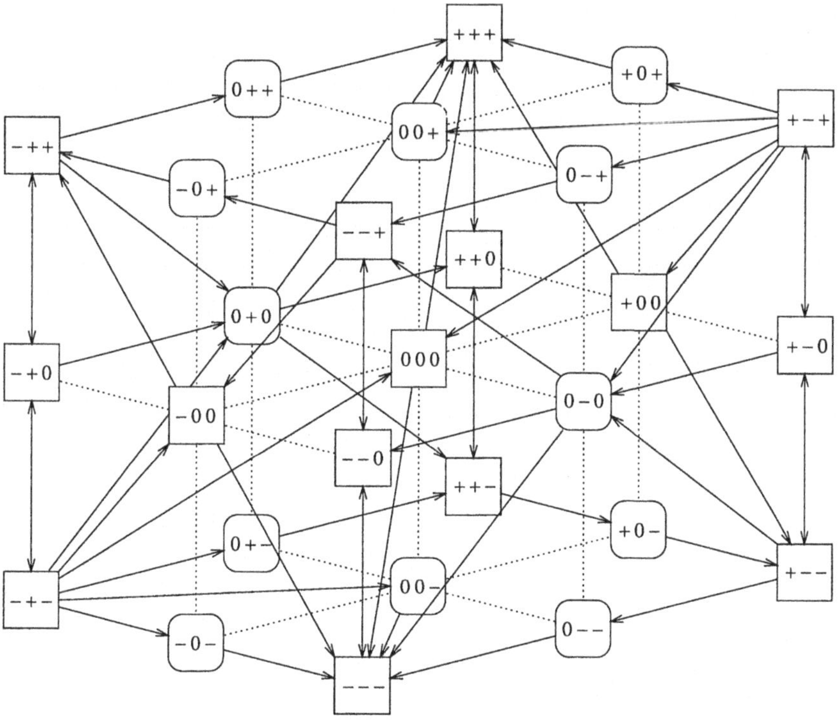

Figure 1.7. Trajectories in s, v, a-mode space

approach these ideal geometrical forms to varying degrees. However carefully you draw a circle, close examination will reveal imperfections in it, minute deviations from the perfect roundness of the ideal Circle that is handled by geometry. On the other hand some circles are better exemplars of the ideal Circle than others, and indeed the criterion for how well drawn a circle is is precisely the extent to which it partakes of the properties of the ideal geometrical circle.[12] Plato's theory of Forms can be regarded as the extension of these rather reasonable geometrical ideas to a wider realm.

Plato claimed that although we cannot perceive forms through the senses, since the senses only give us access to material things, we *can* perceive them through our 'inner sense', the mind. It is presumably this mental dimension that is responsible for the modern use of the word 'idea' to refer not to the eternal Forms in Plato's ideal world, but to elements of our thinking, things in the mind. Whereas Plato would have us evaluate objects in the world by comparing them with the ideal Forms, the latter being the gold standard, as it were, to which the former are answerable, we would nowadays, perhaps under the influence of empiricist philosophy, more readily reverse the priorities, making our ideas (now things in the mind, not in Plato's heaven) answerable to the objects in the world.

This leads us to the picture of an idealization as an imperfect copy of some real-world entity or phenomenon which it is intended to represent. It is imperfect in that its properties do not mirror perfectly the properties of what it represents. Of course all representations are imperfect in this sense, but we do not call all of them idealizations. Our notion of idealization retains some of the elements of the Platonic Forms: whereas the real world is messy, complicated, and obscure, our idealizations are neat, simple, and clear. One might say that idealizations are imperfect because they are too perfect, thereby failing to reflect the inherent imperfections of the world. The justification for using them is that the real world is too complicated for us to grasp without simplifying it and abstracting away many messy details. Idealizations succeed to the extent that the details thereby lost are truly irrelevant to our current concerns; since these details are sometimes relevant and sometimes not, it follows that any idealization will be useful only over the limited range of conditions for which the details are truly irrelevant. Thus use of an idealization must be circumspect: it is a serious error to extend its use beyond

[12] I am reminded here of Vasari's anecdote about Giotto, who, when called upon to demonstrate his artistic prowess before the Pope's emissary, took a clean sheet of paper and drew on it, freehand, a perfect circle.

the range in which it is justified.

At this point we should come down to earth and examine a number of idealizations that have been used in various branches of science, to get a feel for their properties. Physics is a prolific source of idealizations, including

1. An *ideal gas*, which obeys the Gas Law under all conditions. The Gas Law states that for a fixed mass of gas, the quantity

$$\frac{pressure \times volume}{temperature}$$

   is constant. Real gases obey the law fairly accurately over a considerable range, but depart radically from it at extremes of temperature or pressure. The reason for this is that the law is derived on the supposition that the gas consists of many infinitesimal perfectly elastic spherical particles which interact with one another only by collisions. In reality the gas molecules are finite in size, they are neither spherical nor perfectly elastic, and at close range interact by means of intermolecular forces across space. Thus it is at high pressures, when the molecules are packed relatively closely together, that the most significant departures from the Gas Law occur.

2. A *perfectly elastic* wire is one for which there is a fixed relation between load and extension. Many materials behave in very nearly a perfectly elastic fashion for loads up to a certain value (the 'Elastic Limit'), after which they become permanently deformed, unable to revert to their original length on removal of the load. Some materials such as nylon do not behave perfectly elastically even though they may revert to their original length on removal of the load: the load/extension ratio differs according to whether the load is increasing or decreasing.

3. A *perfectly rigid* body is one whose size and shape remains unaltered when it is subjected to any forces. Many bodies approximate to this state, so long as the forces are not too great, but in reality every body changes its shape at least a little bit when subjected to any force.

4. A *perfectly incompressible* fluid is one whose volume remains constant no matter what forces it is subjected to. Liquids in general are pretty nearly incompressible, which forms the basis of all forms of hydraulic engineering; but again in reality the incompressibility is never quite perfect, so this too is an idealization.

In Linguistics, the dominant idealization in the Chomskyan era has been the notion of 'competence' as distinct from 'performance'. As Chomsky puts it,

> Linguistic theory is concerned primarily with an ideal speaker-listener, in a completely homogeneous speech-community, who knows its language perfectly and is unaffected by such grammatically irrelevant conditions as memory limitations, distractions, shifts of attention and interest, and errors (random or characteristic) in applying his knowledge of the language in actual performance. This seems to me to have been the position of the founders of modern general linguistics, and no cogent reason for modifying it has been offered. To study actual linguistic performance, we must consider the interaction of a variety of factors, of which the underlying competence of the speaker-hearer is only one. In this respect, study of language is no different from empirical investigation of other complex phenomena.
>
> (Chomsky 1965: ch. 1)

This particular idealization has been the subject of much criticism, for example by Robinson (1975), Sampson (1980), and Moore and Carling (1982).

The application of mathematics almost always involves some kind of idealization. This is because mathematical concepts are abstract, and the only way we can apply an abstract concept to the real world is by ignoring those aspects of the real world which are not represented in the abstraction. The very act of counting presupposes an idealization by which certain chunks are carved out of reality to serve as countable units. To say that there are twenty sheep in a field is to ignore the fact that each of the sheep is a unique individual differing in countless—but for some purposes irrelevant—ways from each of the other sheep.

Logic, too, involves idealization. The mainstay of modern logic is the first-order predicate calculus, which rests on the idealized notion of a *property* as an all-or-nothing affair; that is, given any property (represented in the formalism by a predicate), logic insists that it be entirely determinate for each object under consideration whether or not it possesses that property. Only to the extent that this idealization holds water can we apply the predicate calculus in real-life situations. Those logics such as Fuzzy Logic which have been specifically devised with the aim of overcoming this 'all-or-nothing' idealization only succeed in compounding the problem by replacing that idealization with a more complex one involving the assignment of precise real numbers to express the degree to which an object possesses a property.

In computer science, the central idealization is the notion of a *virtual machine*: the effect of running a computer program is to enable a piece of general-purpose hardware to simulate the behaviour of an idealized special-purpose machine. Under ideal conditions, the user of the program only gets to see those effects which are implied by the functionality of the latter, the mechanism by which this is achieved being wholly hidden from view. In practice it all too often happens that low-level effects relating to the operating system or the hardware architecture make themselves visible, especially under unusual error conditions unanticipated by the original programmers.

Like all tools, idealization can be a mixed blessing. On the one hand, it is indispensable to science, which is unable to make any progress without the use of simplifying assumptions. On the other hand, an inappropriate idealization can lead to results which bear no relation to reality. The important thing is to understand the limits of the conditions under which an idealization is appropriate. In this book we shall meet many idealizations of various kinds, and I shall sometimes advert explicitly to their conditions of acceptability. Even when I do not, however, it should be borne in mind that I never wish to claim that any particular idealization is universally valid: for of course, idealizations never are.

As a tailpiece to this discussion of idealization, it is worth noting that while idealization usually involves simplification, it can also involve the reverse. Take, for example, the idealization of a coastline as a fractal; here the (presumably) finite geometrical complexity of the real coastline has been replaced by the infinite complexity found in the fractal. What counts as 'complexity' depends, however, on one's point of view. The Mandelbrot set is very complex in the sense that there is an infinite amount of structure to be found in it, but in another sense it is wonderfully simple, since all that is needed to generate it is an algorithm expressible in just a few lines. The distinguishing feature of fractals is their 'complexity in simplicity'. There seems to be a large amount of information in there but it can be reduced to only very little—hence the possibility of using fractals as a basis for data compression.

1.6. Granularity

Another general phenomenon which will be frequently encountered in this book is granularity or grain-size. Normally, granularity is taken to be a property of representations rather than of what is represented, but we can start by distinguishing two notions of granularity, which I shall call 'intrinsic' and 'representational'.

The intrinsic granularity of something is manifested in the various size-scales at which it exhibits significant structure, where this latter notion is to be expanded in terms of heterogeneity, discontinuity, or, more generally, of reduced symmetry. At these scales, our descriptions tend to be in terms of discrete objects, whereas at intermediate scales the natural form of description is in terms of distributions of matter ('stuff' rather than 'things'). This can be illustrated by considering a bowl full of sugar cubes. There is a sequence of scales at which significant structure appears:

- At the scale of around 10 centimetres, the bowl and its contents constitute a single object, sharply distinguished from other such bowls, and other objects of a similar scale.
- At a scale of around a centimetre, the individual sugar cubes come into prominence.
- At a scale of around a millimetre, the constituent grains of each sugar cube appear.
- At a much smaller scale, of around a nanometre, significant structure is manifested by the individual disaccharide molecules.

Although I have referred to appearances in my descriptions of these levels of granularity, the levels are intrinsic because they are determined by the sugar itself: we are not free to choose any scale as a grain-size: at scales of 4 centimetres, or 4 millimetres, or 1 micrometre the content of the sugar bowl does not resolve itself into structured individuals.

This notion of intrinsic granularity is related to Bronowski's notion of 'stratified stability'. Bronowski (1970) pointed out that the universe as a whole exhibits significant structures at a discrete sequence of scales, from clusters of galaxies down through galaxies, stars, planets, everyday objects, molecules, atoms, to subatomic particles, a 'great chain of being' grounded in science rather than theology. For Bronowski, this sequence of scales was not an artefact of the way we humans describe the world, but was rather intrinsic to the nature of the world itself, independently of how we describe it.

The other kind of granularity is representational granularity. This has to do with the fineness of detail afforded by different representations. A familiar example is the grain size of a photographic emulsion, and, more generally, the 'dot' size of a printed picture. Photographic films come in a range of different grain sizes; in general, the finer grain sizes are more desirable because they enable one to capture finer detail in the objects one is photographing, but there is a trade-off between granularity and sensitivity. The price one pays for very fine detail is the necessity for longer exposure times, with the consequent danger of blurring arising from movement of either the camera or the object being photographed. The relatively coarse dot size of photographs in newspapers becomes apparent as soon as one examines them with a lens of even the most modest magnifying power; a curious consequence of this graininess is that a photocopy of a newspaper picture generally gives a clearer image than the original, because the individual dots tend to become blurred together, giving the illusion of a finer level of granularity—though this is, of course, only an illusion, since the process of photocopying cannot possibly increase the amount of veridical detail in the picture.

Another example of representational granularity is the resolving power of an optical instrument such as a microscope or telescope (or indeed, the eye). Novices tend to imagine that the most important aspect of such a device is the magnification; but extra magnification is useless unless it goes hand in hand with extra resolving power. For example, the planet Saturn appears to the naked eye as a bright yellowish pinprick of light. Quite a small telescope can reveal it as a disc surrounded by a ring, one of the most beautiful sights in the heavens. Given sufficient resolving power, a magnification of about twenty times will suffice for this. An inferior telescope, with the same magnification but the resolving power of the human eye, will only show a blurred circle of light, with no hint of the ring. Galileo, who was the first person to observe Saturn telescopically, had a telescope intermediate in quality: he could see that there was something odd about the outline of Saturn, but he was unable to see the strange protuberances on either side of the planetary disc as forming a ring.

A somewhat more abstract, but still visual, notion of granularity is exemplified by the process of *generalization* in cartography. Generalization can be defined as 'the process of reducing the amount of detail in a map in a meaningful way' (Kraak and Ormeling 1996: §5.4). As Kraak and Ormeling point out, 'large-scale maps usually contain more detail than small-scale maps', but 'even at the same scale the level of detail can change'. They note that a map for a reference atlas will hold much more detail than a map of the same area, and to the same scale, in a school atlas. Thus *level of detail* is separable from *scale*: this is the distinction between resolution and magnification again.

Moving away from the visual, we lose our grip on the notion of magnification, but resolution remains in force. The concept of resolution can be applied to any form of representation, referring to the level of detail it affords. This applies both to analogical representations, which represent things by duplicating some aspect of their structure or appearance in a different physical medium, and to symbolic representations, which describe situations by means of more or less complex structures built out of labels that are conventionally associated with various abstractions pertaining to those situations. In connection with the latter kind of representation, we may speak of 'conceptual granularity'. A database which classifies living things only as birds, mammals, fish, insects, and so on, achieves a coarser level of conceptual granularity than one which further classifies birds as raptors, songbirds, and so on, and similarly with the other broad classes; a finer level of granularity still is obtained if individual species can be named.

Hobbs (1985) treats conceptual granularity in the context of knowledge representation using the first-order predicate calculus. Given any first-order theory, one can move to a simpler theory having coarser granularity by taking a subset **R** of the predicates in the theory and quotienting out an equivalence relation called 'indistinguishability', defined by the second-order rule:[13]

$$x \sim y \quad \triangleq \quad \forall P \in \mathbf{R}(P(x) \leftrightarrow P(y)).$$

The predicates in **R** are chosen to be those that are relevant to the knowledge-representation task being undertaken; domain elements are indistinguishable if no relevant predicate distinguishes them. The domain of the simplified theory is now taken to be the set of equivalence classes of the original domain that are determined by the indistinguishability relation.

Hobbs goes on to discuss cases where the indistinguishability is given *a priori* rather than being derived from a set of relevant predicates. These cases are strongly related to the process of idealization, to which Hobbs makes an explicit connection. Idealization can be viewed as a process of filtering out unwanted details, in effect a kind of coarsening of the conceptual granularity—though an idealized picture differs from a merely grainy one in that where the latter has blurred outlines, the former presents them as quite sharp. It can also have the effect of reducing a continuous scale of variation to a discrete one, hence linking in with the earlier discussion of quantitative versus qualitative state spaces.

Granularity will be important to us throughout this book because many—perhaps most—of the spatial and temporal attributes I shall discuss are sensitive to the level of granularity at which the objects of those attributes are

[13] In this book, formulae of the form '$X \triangleq Y$' are to be read as 'X is defined to be equivalent to Y'. This and other symbols introduced throughout this book are collected together in the Index of Special Symbols on p. 394.

represented. In the ensuing chapters I shall point out many examples of this granularity-sensitivity, as for instance in relation to connectivity (§4.2), shape (§4.7), process (§5.2), and continuity (§7.1).

2

Space

2.1. Places and spaces

It is customary to distinguish between *absolute* and *relational* models of space. In an absolute model, space is regarded as having existence and properties independently of the existence of any object; and the objects acquire their spatial properties and relations in consequence of their being located in this pre-existing space. In a relational model, on the other hand, objects come first, and space exists only in so far as the objects have spatial properties and relations: space is, in other words, a logical construct from objects and their properties.[1]

Objects are related to space by way of the notion of position. Absolute space provides a copious set of *places* to serve as the possible positions of objects, each place being a portion of space capable of being occupied by an object of appropriate size and shape. Relational space is no less abundant in its positional possibilities, the difference being that here the positions are always defined relative to the positions of existing objects rather than given as independent entities in their own right.

A potential position is a place. What kinds of place should we recognize? In three-dimensional space, the place of a material object is an extended region of space having volume and shape equal to those of the object itself. It is a 'chunk' of space. But as we shall see in Chapter 3, material objects are not the only types of objects we may wish to consider. Amongst others are *parasitic* features of material objects such as edges, corners, and surfaces, which have no volume in themselves but depend for their existence on objects which have volume. These too have positions, but since they lack volume, their positions must lack volume too. The corner of a cube, for example (here we assume an idealized 'perfect cube') has no extension at all, and its position is therefore a point in space; likewise, an edge of the cube has extension in

[1] An exactly parallel situation exists with respect to time and events, as discussed in §5.1.

only one dimension, so its position is a line. The elements of space we may wish to refer to are therefore quite varied: as well as three-dimensional regions with volume, there are two-dimensional surfaces having area but no volume, one-dimensional lines having length but neither area nor volume, and dimensionless points lacking length, area, and volume. Depending on how we model space, we may also wish to admit elements of mixed dimension, for example a spherical volume adorned with a planar equatorial flange or linear spikes. We may also wish to allow spatial elements of fractional dimensions, as exemplified by the many kinds of fractals discovered in recent years.

Clearly the three dimensions of space, much more than the single dimension of time, allow the existence of a riotous profusion of different kinds of spatial element. Any formal account of spatial change must in some degree systematize this profusion, providing a taxonomy and a set of regulatory principles to govern the spatial representations we can avail ourselves of.

In this chapter I shall adopt an absolute view of space; the objects which inhabit this absolute space will be discussed in the next chapter. This order of proceedings is for convenience only; it should not be understood as implying any commitment to a view of absolute space as philosophically 'correct'. I shall survey a number of different approaches to the description and representation of space, bearing in mind the particular needs that will arise in the context of qualitatively describing the phenomena of spatial change.

2.2. Axiomatic geometry

The earliest detailed and systematic study of space that we know of is the geometry of ancient Greece. Herodotus suggests, in Book II of his *Histories*, that this science had its origins in Egypt, where techniques of measurement and computation were developed to enable property owners to restore field boundaries that had been obliterated by the yearly flooding of the Nile. The geometric principles thereby developed were refined by the Greeks into a vast edifice of pure mathematics in which the practical concerns of the Egyptian surveyors played no further part. The foundations of this edifice were systematically expounded by Euclid in his great 13-volume work, *The Elements*, which was revered well into the twentieth century as one of the central texts of mathematics. Its influence on the development of western culture has been profound.

The Elements represents the first appearance of the axiomatic method. At the start of the enterprise a number of postulates called 'axioms' are asserted as unquestionably true, together with a number of rules for deriving new propositions from old ('rules of inference'). The rules of inference are used for deriving from the axioms, in a formal and systematic manner, a series of theorems, propositions whose truth is guaranteed so long as the axioms are true. The power of this method is manifest in both the quantity and quality of the theorems that Euclid was able to derive by its means, though perhaps paradoxically it must be said that the Greeks themselves, and most of their successors (right up until the close of the nineteenth century), failed to appreciate the true nature of the axiomatic approach—as will be discussed in more detail in §2.2.3.

2.2.1. Euclidean geometry

Euclid's geometry handles spatial entities such as points, lines, surfaces, solids, figures, and angles, and spatial properties and relations such as length, area, volume, congruence, similarity, and parallelism. Although he purports to define the basic spatial entities, his definitions would be of little help to anyone who did not already have a good idea (derived perhaps from intuition and common usage) of what they were, and, moreover, the definitions play no further part in the development of the theory. None the less they are important because they serve to link the otherwise abstract axiomatics to our intuitive understanding of space.

Amongst other things, Euclid tells us that

- A point is that which has no part.
- A line is breadthless length.
- The extremities of a line are points.
- A surface is that which has length and breadth only.
- The extremities of a surface are lines.
- A boundary is that which is an extremity of anything.
- A figure is that which is contained by any boundary or boundaries.

The concepts introduced by these definitions are idealizations (cf. §1.5). If asked to point out an example of a point in the physical world, one might perhaps point to a dot on a piece of paper, such as the full stop at the end of this sentence. This 'point' is actually an approximately circular patch having a diameter of about half a millimetre. Through a hand-lens one can clearly distinguish its left half, say, from its right, so it is scarcely true that it 'has no part'. The form of idealization required to get from a dot of this kind to a Euclidean point is a *limiting process*: we imagine a series of ever smaller physical dots and say that 'in the limit', as the diameter decreases to zero, we arrive at a true point. This is paradoxical, because so long as the dot is still a dot, with a diameter greater than zero, it is not a Euclidean point; and once the diameter has become zero, there is nothing left at all.

An alternative source of the point idealization is in some ways more satisfactory. Figure 2.1 shows the flag of the Czech Republic. It contains white, red, and blue portions, which meet at a common apex. This apex occupies no area and has no parts. In this respect it is a better representation of the idealized Euclidean point than are the dots discussed in the previous paragraph. Unfortunately, this representation is not entirely without problems either. On close examination, it will be found impossible to locate the meeting-point of the three colours precisely. This is because the description of portions of the flag as 'white', 'red', or 'blue' cannot be taken to arbitrarily fine granularity. Below the molecular level we can no longer regard the flag as a uniform flat expanse which is everywhere coloured. Thus the representation of a point as the meeting-point of colour edges is also something of an idealization. Later, in §2.4, I shall discuss some of the implications of choosing one or the other of these two routes to the idealized Euclidean point.

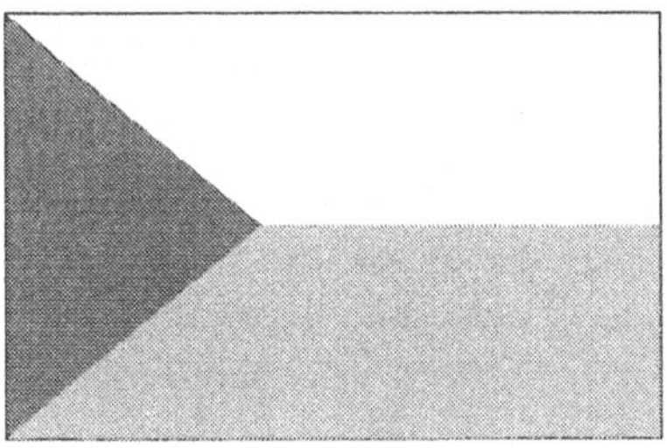

Figure 2.1. Three coloured areas meeting at a point

A similar choice exists as regards Euclid's other terms. A Euclidean line can be represented as a narrow ribbon-like trace, as when we use a pencil or pen to draw a line on a piece of paper. Alternatively, it may be represented as an edge, where an expanse of one colour meets an expanse of another colour (there are three examples of such edges in Figure 2.1). Similarly, a surface may be represented as a thin membrane; or as the interface between a volume of space occupied by one kind of matter and a volume occupied by a different kind, for example the surface of a lake, where the water meets the air. In each case the first kind of representation clearly lacks the property ascribed by Euclid (a pencil line is not breadthless, a membrane has thickness as well as length and breadth), whereas the second kind comes closer to Euclid's conception.

Euclid's axioms consist of five 'common notions', which handle the (not specifically geometrical) notions of 'part', 'whole', and 'equal', as well as five specifically geometrical axioms. An example of Euclid's common notions is the rule that 'things which are equal to a third thing are equal to each other'. In modern logical notation this rule would be written as

$$\forall x \forall y \forall z (x = z \wedge y = z \rightarrow x = y),$$

a combination of the symmetry and transitivity properties for equality.

Some of the axioms take the form of *constructions*, which describe the conditions under which different geometrical figures may be constructed. For example, given a point and a distance that have already been constructed, we can construct a circle, of radius equal to the given distance, centred on the point. The possible constructions in Euclidean geometry are precisely those which can be performed using an uncalibrated ruler (a 'straight-edge') and a compass, with the proviso that the span of the compass can only be held fixed so long as one of its points remains in the same position.

Many of the theorems, or 'propositions', assert the possibility of some geometrical construction. An example is the very first proposition in the *Elements*, which is *On a given finite straight line to construct an equilateral triangle.* The proof shows how to perform this construction, and then shows that the resulting triangle is indeed equilateral. Some theorems refer, not to new constructions, but to properties of existing constructions; new constructions may be involved in their proofs. An example is the most famous of all propositions in the *Elements*, Proposition 47: *In right-angled triangles the square on the side subtending the right angle is equal to the squares on the sides containing the right angle.*

2.2.2. Non-Euclidean geometry

Amongst the five specifically geometrical axioms of the *Elements*, the fifth has always been a source of unease. Euclid himself delayed invoking it for as long as possible, perhaps sensing that this axiom was considerably less self-evident than the other four. As Euclid formulated it, the fifth axiom states that:

> if a straight line falling on two straight lines make the interior angles on the same side less than two right angles, the two straight lines, if produced indefinitely, meet on that side on which are the angles less than the two right angles.

Quite apart from the matter of how self-evident this axiom may be, it is noteworthy that it is much *longer* than the other axioms; in the original Greek, the number of words (excluding the initial 'Let it be postulated' and the linking 'and's) in the five axioms are 9, 8, 6, 7, and 34 respectively: the fifth axiom is longer than the other four put together!

Euclid could have replaced this fifth axiom by something a little simpler. Given the other axioms, it comes out as equivalent to *Playfair's Axiom:*

> Given a line L and a point P, there is a unique line through P parallel to L.

In other words, of all the lines parallel to L, exactly one passes through P, and of all the lines through P, exactly one is parallel to L. If this *seems* obvious, it is partly because of our long habituation to Euclidean geometry as *the* geometry most frequently invoked in actual use—though there is also a deeper reason to do with the fact that at small scales the non-Euclidean alternatives to Euclidean geometry do closely approximate to Euclidean geometry in ways that can be made quite precise.

Whether or not an axiom concerning lines, points, and circles is self-evident must depend on our intuitive understanding of these terms. We are accustomed to thinking of a straight line as the shortest distance between two points. Suppose, then, that we confine ourselves to the surface of a sphere. Between two points on the sphere, the shortest path *on the surface of the sphere* is a great circle, the intersection of the spherical surface with a diameter plane (for example, on the earth, the equator, or any meridian). What happens if we interpret 'line' to mean a great circle?

If we interpret 'point' in the usual way, Euclid's first axiom becomes false: it is not true that any two points determine a unique line, since two *antipodal* points (lying at opposite ends of a diameter) lie on an infinite number of distinct great circles—indeed, any great circle through one must also pass through the other. In order to rectify this, we reinterpret 'point' to mean an antipodal pair of ordinary points, in effect counting the two members of such a pair as one. This in turn means that we must reinterpret 'circle' to mean a pair of circles such that each point on one is antipodal to a point on the other (for example, on the earth, two lines of latitude at equal distance from the equator, say 50°N and 50°S).

With these adjustments, Euclid's first four axioms are satisfied;[2] but the fifth is not. In this 'spherical' geometry, the phenomenon of parallelism is non-existent since any two 'lines' (great circles) meet at a 'point' (antipodal point-pair). This means that the geometry of the surface of a sphere, when described in these terms, is non-Euclidean.

Another form of non-Euclidean geometry negates the fifth axiom in the opposite way: instead of there being *no* lines through P parallel to L, as in spherical geometry, we now postulate *infinitely many* lines. If this is the only adjustment we make to Euclidean geometry, we obtain a system called 'hyperbolic' geometry. The easiest way to visualize it is through a construction originally due to Beltrami, but now more generally associated with the name of Poincaré. The 'space' of the geometry is the interior of a circular disc in the Euclidean plane. A 'point' is interpreted as any ordinary point in this space, but a 'line' is now a circular arc meeting the edge of the disc at right angles. In Figure 2.2 we see, in the left-hand illustration, two lines L_1 and L_2 meeting at a point P, and a line L which does not meet either L_1 or L_2 and is therefore parallel to both of them. This shows that in hyperbolic geometry the relation 'parallel' is not transitive, and also illustrates the failure of the fifth axiom: L_1 and L_2 are two distinct 'lines' parallel to L and passing through P. In fact, any 'line' enclosed between the two dotted lines will have this property. Although the space appears to be finite, we must understand 'distance' along a 'line' to increase exponentially as we approach the bounding circle, so that that boundary, which is not itself part of the space, is infinitely distant from every point in the space.

A number of familiar theorems of Euclidean geometry turn out to be equivalent to the fifth axiom, and therefore false in non-Euclidean geometries. Two such theorems are:

- The locus of points at a fixed distance from a given straight line is a pair of straight lines parallel to the given line.

[2] The second axiom, which says that a line can be extended indefinitely, needs to be phrased carefully to avoid the implication that a line must be of infinite length—it is enough that it has no end-points, which is certainly true of a great circle.

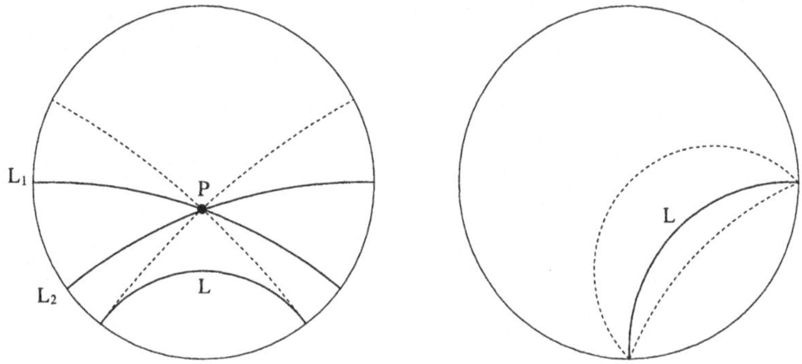

Figure 2.2. 'Lines' in hyperbolic geometry

- The interior angles of a triangle together make up two right angles.

On the sphere the locus of points at a fixed distance from a straight line is a *circle*: for example, the points at a given fixed distance from the equator form two lines of latitude. These are not straight lines, even in the sense given to this term in spherical geometry. In hyperbolic geometry, as we have seen, a line can be represented by a circular arc cutting the boundary circle at right angles. The locus of points equidistant from this line turns out to be a pair of circular arcs, distinct from the line, and meeting the line at the boundary, as illustrated by the dashed arcs on either side of L in the right-hand illustration of Figure 2.2. These do not count as straight lines since they are not perpendicular to the boundary.

Again, on the sphere, the angle sum of a triangle depends on its size. For small triangles it is a little greater than 180°, but for larger triangles it is greater, and the amount by which it exceeds two right angles is proportional to the area. For example, the triangle formed by London, Edinburgh, and Dublin has an angle sum of 180.11°, whereas the much larger triangle formed by London, New York, and Rio de Janeiro has an angle sum of about 236°. Figure 2.3 shows an equilateral triangle whose area is one-eighth that of the surface area of the sphere, whose angles are all right angles, making the angle sum *three* right angles, not two.[3]

Although non-Euclidean geometries have many features that are strongly counterintuitive to anyone whose exposure to geometry has only been of the

[3] Strictly speaking, the triangle consists of two parts, the triangle shown in the figure, and its antipodal image; so its area is really a quarter of the sphere's surface, not an eighth.

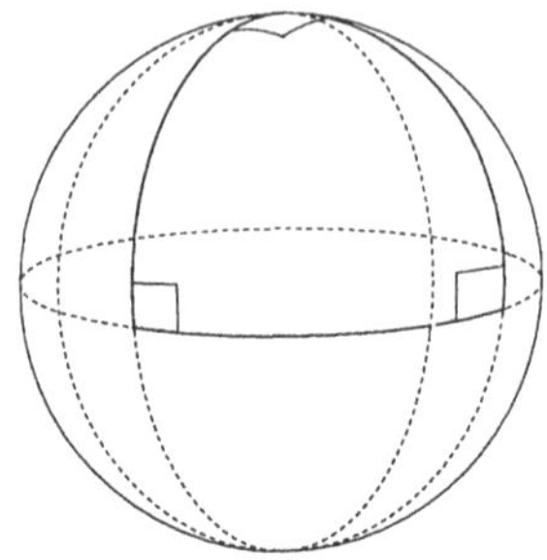

Figure 2.3. An equilateral triangle with three right angles on the surface of the sphere

Euclidean kind, these are well worked-out and consistent mathematical theories. We should take them seriously, especially in view of the well-confirmed General Theory of Relativity, according to which the space we actually live in, our universe, has a geometry which is fundamentally non-Euclidean.

2.2.3. Axiom systems as uninterpreted calculi

The advent of non-Euclidean geometry in the early years of the nineteenth century brought in its train a revolution in our understanding of the axiomatic method. This was first made fully explicit towards the end of the century by David Hilbert, the foremost German mathematician of the day. Hilbert liked to say:

> One must be able to say at all times—instead of points, straight lines, and planes—tables, chairs, and beer mugs.
>
> (Reid 1996)

What he meant by this was that an axiom system does not in itself tell us how the terms occurring in it should be interpreted, except in so far as any allowable interpretation must be such as to make the axioms and theorems of the system come out true. Any such interpretation of the terms provides a *model* for the axiom system, and the system may be understood as saying things about any of its models.

More generally, any consistent *theory*, in the sense of a set of statements regardless of the means by which they are generated, may be said to have different models according as the terms in the theory are interpreted differently, subject only to the constraint that they are interpreted in such a way as to make the statements of the theory come out true.

In the next section I shall take a brief look at what may be fittingly called the *standard model* of Euclidean geometry, by which points are interpreted as tuples of real numbers (co-ordinates) and lines are interpreted as linear equations over those tuples. This is Cartesian geometry, which since the seventeenth century has been one of the cornerstones of mathematics, and which in some form or other is doubtless familiar to every reader of this book.

2.2.4. Co-ordinate geometry as a model for Euclidean geometry

In two-dimensional Cartesian geometry a point is represented by an ordered pair (x, y) of real numbers, and every such pair represents exactly one point. A straight line is represented by an equation of the form $ax + by + c = 0$, where a, b, and c are again real numbers; with the condition that the equations $ax+by+c=0$ and $a'x+b'y+c'=0$ represent the same line just so long as $ac' = a'c$ and $bc' = b'c$. The condition for a point to lie on a line is simply that its co-ordinates satisfy the equation. Given two distinct lines $ax+by+c=0$ and $a'x + b'y + c' = 0$, the condition that they meet is that there is some pair (x, y) satisfying both equations. The condition for this is that $ab' \neq a'b$. Thus parallel lines can be characterized as distinct lines for which $ab' = a'b$.

With points, lines, incidence, and parallelism so defined (and analogous definitions can be given for three-dimensional space), all of Euclid's axioms are satisfied. Consider the fifth axiom, for example. The lines parallel to $ax+by+c=0$ can all be expressed in the form $ax+by+c'=0$ for different values of c'. Given a point (d, e), there is a unique value of c' such that the line $ax + by + c' = 0$ passes through that point, namely $c' = -(ad + be)$. This verifies Playfair's axiom, and hence also Euclid's fifth.

Note that Euclid's axioms refer to circles, which in turn are defined in terms of distance: a circle consists of all those points whose distance from a given point (the centre of the circle) is equal to a fixed amount (the radius of the circle). In Cartesian geometry the distance between points $P_1 \equiv (x_1, y_1)$ and $P_2 \equiv (x_2, y_2)$ is defined as the quantity

$$d(P_1, P_2) \triangleq \sqrt{(x_1 - x_2)^2 + (y_1 - y_2)^2},$$

and hence the equation of the circle of radius r and centred at (x_0, y_0) is $(x - x_0)^2 + (y - y_0)^2 = r^2$.

The great power of Cartesian co-ordinates is that they effectively reduce geometric questions to algebraic ones. This power is what enabled the mathematicians of the eighteenth century and after—especially when abetted by the invention of the differential and integral calculi—to solve problems in Euclidean geometry that were beyond the reach of Euclid's axioms. Much of our everyday mathematics now would be unthinkable without Cartesian geometry, which is therefore rightly included, in some form or other, in all mathematical syllabuses. But since it is essentially a *quantitative* technique, grounded in the principle of describing all geometrical phenomena in terms of equations satisfied by the exact numerical co-ordinates of certain points, it will not play a prominent part in the rest of this book.

The same remarks apply to other numerical co-ordinate systems, such as polar co-ordinates, in which the position of a point P is specified in terms of its distance r from a fixed origin O together with the anticlockwise angle between a fixed line through O and the line OP. This system can be generalized to three dimensions in two different ways, giving us spherical polar co-ordinates and cylindrical polar co-ordinates. The former, when restricted to the surface of a fixed sphere centred on the origin, corresponds exactly to the system of geographical latitude and longitude.

2.3. Point-set topology

Towards the end of the nineteenth century, a number of developments in mathematics, including the rigorous treatment of continuity, the characterization of the real numbers as limits of convergent sequences of rationals, the introduction of measure theory, and Cantor's investigations into the cardinality of infinite sets, all came together to produce a unified conception of mathematical space in terms of *sets of points*. In this conception, the properties of spatial entities, of whatever dimensionality, are studied in terms of the properties of sets of points corresponding to those entities. Quantitative properties are well handled by co-ordinate geometry and its derivatives such as the calculus. By contrast, another mathematical discipline, *topology*, is more concerned with investigating a collection of properties that may be described as qualitative rather than quantitative. In particular, it explores the general notions of limits and continuity in a way that is not tied to the quantitative treatment of these notions in the calculus. In its most general form topology is a highly abstract discipline that can serve as a powerful tool in many branches of mathematics. Point-set topology specifically handles these topological notions as they apply to *sets*; the elements of the sets are conventionally called 'points' in this context, though they do not necessarily bear any resemblance to points as these are understood in geometry.

2.3.1. Open and closed sets

Perhaps the easiest way to introduce the essential concepts of point-set topology is by looking at how they are instantiated in the familiar case of Cartesian geometry. For simplicity, I begin with a two-dimensional Cartesian space, which may be represented by the set of real number pairs with a distance function d defined as described above. Individual real number pairs are the 'points' of this space, while the 'regions' are sets of such points. While we may not wish to regard an arbitrary set of points as a region, it is reasonable to include amongst the regions of two-dimensional space the *circular discs*, for example the set of all points within a given radius r of a given point a. This can be understood in two different ways, depending on whether 'within' is interpreted in an *inclusive* sense, as including those points whose distance from a is exactly r, or in an *exclusive* sense, as excluding them. The former interpretation leads to the concept of an 'open' disc, in this case the set

$$B(a, r) \triangleq \{x \in \mathbb{R}^2 \mid d(x, a) < r\},$$

whereas the latter leads us to a 'closed' disc,

$$cl(B(a, r)) \triangleq \{x \in \mathbb{R}^2 \mid d(x, a) \leq r\}.$$

The relation between these two discs is that the closed disc consists of the open disc together with its boundary, that is, the set of points whose distance from a is exactly r.

If you are situated in an open disc, and you want to move outside the disc, then you have to cross over a region which is still inside the disc. For example, if you are at the point (9,0) in the open disc $B((0, 0), 10)$, then the nearest point which is not also in the disc is (10,0); to get to it you have to follow a path of length at least 1 which lies entirely inside the disc. With a closed disc it is different. You can be situated on the boundary of the disc, and in that case there is no positive lower limit to the distance you have to traverse in order to reach points outside the disc. For the disc $cl(B((0, 0), 10))$, if you are located at the point (10,0), then there is no lower limit to the distance separating you from points outside the disc: for any number $\epsilon > 0$ we can find points outside the disc, such as $(10 + \epsilon, 0)$, which are that distance away from you. The number ϵ can be made as small as you like.

A useful concept to bring in at this point is that of a 'neighbourhood'. In the present context, a neighbourhood of a point a is any set of points which contains some open disc $B(a, r)$. We can then define open and closed *sets* as follows:

- A set of points is *open* if every point in the set has a neighbourhood contained entirely within the set.
- A set of points is *closed* if its complement is open.

From these definitions we can prove that

- *The union of arbitrarily many open sets is open.*
  For let $\mathcal{S}$ be any collection of open sets and let x be any point in their union. Then $x \in S$ for some $S \in \mathcal{S}$. Since S is open, some neighbourhood of x is contained in S, and hence also in $\bigcup \mathcal{S}$ (since S is a subset of $\bigcup \mathcal{S}$). Hence $\bigcup \mathcal{S}$ is open.
- *The intersection of arbitrarily many closed sets is closed.*
  The complement of $\bigcap \mathcal{S}$ is the union of the complements of the elements of $\mathcal{S}$. These sets are open, so their union is open. Hence $\bigcap \mathcal{S}$ itself is closed.
- *The empty set and the universal set (which contains every point in the space under consideration) are both open.*
  It is vacuously true that every point in the empty set has a neighbourhood in the set (since there are no points in the empty set); further, any neighbourhood of a point in the universal set (that is, of any point whatever) must be entirely in the universal set.
- *The empty set and the universal set are both closed.*
  This follows from the last result since the universal set and the empty set are mutually complementary.

To prepare for the remaining two results we may observe that the intersection of any two neighbourhoods of a given point is itself a neighbourhood of that point. This follows from the fact that the open discs centred on a point are *nested* in the sense that given any two such discs, one contains the other. We can now prove that

- *The intersection of two open sets is open.* Let S_1 and S_2 be open, and let $x \in S_1 \cap S_2$. Since x is in S_1, it has a neighbourhood $N_1 \subseteq S_1$, and since it is in S_2 it has a neighbourhood $N_2 \subseteq S_2$. By the property noted above, $N_1 \cap N_2$ (which is a subset of $S_1 \cap S_2$) is also a neighbourhood of x, so x has a neighbourhood contained in $S_1 \cap S_2$. Hence $S_1 \cap S_2$ is open.
- *The union of two closed sets is closed.* Let S_1 and S_2 be closed. Then their complements S_1' and S_2' are open, so by the previous result the intersection $S_1' \cap S_2'$ is also open. The complement of this set, which is $S_1 \cup S_2$, must therefore be closed.

Note that these two results can be straightforwardly generalized to the intersection of finitely many open sets and the union of finitely many closed sets, but they cannot be generalized beyond this to infinite collections of sets. For example, in $\mathbb{R}^2$, the infinite sequence of sets

$$B(a, r+1), B(a, r+\tfrac{1}{2}), B(a, r+\tfrac{1}{3}), B(a, r+\tfrac{1}{4}), \ldots$$

are all open, but their intersection is the *closed* set $cl(B(a,r))$. Likewise, the sets

$$cl(B(a, \tfrac{1}{2})), cl(B(a, \tfrac{2}{3})), cl(B(a, \tfrac{3}{4})), cl(B(a, \tfrac{4}{5})), \ldots$$

are all closed, but their union is the open set $B(a, 1)$.

Although I have defined neighbourhoods in a very specific way in terms of the open discs $B(a, r)$, the proofs I have just given of the properties of open and closed sets only rely on certain rather general features of neighbourhoods. They do not rely on the specific interpretation I have given to the term 'neighbourhood' in two-dimensional Cartesian space. More generally, any interpretation of that term which satisfies certain minimum conditions can form a basis for the definition of open and closed sets sufficient to support the proofs given above.

The conditions we require for neighbourhoods are that

1. If N is a neighbourhood of x then $x \in N$.
2. If N_1 is a neighbourhood of x and $N_1 \subseteq N_2$ then N_2 is a neighbourhood of x.
3. If N_1 and N_2 are neighbourhoods of x then so is $N_1 \cap N_2$.
4. If N is a neighbourhood of x then there is a neighbourhood $N' \subseteq N$ of x which is a neighbourhood of each of its points.

The fourth condition ensures that every neighbourhood of a point must contain an open neighbourhood of that point. A common approach to topology is to start with the notion of open set as primitive, and then define a neighbourhood of a point x as any set N such that there is an open set O for which $x \in O \subseteq N$. Condition (4) ensures that if we start with neighbourhoods, define open sets as we did above, and then apply the definition of neighbourhood in terms of open sets, then we recover the neighbourhoods we started out with.

However we define the neighbourhoods of a point, so long as conditions (1)–(4) are satisfied, we can go on to define open and closed sets and these sets will have the properties enumerated above. Any such system, consisting of a set of points together with a specification of which sets of points are to count as neighbourhoods of each given point, is referred to as a *topological space*.

The notions of open and closed sets enable us to define a number of further topological ideas which are closely connected with some of our qualitative intuitions about space. The most important of these ideas are the following:

- The interior $int(S)$ of a set S is the union of all open sets contained in S. As a union of open sets, it is itself open, and since it contains every open set contained in S, it is the *largest* (that is, most inclusive) open set contained in S.
- The closure $cl(S)$ of a set S is the intersection of all closed sets containing S. As an intersection of closed sets, it is itself closed, and since it is contained in every closed set containing S, it is the *smallest* (least inclusive) closed set containing S.
- The exterior of a set S is the union of all open sets that are disjoint from S. It is, in fact, both the interior of the complement of S and the complement of the closure of S.
- The boundary ∂S of a set S is the set of points in $cl(S)$ but not in $int(S)$, that is, $\partial S = cl(S) \setminus int(S)$.

It follows from these definitions that a set is open if and only if it is its own interior, that a set is closed if and only if it is its own closure, and that an open set is disjoint from its boundary whereas a closed set contains its boundary. Moreover, the boundary of any set is necessarily closed, and is its own boundary; the interior of the boundary is empty. In a rich topology like that of two-dimensional Cartesian space, sets with empty interior are insubstantial things, lacking area.

- A set S is connected if whenever it is a subset of the union of two disjoint open sets, it is a subset of just one of them; that is, if for any open sets S_1 and S_2,

$$(S \subseteq S_1 \cup S_2) \wedge (S_1 \cap S_2 = \emptyset) \Rightarrow (S \subseteq S_1) \vee (S \subseteq S_2).$$

An arbitrary set of points S is likely to be neither open nor closed, but since we have $int(S) \subseteq S \subseteq cl(S)$ the differences between it and the 'nearest' closed and open sets are contained in the boundary ∂S—and hence insubstantial in the normal two-dimensional Cartesian topology.

2.3.2. Modelling regions as sets of points

Topological ideas, being of a highly general character, have found wide application throughout mathematics, but for my present purpose I am only interested in their usefulness for modelling our intuitions concerning physical space. Narrowing the focus to the Cartesian spaces $\mathbb{R}^2$ and $\mathbb{R}^3$, we find that whereas topology will handle arbitrary sets of points in these spaces, our spatial intuitions are much more limited. Of key importance is the idea of a *region*, but it is not a straightforward task to define exactly what should be meant by this term.

In the Cartesian plane, which sets of points should we count as regions? It is not hard to specify mathematically any number of sets of points which we should certainly wish to exclude from regionhood; some examples are:

- Those points, contained in a circular disc, whose co-ordinates are rational numbers.
- Those points, contained in a circular disc, whose co-ordinates, when expressed in base 3, lack the digit '1'.
- The Mandelbrot set.

Some more doubtful cases, which we may or may not wish to include as regions, are

- An open circular disc with the points along one of its radii removed.
- An open circular disc together with half the points on its circumference.
- Two closed circular discs meeting at a single point on the circumference of each.

These are illustrated in Figure 2.4, in which, following the usual convention, the boundary of a set is drawn as a solid line if it is part of the set and as a dashed line if it is not.

Notwithstanding my earlier remarks concerning the positions of edges, corners, and the like, it seems not unreasonable that, in the plane, a 'first-class' region should have an area (and likewise, in three-dimensional space, a volume). Thus a first requirement for a set of points to model a region of space is that it should have non-empty interior. Thus the boundary of a set

Figure 2.4. Three point-sets: are they regions or not?

of points will not itself count as a region. The simplest way to ensure that a set of points has non-empty interior is to insist that it is open. This eliminates 'regions' with isolated points or bits of boundary, but it still lets in examples like the open disc with a missing radius (as in the leftmost illustration in Figure 2.4). The boundary of this set includes the missing radius. Now, to quote Smith (1993), 'It seems to be an important mark of the pre-analytic concept of boundary ... that a boundary of y should divide y from something else'. The trouble with our example is that the missing radius, which is indeed part of the boundary, does not divide the set from something else; it only divides one part of the set from another part.

The idea that the boundary of a region is the interface between the region itself and the outside world suggests, as a further restriction on which point-sets are to count as regions, the requirement that every boundary point is in contact with both the interior and the exterior of the region. More exactly, every neighbourhood of a boundary point of a region should overlap both the interior and the exterior of the region. If the region is an open set, then this condition turns out to be equivalent to the following simply expressed requirement: the set is the interior of its closure: $S = int(cl(S))$. A set with this property is called a 'regular' open set.

The open disc with a missing radius, although it is an open set, is not regular open. Its closure is the closed disc, with no radius missing, so the interior of its closure is the open disc, again with no radius missing. Thus the set is not equal to the interior of its closure, so it is not regular open.

A number of authors have suggested regular open sets as an appropriate point-set model for spatial regions. The earliest example I am aware of is Cartwright (1975), who, while applying the word 'region' to arbitrary point-sets, is really interested in those regions which can act as the positions of material objects ('receptacles'), and these he identifies with regular open sets. See also Forrest 1996. An alternative possibility is to use regular *closed* sets, that is, sets which are equal to the closure of their interiors. There is a one-to-one correspondence between regular closed and regular open sets: the closure of a regular open set is regular closed, and the interior of a regular closed set is regular open. Regular closed sets are used by Davis (1990). Another possibility is to allow *both* regular open and regular closed sets, as proposed by Asher and Vieu (1995), whose theory is discussed below (§2.5.2).

Regular sets, whether of the open or closed variety, are not without their problems, however. Although the restriction to such sets serves to exclude many 'pathological' point-sets which one would not want to regard as *bona fide* regions, it does not exclude them all. Pratt and Lemon (1997) point out that the boundary of a regular region can include part of the curve $y = \sin \frac{1}{x}$ on the interval $(0, 1)$; this curve is pathological in the sense that it has an infinite series of maxima and minima in the vicinity of $x = 0$, and the region inherits this pathological character—for example, although the region is finite, a straight line can cross its boundary in infinitely many places. (See p. 308 for further discussion of this curve.)

For this reason, Pratt and Lemon restrict their regions still further to *polygons*. More precisely, they define a *basic polygon* to be the intersection of finitely many open half-planes in $\mathbb{R}^2$, and a polygon to be the regular union of finitely many basic polygons. By the 'regular union' is meant the interior of the closure of the set-theoretic union. This is to prevent the formation of, for example, an open rectangle with a line of points missing across the middle, such as would result from the union of two adjacent open squares: taking the regular union incorporates the 'missing' line into the set. An alternative model, with the same first-order theory, is based on $\mathbb{Q}^2$ rather than $\mathbb{R}^2$. I shall return to Pratt and Lemon's polygonal models later, in the discussion of mereotopology (see §2.5.3).

2.3.3. Regions of different dimensions

The spatial concepts we handle in everyday life include entities which, in idealization at least, are extended in zero, one, two, or three dimensions. To describe the room I am sitting in, I might wish to refer to the whole interior space of the room, which is a three-dimensional region, or one of the walls, which, if understood in the sense of a surface, is a two-dimensional region, or the line along which the wall meets the floor, which is a one-dimensional entity ('region' hardly seems an appropriate word in this case), or finally the corner where two adjacent walls and the floor meet, which is a point or zero-dimensional entity.

In the context of three-dimensional space, then, we may wish to consider spatial elements extended in 0, 1, 2, or 3 dimensions. A useful concept in this connection is that of 'co-dimension': we say that an m-dimensional entity embedded in an n-dimensional space has co-dimension $n - m$. For spaces of three or fewer dimensions, this gives rise to ten different possible combinations of dimension and co-dimension. These are illustrated in Figure 2.5.

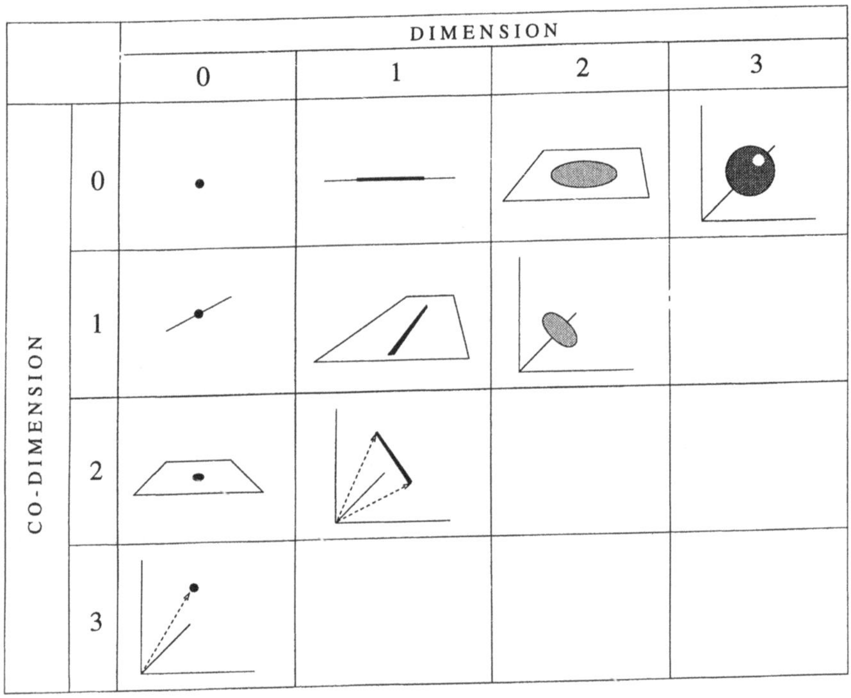

Figure 2.5. Dimension and co-dimension: the ten possibilities in space of three or fewer dimensions

Along the top row of the figure are shown regions of co-dimension zero, regions of the same dimension as the space they are embedded in: from right to left, volumes in three-dimensional space, areas in the plane, line segments on a line, and both last and least, a single point constituting the whole of its space. The embedding space need not be Euclidean: areas on the surface of the Earth have co-dimension zero too, so long as they are considered from an earth-bound perspective, as in most geographical contexts (Egenhofer and

Mark 1995). From the point of view of three-dimensional space, they have co-dimension 1, and thus belong in the second row of the table.

Here are some ways in which co-dimension is important:

- *Boundaries.* In Euclidean space, if two equidimensional regions of co-dimension zero have identical boundaries then they are either equal or complementary; but there are infinitely many distinct equidimensional regions of positive co-dimension having identical boundaries. Thus a given circle in a plane is the boundary of just two regions: the interior of the circle and the exterior of the circle. The same circle in three-dimensional space is the boundary of many regions, including for example the two portions into which the circle divides the surface of any sphere in whose surface it is located. This is important in connection with measuring the separation between regions—see §7.2.
- *Connection.* The possible modes of connection for two regions depends on their co-dimension. If both regions are of co-dimension zero then there are essentially just the eight modes of connection portrayed in Figure 2.9 (p. 77). If one or both regions have positive co-dimension then many more possibilities open up—see §2.5.4.
- *Distance and overlap.* If a region is displaced by a small amount, then the resulting region must overlap with the first if they are of co-dimension zero, but need not do so if they are of positive co-dimension. This is illustrated in Figure 2.6. This observation will be relevant to the discussion of separation measures on regions in §7.3.5.

(*a*)

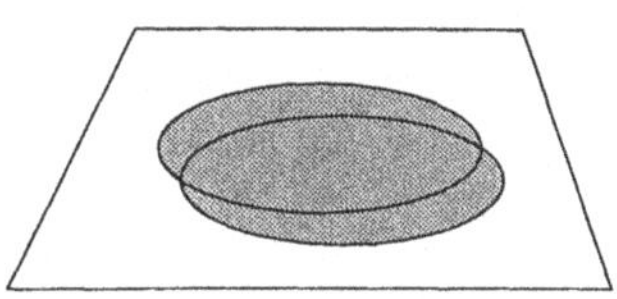

(*b*)

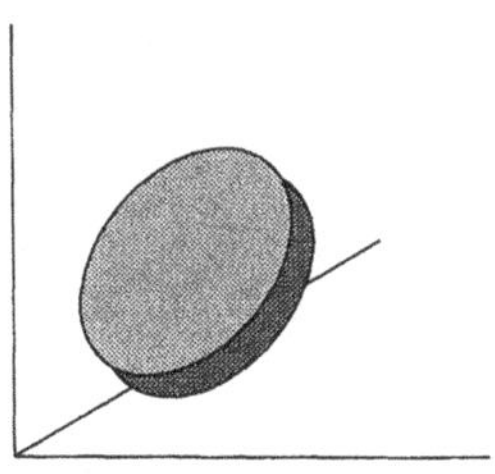

Figure 2.6. Small displacements: (*a*) A region of co-dimension zero overlaps a slightly displaced copy of itself, (*b*) a region of positive co-dimension can be slightly displaced so that it no longer overlaps its original position

How should these different dimensions be handled within a uniform topological framework? To see the problem, suppose we were to model the interior of the room as a set of points, say the interior of the cuboid specified in $\mathbb{R}^3$ by the equations:

$$0 < x < l,\ 0 < y < w,\ 0 < z < h$$

where l, w, and h are the length, width, and height of the room respectively. This is a regular open set in the usual topology on $\mathbb{R}^3$; even better, it is a polyhedron, which is the natural generalization to $\mathbb{R}^3$ of the polygons chosen by Pratt and Lemon as their models for two-dimensional spatial regions.

The walls can similarly be specified as sets of points satisfying various equations. One of the walls has the equations:

$$x = 0,\ 0 < y < w,\ 0 < z < h.$$

This is part of the topological boundary of the cuboid. In the $\mathbb{R}^3$ topology it is *not* an open set: in fact it has empty interior and is therefore about as far from being an open set as it is possible to be. It is closed, but it is not regular closed. On the other hand, it *is* a polygon. But Pratt and Lemon's polygons are regular open, and this is not.

The source of the difficulty is that point-sets cannot be regarded as open or closed when taken out of context. Only in the context of a given topological space can a set be regarded as open or closed, and the question is what is the appropriate topological space to consider. In saying that the wall specified above is not open, we are implicitly invoking the topology appropriate to the whole three-dimensional space. If we want both to accept the wall specified by the equations above as a bona fide spatial entity, and we also want all bona fide spatial entities to be, say, regular open sets, then we have to evaluate the wall in an appropriate context, a context in which it *is* regular open.

An obvious approach is to note that the wall is a part of the surface of the room, which is the boundary, in $\mathbb{R}^3$, of the point-set used to model the room. We would like to say that the wall is a regular open set in a topology defined on the point-sets contained in that surface. How do we specify that topology? There is a ready-made mathematical answer in the notion of a subspace topology.

Given a topology on a space Σ, and a subset $\Pi \subset \Sigma$, we can induce a 'subspace topology' on Π in a natural way as follows. For any point $p \in \Pi$, the neighbourhoods of p in the subspace topology are taken to be all the intersections of the form $N \cap \Pi$, where N is a neighbourhood of p in the original topology on Σ. It follows from this definition of the subspace topology that the open sets in the subspace topology on Π are simply the intersections of Π with the open sets in Σ.

Two examples are shown in Figure 2.7. The cuboid represents a region in three-dimensional space. Two points on the surface of the cuboid are shown, each with one of its neighbourhoods, an open spherical region centred on the point. The intersections of these neighbourhoods with the surface of the cuboid are shown shaded: these are neighbourhoods of the points in the subspace topology on the surface. In the case of the cuboid on the top face, the neighbourhood shown is simply a circle (appearing foreshortened in the diagram as an ellipse), since the original spherical neighbourhood only intersects the surface of the cuboid in the top face. For the point lying on an edge of the cuboid, the spherical neighbourhood intersects the surface in two semicircles, one on each of the faces adjacent to that edge. Other cases can result in still more complex regions, but they all count as neighbourhoods on the surface, and can serve as a basis for the topology on the surface.

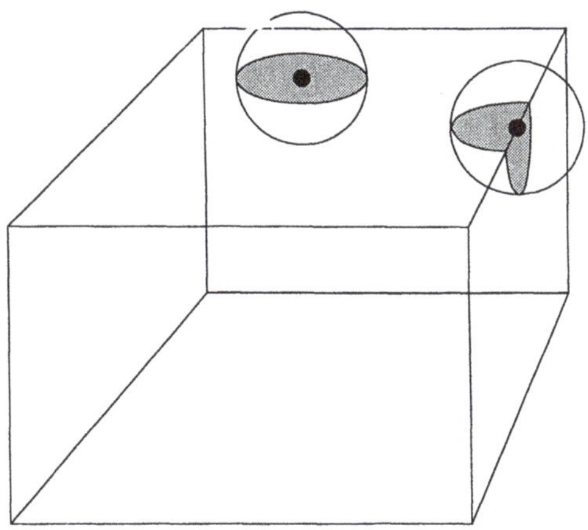

Figure 2.7. Neighbourhoods in a subspace topology

The network of edges of the cube can be regarded as a one-dimensional space, and since the second point shown in Figure 2.7 lies on an edge, it belongs to this space. The original spherical neighbourhood of the point intersects this network of edges in an open line segment, a diameter of the sphere. This line segment is a neighbourhood of the point in the subspace topology induced on the network of edges.

I am now in a position to give a uniform account of regions in different dimensions. For simplicity, regions will be modelled as regular open sets, even though, as already demonstrated, the regular open sets include some objects we might hesitate to accord the status of region to. Our fundamental space is $\mathbb{R}^3$. A three-dimensional region is a regular open set in the usual topology on this space. I shall use $\mathcal{R}_3$ to denote the set of three-dimensional regions. For any such region, its topological boundary, or surface, is a two-dimensional space, with a subspace topology. A two-dimensional region is any regular open set in the subspace topology on any of these surfaces. The set of two-dimensional regions will be denoted $\mathcal{R}_2$. The boundary of a two-dimensional region is a one-dimensional space, and it too has a subspace topology (we can regard this as induced from the three-dimensional topology or from the subspace topology on the surface: it makes no difference). A one-dimensional 'region' is any regular open set in the subspace topology on any of these one-dimensional spaces: such regions form the set $\mathcal{R}_1$. The boundary of a one-dimensional region is simply a set of isolated points; these are zero-dimensional entities, belonging to the set $\mathcal{R}_0$.

Despite the attractive simplicity of this scheme, it is not without its problems. Although the definitions of $\mathcal{R}_2$ and $\mathcal{R}_1$ may seem natural and well motivated, they have some consequences that are perhaps counter-intuitive. Consider, for example, the illustrations (*a*) and (*b*) in Figure 2.8. Both figures can be constructed out of two straight line segments, but the one in (*a*) corresponds to an element of $\mathcal{R}_1$ whereas the one in (*b*) does not. To see why this is, refer to (*d*) and (*e*) in the same figure. The two shaded squares in (*d*) represent an element of $\mathcal{R}_2$, say Z; the dotted circle represents an element of $\mathcal{R}_3$, say Y (it should be interpreted as a sphere rather than a circle, intersecting the picture-plane to which the rest of the figure is confined). The crossed lines of (*a*) are revealed as the intersection $Y \cap \partial Z$, which is in $\mathcal{R}_1$.

Turning now to (*b*), we may note that although this figure can also be regarded as a subset of ∂Z, it is not open—and hence *a fortiori* not regular open—in the corresponding subspace topology. Perhaps we have chosen the wrong Z; but it is easy to see that *no* element of $\mathcal{R}_2$ has a boundary in whose subspace topology the figure in (*b*) is an open set. The shaded area in (*e*) represents a different element of $\mathcal{R}_2$; the small dotted circle shows how any neighbourhood of its centre point must include stretches of all four boundary segments radiating from that point. In the subspace topology on ∂Z, the interior of the closure of the figure in (*b*) is missing the point at which the line segments meet. The resulting set 'looks' non-regular to our intuition.

Similar phenomena occur in three dimensions too.

Parts (*c*) and (*f*) of Figure 2.8 illustrate another oddity. The straight line segment in (*c*) is part of the boundary of both the shaded area in (*d*) and the shaded area in (*f*). However, in the subspace topology on the latter boundary, it is regular open, whereas on the subspace topology on the former, it is not, for the same sort of reason as the figure in (*b*) is not regular open in that topology either. Thus to show that a one-dimensional figure X is in $\mathcal{R}_1$, one must find an element Z of $\mathcal{R}_2$ such that X is regular open in the subspace topology on ∂Z. To show that X is *not* in $\mathcal{R}_1$ it is not enough to find a $Z \in \mathcal{R}_2$ such that X lies within, but is not regular open in, ∂Z; rather, one must show that X is not regular open in ∂Z for *any* $Z \in \mathcal{R}_2$. Similar remarks apply to the other dimensions.

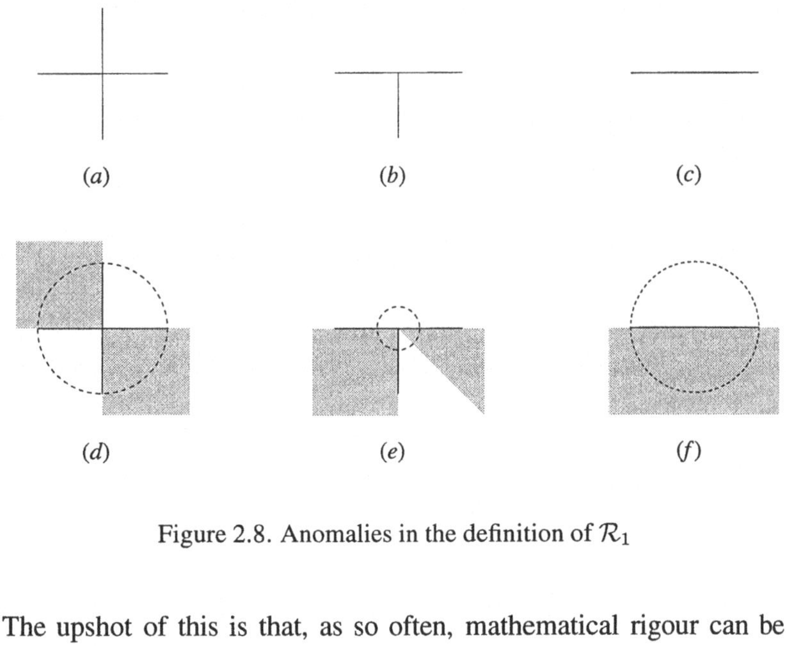

(*a*) (*b*) (*c*) (*d*) (*e*) (*f*)

Figure 2.8. Anomalies in the definition of $\mathcal{R}_1$

The upshot of this is that, as so often, mathematical rigour can be the enemy of intuition. The problems arise particularly in the case of regions of *positive co-dimension*, that is, regions of lower dimension than the space in which they are regarded as being embedded. I do not believe there is an easy solution to the problem of characterizing regions of positive co-dimension in a way that is both soundly based in topological theory and conformable to the demands of human spatial intuition; there is surely nothing final about our provisionally adopted solution.

2.4. Topology without points

At several places during the preceding sections of this chapter there has appeared, just below the surface as it were, a tension between two different kinds of construction by which space might be modelled. On the one hand we have what might be called the *bottom-up* approach, by which the primitive spatial entities are taken to be points, with more complex entities such as lines, surfaces, and solids being built up out of these. On the other hand, we have the *top-down* approach, whereby the primitive spatial entities are solids, the chunks of three-dimensional space capable of being occupied by extended material bodies; following this approach, we must regard surfaces, lines, and points as derivative entities dependent for their existence on solids.

This tension was implicit in §2.2.1, where I referred to the two kinds of idealization which might be represented by Euclidean points, lines, and surfaces. If we adopt the idealization according to which a point is an 'infinitely small dot', then we are, in effect, regarding a point as in some sense a *constituent* of space. Under this conception, the point is a positive presence, a constituent part of any line, surface, or solid within which it lies. In the pre-ideal situation where a point is, say, a little cube measuring one millimetre along each side, we can construct a volume of one cubic metre out of exactly one billion points. The smaller we make the points, the more of them we need to construct this volume. In the limit we are led to the idea of a volume constructed out of infinitely many dimensionless points. This is paradoxical, since it is difficult to see how a finite volume can be constructed out of parts which individually have no volume, however many of them there are. One would have to regard volume as an 'emergent' property, although to say this is merely to label the problem, not to solve it.

An alternative approach, seen by many as closer to human experience, is to adopt the other idealization, according to which solids come first, and surfaces, lines, and points are regarded as *interfaces* between contiguous solids. As we saw in the case of the Czech flag (Figure 2.1), there is no temptation on this view to say that any higher-dimensional entity is constructed out of entities of lower dimension. We can still say that a point lies in a line, surface, or solid, that a line lies in a surface or solid, and that a surface lies in a solid; we can also say of any of the lower-dimensional entities that it *bounds* one of the higher-dimensional ones, in other words that it forms part of or lies within its boundary. What we cannot do, on this conception, is to say that a lower-dimensional entity is a *part* of any higher-dimensional entity. Here we are assuming a particular conception of what we mean by a 'part': a part, in this sense, is a *contributory* part, something that can in principle be removed from what it is part of to leave some other part as remainder. We might still be able to characterize lines, surfaces, and solids in terms of sets of points: that is, we can uniquely specify any of these entities by specifying the set of points lying in them. But that is not to say that any of these entities *is* a set of points; just as, while we can uniquely specify a triangular area in the plane by specifying just three points, no one would want to claim that the triangular area *is* the set whose members are just those three points.

The top-down construction of space is not without problems of its own, however. It was glibly suggested above that lower-dimensional entities are interfaces between higher-dimensional ones. While we may have a clear enough intuitive understanding of what is meant by this, it is not entirely straightforward to put this idea on a logically more rigorous footing. What exactly is an interface? We cannot be given interfaces for free, along with the solids, since that would be to negate the idea of a top-down construction of space. We need somehow to *derive* the interfaces from the solids. In the next few sections I shall look at some ways in which this might be done.

2.4.1. Points constructed as sets of solids

The method most commonly proposed for constructing lower-dimensional entities from higher is to specify an entity of the former type by means of a *set* of entities of the latter type. Thus a point may be defined as a set of solids: typically the set of solids that we want to be able to speak of as *containing* the point, or some distinguished subset of this set. In the remainder of this section I shall informally review two ways that have been suggested for doing this, then in the next two sections I shall attack the problem with rather more formal precision.

De Laguna's abstractive elements

The system proposed by de Laguna (1922) posits a basic ontology of *solids*, which are to be understood as spatial regions, the spaces occupied by physical solids. Laguna borrows from Whitehead the notion of an 'abstractive set', which is a set of solids that is (*a*) nested (that is, of any two members of the set, one is contained in the other), and (*b*) such that there is no solid contained by every member of the set. The 'abstractive element' corresponding to a given abstractive set is now defined as the set of all solids containing at least one member of the set (and therefore infinitely many of them). Laguna next defines a *point* to be an abstractive element which is not a proper subset of any other abstractive element, in other words a *maximal* abstractive element. In effect, a point is identified with the set of solids that would ordinarily be said to contain the point, though the construction using abstractive elements enables this set to be characterized without reference to the point itself, thereby avoiding circularity. In a similar way, lines and surfaces (and also, it should be added, 'hybrid' objects of mixed dimension, such as a planar object with a linear spike) can be specified as various kinds of non-maximal abstractive elements.

How is this construction related to the bottom-up construction of solids from points? Suppose we regard a solid as a bounded regular open set in $\mathbb{R}^3$. Then an abstractive set is a collection of bounded regular open sets with the properties that (*a*) given any two of them, one contains the closure of the other, and (*b*) the intersection of all members of the collection does not itself contain any non-empty open set. It can be proved that, in $\mathbb{R}^3$, any such collection must have non-empty intersection, and this will have empty interior because of condition (*b*). If the abstractive element is maximal, the intersection will consist of a single point. Note that our condition (*a*) does not exactly correspond to Laguna's condition (*a*) as presented in the last paragraph. This suggests that the Laguna construction should be corrected so that (*a*) reads: of any two members of the set one is contained *together with its boundary* in the interior of the other. Thus corrected, the Laguna construction can be modelled in the point-set topology in $\mathbb{R}^3$, with satisfactory results. But the *purpose* of the construction is to show how points can be defined in terms of solids that are not themselves conceived of as constructed out of points; thus even if it had no model in point-set topology this would not in itself discredit it. The fact of its having such a model, however, may cast doubt on its usefulness, on the grounds that we might as well work within point-set topology from the start.

Tarski's spheres

Tarski (1956) takes as his primitive elements *spheres*, together with two mereological postulates regarding the part–whole relation. On this basis, in a series of steps defining various notions to do with tangency, he defines what it is for two spheres to be concentric. He is then able to define a point as 'the class of all spheres which are concentric with a given sphere' and a solid as 'an arbitrary sum of spheres'. Adjoining certain further axioms, including the restriction of solids to those for which the class of all interior points is a non-empty regular open set, he arrives at an axiom system for the 'geometry of solids'.

Both Nicod (1930) and Tarski point out that the geometry of solids can be modelled within ordinary Euclidean point-set geometry, and vice versa. It follows that, from a formal point of view, the choice between them is a matter of taste and convenience rather than one of substance. We shall meet a similar situation in Chapter 5 regarding the equivalence of instant-based and interval-based theories of time.

2.4.2. Regions as primitive entities

Any definition of lower-dimensional entities in terms of higher-dimensional ones must be premised on the idea that we can develop a theory of higher-dimensional entities—which we may as well call regions—without taking the standard mathematical route of regarding such entities as aggregates (modelled as sets) of the lowest-dimensional type of entity, namely points. This raises an obvious question: if regions are not sets of points, what are they? The author of a formal theory of regions is actually under no obligation to answer this question in a rigorous way. Regions can simply be primitive elements of the theory, specified purely in terms of the properties ascribed to them by the theory. For the rest, it is perhaps sufficient to say something about their being 'chunks of space' and leave it to the consumers of the theory to form suitable intuitions, based on their own spatial experience, of what kind of thing is meant. (Something like this is essentially what Euclid was doing, even if he did not think of it in quite this way, in his sequence of 'definitions' of points, lines, and surfaces.)

Whereas in the bottom-up approach the properties ascribed to regions arise as consequences of the properties of the primitive non-regional elements in terms of which they are defined, in the top-down approach the properties of regions have to be postulated explicitly. There is room for discussion about just what properties we postulate, just as in the other approach there is room for discussion about just how we define regions in terms of points. For regions, some of the important issues are:

- What counts as a region? In particular, should we admit the existence of a 'universal' region U, comprising the whole of space, or of a 'null' region Λ comprising no space?
- What relations between regions should we consider? Obvious choices

include 'part', 'overlap', 'connection', and of course 'equality'. We need to decide which relations are to be postulated as primitives, and how the other relations are defined in terms of them. For the primitive relations we need axioms to determine their properties (such as reflexive or irreflexive, symmetric, asymmetric or antisymmetric, transitive).

- What operations on regions should we consider? Again, there are some obvious choices, such as 'fusion', 'greatest common part', 'difference', and 'complement'. Some of these may be taken as primitive, others defined, and, as with the relations, we need axioms to establish properties such as commutativity, associativity, and idempotence, as appropriate.

Not all the choices involved in the above list are independent of one another, but, even so, there is a wide range of options, both as regards which regions, relations, and operations we admit and the properties we ascribe to these. Any definite set of choices is tantamount to selecting an abstract algebraic structure to be the formal expression of our theory of regions. Mathematics is replete with algebraic structures of many kinds (groups, rings, fields, vector spaces, and so on), and on one view the essence of computer programming is the appropriate selection and construction of algebraic structures whose properties are such as to enable them to model different aspects of the real world. We are thus working in a well-trodden area, and almost all of the structures I shall discuss below are existing mathematical theories taken 'off the shelf', their primitive terms being interpreted as referring to regions and various relations or operations on regions.

A widely used example of such a theory is Boolean algebra. A Boolean algebra is an abstract mathematical structure defined on some set of individual elements $\mathcal{R}$—in our interpretation, the elements of $\mathcal{R}$ will be understood to be regions of space. Amongst these individuals are a universal element 1, representing in our interpretation the 'universal region', covering the whole of space, and a null element 0. This latter is hard to interpret as a region: we must regard it as a convenient fiction introduced in order to simplify the technical operations of the algebra. The elements of $\mathcal{R}$ can be acted on by three primitive operations. Two of these, which we shall denote $+$ and $\cdot$, are binary operations; in the context of Boolean algebra, they are generally referred to by the names 'join' and 'meet'. When the individuals of the theory are interpreted as regions, $X + Y$ is understood as the region resulting from 'putting together' X and Y—it is a region containing both X and Y as parts, with nothing left over, their 'fusion'; and $X \cdot Y$ is interpreted as the overlap of X and Y, the greatest common part of X and Y. If X and Y have no common parts, their overlap will be the null region. The third primitive operation is complementation, which acts on a element X to produce an element $-X$ which is in a precise sense complementary to X. In the regions interpretation, $-X$ is the whole of space apart from X: what is left when X is, as it were, removed.

Table 2.1. *Axioms for a Boolean algebra*

Commutativity	$X+Y=Y+X$ $X\cdot Y=Y\cdot X$
Associativity	$X+(Y+Z)=(X+Y)+Z$ $X\cdot(Y\cdot Z)=(X\cdot Y)\cdot Z$
Distributivity	$X\cdot(Y+Z)=(X\cdot Y)+(X\cdot Z)$ $X+(Y\cdot Z)=(X+Y)\cdot(X+Z)$
Absorption	$X+(X\cdot Y)=X$ $X\cdot(X+Y)=X$
Complementation	$X+(-X)=1$ $X\cdot(-X)=0$

The Boolean operations are stipulated to obey the set of axioms listed in Table 2.1. Various other operations and relations can be defined in terms of these. An important new operation is the difference operation defined by

$$X-Y \triangleq X\cdot(-Y);$$

the difference $X-Y$ represents that part of region X which does not overlap Y. From this, in turn, we can define the symmetric difference

$$X \triangle Y \triangleq (X-Y)+(Y-X),$$

which consists of the parts of X which do not overlap Y, together with the parts of Y which do not overlap X. The symmetric difference represents the extent to which two regions differ from one another: it is null just when the two regions are equal.

Of the relations, perhaps the most important is an ordering relation $\leq$ defined on $\mathcal{R}$ by the rule:

$$X \leq Y \triangleq X\cdot Y=X.$$

(The definiens here can be replaced by the provably equivalent formula $X + Y = Y$.) In the regions interpretation, this can be understood as saying that region X is part of region Y. It is easily proved that $\leq$ is a weak partial order, that is, it is reflexive, transitive, and antisymmetric, as befits a part–whole relation.

An 'optional extra' for a Boolean algebra is to generalize the binary operations $+$ and $\cdot$ so that they can be applied to an arbitrary number of elements, and not just to two at a time. If S is a set (possibly infinite) of elements of a Boolean algebra, interpreted as a set of regions, then $\sum S$ is the fusion of all of them, while $\prod S$ is their greatest common part. In a general Boolean algebra, these operations may not be defined for all sets of elements; if they are, then the Boolean algebra is said to be *complete*.

As an abstract axiomatic theory, Boolean algebra has many different interpretations, of which the two best known are *set theory* and *propositional logic*. In the former, the individuals are interpreted as sets: say all the subsets of some universal set U. Then 0 is the empty set ($\emptyset$), and $+$, $\cdot$, and $-$ are the operations of union ($\cup$), intersection ($\cap$), and complementation respectively. The derived relation $\leq$ is interpreted as the subset relation ($\subseteq$). Under this interpretation, all the axioms are satisfied, thus verifying that set theory is indeed a model for Boolean algebra.

In the propositional logic interpretation, the individuals are *statements* which may be true or false,[4] the universal element is a statement which is always true (a tautology), the null element is a statement which is always false (a contradiction). The operations $+$, $\cdot$, and $-$ are interpreted as truth-functions of statements: disjunction ($\vee$), conjunction ($\wedge$), and negation ($\neg$), respectively; and $X \leq Y$ means that X logically implies Y (that is, the conditional $X \rightarrow Y$ is a tautology). Once again, the axioms are all satisfied, so propositional logic is a model for Boolean algebra.

For us, a particularly interesting interpretation of Boolean algebra is in terms of regular open sets in a topological space. The individuals are the regular open sets, with 1 being the universal set and 0 the empty set. The operator $\cdot$ is interpreted as intersection, just as in the Set Theory interpretation, but now $X + Y$ is the regular union $int(cl(X \cup Y))$, and $-X$ is the interior of the complement $int(U \setminus X)$. It can be verified that the axioms are satisfied by this interpretation—for details, and more information on Boolean algebras generally, see Koppelberg 1989.

[4] More exactly, they should be taken to be equivalence classes of statements under the relation of logical equivalence.

If regions are interpreted as open sets more generally, without the requirement of regularity, then we no longer have a Boolean algebra. The reason for this is that since the union of open sets is again open, we have to interpret $+$ as ordinary union rather than regular union; on the other hand we still have to interpret $-X$ as $int(U \setminus X)$, since the set-theoretic complement of an open set is not in general open. We now find that $X + (-X)$ is interpreted as $X \cup int(U \setminus X)$, which in general is not equal to U. Thus the first complementation law does not hold, so the structure we have defined is not a Boolean algebra. Instead, it belongs to a weaker class of structures known as Heyting algebras in which the complementation law $X \cdot (-X) = 0$ is satisfied but $X + (-X) = 1$ is not (Johnstone 1982; Stell and Worboys 1997).

In basing a theory of regions on Boolean algebra one is not thereby committing oneself to any particular model or class of models. I might claim that regions do in fact obey the axioms of a Boolean algebra under the operations of fusion, greatest common part, and complementation, while repudiating the idea that regions must be identified with regular open sets. I might, for example, want to ascribe other properties to regions, which are incompatible with their interpretation as regular open sets. On the other hand, if the only properties I want to ascribe to regions are just those that are expressed by the Boolean algebra, then presumably *any* model for the Boolean algebra will be an equally good model, however intuitively implausible, for the regions domain.

2.4.3. Construction of points as sets of regions

In order to define points in terms of regions, we need to have some idea of how we want the points we are going to construct to be related to the regions we are going to construct them from. A possible approach to this problem is to consider how points are related to regions in the standard mathematical approach, in which regions are constructed as sets of points. In the latter approach, regions are modelled as some kind of set: in order of increasing fussiness as to what can count as a region, we have considered general sets, open sets, and regular open sets (the latter two types presupposing, of course, the existence of a topology on the set of points). I shall refer to regions of these three kinds as S-regions, OS-regions, and ROS-regions respectively.

The first important property to consider is the following:

(a) A point is uniquely characterized by the set of regions it belongs to.

This is trivially true for S-regions, since for a given point p there is no other point belonging to the set $\{p\}$. For OS-regions, the property does not *have* to hold, but the condition for it to do so is rather weak, namely, that whenever the open sets containing point p are exactly the same as the open sets containing q, then p and q are the same point. A topology satisfies this condition so long as it has the property that of any two distinct points, at least one is in an open set not containing the other. A topology with this property is known as a T_0 topology, and almost all topologies of interest are of this kind—as is, in particular, the standard topology on Euclidean space. For ROS-regions, the condition on the topology for the property (*a*) to hold is that, of any distinct points, at least one is in a regular open set not containing the other. This certainly holds for any topology with the property that for any two distinct points we can find disjoint open sets each containing one of the points. A topology with this property is called a Hausdorff topology, and again, most topologies of interest to mathematicians are of this kind.[5]

Property (*a*) suggests that the set of regions containing a point can be taken as representative of the point itself. We may now consider some further properties that these sets of regions might be supposed to have. In the following, the term 'p-region' refers to a region containing the point p.

(*b*) The null region, if it exists, is not a p-region.

(*c*) The intersection of any two p-regions is a p-region.

(*d*) Any region having a p-region as a subset is itself a p-region.

(*e*) Of a region and its complement, exactly one is a p-region.

(*f*) If the union of a collection of regions is a p-region, then at least one member of the collection is a p-region.

For S-regions, all of these properties hold. To determine whether the properties hold for OS- and ROS-regions, we need to make clear what is to be understood by the operations of complementation and union in these cases, since the set-theoretic complement of an open set is not in general open, and the set-theoretic union of two regular open sets is not in general regular open. For OS-regions, union presents no problems, since the union of open sets is always open; but we must replace complement here by 'open complement', by which is meant the interior of the set-theoretic complement. This replacement will also work for ROS-regions (since the open complement of a regular

[5] Note, however, that later on, in §2.6.2, I shall make essential use of non-Hausdorff topologies—but this will be in a context where the present discussion has no relevance.

open set is itself regular open), but in this case we must also interpret union to mean the regular union $int(cl(S_1 \cup S_2))$ (see §2.3.2). For OS-regions, all the properties except (*e*) hold; (*e*) fails because a point on the boundary of a region is neither in the region itself nor in its open complement. This applies equally to ROS-regions, for which a further problem arises with respect to (*f*), since a point in a shared boundary segment of two regular open sets is in their regular union though not in either set on its own.

The above discussion assumed that points were given as primitive entities and regions were certain sets of points. But our main concern in this section is just the reverse of this procedure, namely to construct points as sets of regions, the latter being taken as primitive. A natural way to proceed is to take some selection from properties (*a*) to (*f*) as constraining the kinds of sets of regions that are to be identified with points.

Let us work within the framework of a Boolean algebra whose elements are regions. In this context, the properties (*a*) to (*f*) can be rephrased as follows:

(*a*) A point is a set $\mathcal{P}$ of regions.

(*b*) A point does not contain the null region: $0 \notin \mathcal{P}$.

(*c*) If two regions belong to a point, so does their greatest common part: $X \in \mathcal{P} \wedge Y \in \mathcal{P} \to X \cdot Y \in \mathcal{P}$.

(*d*) If any part of a region belongs to a point, so does the whole: $X \in \mathcal{P} \wedge X \leq Y \to Y \in \mathcal{P}$.

(*e*) Of a region and its complement, exactly one[6] belongs to any given point: $X \in \mathcal{P} \oplus -X \in \mathcal{P}$.

(*f*) If the fusion of a set of regions belongs to a point, so does at least one of those regions: $\sum \mathcal{S} \in \mathcal{P} \to \exists X \in \mathcal{S} : X \in \mathcal{P}$.

A set of regions satisfying properties (*b*)–(*d*) is called a 'filter'. If in addition it satisfies (*e*), it is called a prime filter or 'ultrafilter'. If further it satisfies (*f*), it is called a complete ultrafilter.[7] Depending on precisely which selection of

[6] In the presence of (*b*) and (*c*) this could be weakened to 'at least one', since $X \in \mathcal{P} \wedge -X \in \mathcal{P}$ would imply, by (*c*), that $0 \in \mathcal{P}$, contradicting (*b*).

[7] In this case, we no longer have to state (*e*) explicitly, since it is implied by (*d*) and (*f*): by (*d*), $X + (-X) = 1 \in \mathcal{P}$, so by (*f*), either $X \in \mathcal{P}$ or $-X \in \mathcal{P}$. This does not contradict the earlier observation that for OS-regions we have (*f*) but not (*e*), since, as we have seen, the OS-regions do not, in fact, form a Boolean algebra.

properties we choose out of *(b)*–*(f)*, we will obtain rather different conceptions of points. For a given universe of regions $\mathcal{R}$, I shall write $Points(\mathcal{R})$ to denote the set of points constructible from the regions of $\mathcal{R}$ under the particular conception we have chosen,

It is natural to consider whether, having constructed points out of regions, we can then go on to characterize a region in terms of an associated set of points, thereby coming round full circle back to the point-set perspective. Conversely, if we start with a model for regions as point-sets, and then construct new 'points' as sets of regions, will the 'points' constructed from regions correspond to the points out of which the regions themselves were originally constructed?

For convenience, let us denote by $PS(X)$ the set of points to which the region X belongs:

$$PS(X) \triangleq \{\mathcal{P} \in Points(\mathcal{R}) \mid X \in \mathcal{P}\}.$$

Since $PS(X)$ is a set of points, it is therefore a *set of sets of regions.*

If we assume *(b)*–*(e)* above, we can straightforwardly prove the following correspondences between Boolean operations on regions and set-theoretic operations on point-sets:

- $PS(0) = \emptyset$.
- $PS(1) = Points(\mathcal{R})$.
- If $X \leq Y$ then $PS(X) \subseteq PS(Y)$.
- $PS(X \cdot Y) = PS(X) \cap PS(Y)$.
- $PS(X + Y) = PS(X) \cup PS(Y)$.
- $PS(-X) = Points(\mathcal{R}) \setminus PS(X)$.

What is less straightforward to prove is that if $PS(X) = PS(Y)$ then $X = Y$. It *is* straightforward to prove that $PS(X) = PS(Y)$ if and only if $PS(X \bigtriangleup Y) = \emptyset$. From this it can be seen that in order for every point-set to characterize a unique region, it is sufficient that every non-null region belong to at least one point; then the symmetric difference of any two distinct regions will belong to at least one point, so the point-sets of the two regions will be distinct. What we need to prove, then, is that any non-null region X belongs to at least one set of regions satisfying *(b)*–*(e)* above.

It is easy enough to produce a set of regions, containing X, and satisfying (*b*)–(*d*) (a filter); the following set will do:

$$F(X) = \{Y \in \mathcal{R} \mid X \leq Y\}.$$

We now need to extend this filter to an ultrafilter, in other words a set of regions satisfying (*b*)–(*e*). If we assume that $\mathcal{R}$ is denumerable (which means that it can be put into exhaustive one-to-one correspondence with the set of natural numbers), then this can be done as follows: let the regions in $\mathcal{R}$ be $R_0, R_1, R_2, R_3, R_4, \ldots$. We shall generate a nested sequence of filters $F_0 \subseteq F_1 \subseteq F_2 \subseteq F_3 \subseteq \cdots$ as follows. Let F_0 be $F(X)$ as given above. Then for each i, if $-R_i$ is a member of F_i, let F_{i+1} be equal to F_i; otherwise, let F_{i+1} be the smallest filter containing both F_i and R_i. In either case we have $F_i \subseteq F_{i+1}$ and $0 \notin F_{i+1}$, and the union of all the F_i is now the ultrafilter we require. It must be an ultrafilter since for any region R_i, either it or its complement (but not both) gets to be in F_{i+1}; and it extends $F(X)$ since this was the first element of the sequence.

Essentially the same construction works when $\mathcal{R}$ is non-denumerable, but in this case it depends on an important principle of proof called the *Axiom of Choice*. This is a principle that is frequently invoked in mathematical proofs, but which cannot itself be derived from the basic axioms of set theory on which such proofs typically depend. Instead, it has to be assumed as a primitive axiom in its own right; its position with respect to the other axioms of set theory is somewhat analogous to the position of Euclid's parallel postulate with respect to the other axioms of geometry. In both cases, the additional principle seems intuitively much less obvious than the others, yet it cannot be derived from them. The Axiom of Choice states that, given any collection of sets, one may always assume the existence of a further set which contains one member from each set in the original collection. The special case in which our collection contains finitely many finite sets *can* be proved from the other axioms, but the more interesting cases all involve infinite sets, and that is where the other axioms prove inadequate and our intuitions become much less reliable. I omit the details here, but in brief, the proof we require generalizes the inductive construction described above to a *transfinite induction*, the possibility of which depends on a principle called the *Well-ordering Principle* which is provably equivalent to the Axiom of Choice.

Given that every non-null region belongs to some point, it then follows that a region is uniquely characterized by the set of points to which it belongs, and, moreover, the Boolean operations on regions correspond in a systematic way to the set-theoretic operations on their point-sets. This kind of result can be taken in two diametrically opposed ways. On the one hand, it can be taken as legitimizing the approach in which regions are taken as primitive elements of a Boolean algebra, by showing that the standard set-theoretical type of model for space can be generated from it; on the other hand, it can be taken as showing that, in spite of any philosophical qualms one might have about founding a theory of regions (which are more or less tangible entities, as the possible positions of actual objects) on a theory of points (which are a step further removed from actuality), from a technical point of view one might as well do so if it enables one to exploit the vast body of existing theory in relation to point-sets rather than embarking on the comparatively little-known path of a regions-only approach.

2.5. Mereotopology

Mereotopology refers to those systems which combine the disciplines of mereology and topology. Mereology is the theory of parts and wholes. In some ways it can be likened to 'Boolean algebra without a null element'. Typically it is presented as an axiomatic theory, based on a single primitive relation meaning 'X is part of Y'. The entities whose parts form the subject-matter of mereology may be very various, from material bodies, to abstract entities such as texts, to regions of space. Different sets of axioms may be appropriate for different types of entities. In this chapter I shall be concerned with mereology only in so far as it is applied to the description of spatial regions. Mereology originated in the work of the Polish logician Stanislaw Leśniewski (1886–1939); the mereological 'Calculus of Individuals' of Leonard and Goodman has also been influential (Leonard and Goodman 1940; Goodman 1951). A useful detailed survey of the field is Simons 1987; see also chapter 3 of Casati and Varzi 1999. Mereology cannot in itself provide a basis for topology. In particular, the notion of 'connection', which is the key topological notion for the qualitative description of space, cannot be defined in terms of the part–whole relation alone.[8] This is why topological notions have to be added to mereology to provide an adequate qualitative theory of space. As we shall see below, it turns out that in some cases mere-

[8] This does not contradict the fact, mentioned in §2.4.1, that Tarski was able to define internal and external tangency for spheres purely in terms of the proper-part relation; here he was tacitly exploiting special properties possessed by spheres but not by arbitrary regions. These properties are themselves metric in character, and hence can support topological notions.

ology can be subsumed within topology, since for these cases the part–whole relation can be defined in terms of connection.

2.5.1. Classical extensional mereology

Mereology uses an ontology of unspecified 'individuals' with a single primitive relation 'X is a proper part of Y'. This relation is specified to be irreflexive and transitive, and hence also asymmetric. It is sometimes more convenient to use the reflexive closure of this relation, 'X is a part of Y', which each individual bears to itself as well as to any individual it is a proper part of; this relation resembles the relation $\leq$ of Boolean algebra, but cannot be simply identified with it since we are working in a different system. Mereology is *not* a Boolean algebra, since it has no place for a null individual.

We can use the part relation to define a further relation 'X overlaps Y', meaning that there is an individual Z such that Z is a part of both X and Y.[9] I shall use the following notations:

$x \sqsubset y$	x is a proper part of y
$x \sqsubseteq y$	x is a part of y
$x \circ y$	x overlaps y

Part and overlap are defined in terms of proper part by the formulae:

$$x \sqsubseteq y \quad \triangleq \quad x \sqsubset y \vee x = y,$$
$$x \circ y \quad \triangleq \quad \exists z(z \sqsubseteq x \wedge z \sqsubseteq y).$$

What Simons calls Classical Extensional Mereology makes use of four axioms in addition to those of first-order logic, namely

- The proper-part relation is asymmetric:

Ass $\qquad x \sqsubset y \to \neg(y \sqsubset x).$

- The proper-part relation is transitive:

Trans $\qquad x \sqsubset y \wedge y \sqsubset z \to x \sqsubset z$

[9] In the Calculus of Individuals (Leonard and Goodman 1940), 'overlap' is taken as primitive, and 'part' defined in terms of it as follows:

$$x \sqsubseteq y \triangleq \forall z(z \circ x \to z \circ y).$$

- The *Weak Supplementation Principle*. If x is a proper part of y then y has a proper part which does not overlap x:

  WSP $$x \sqsubset y \rightarrow \exists z(z \sqsubset y \wedge \neg(z \circ x))$$

- The *General Sum Principle* (here Φ is an arbitrary predicate, so this axiom is actually an axiom schema):

  GSP $$(\exists x \Phi(x)) \rightarrow \exists x \forall y(y \circ x \leftrightarrow \exists z(\Phi(z) \wedge y \circ z))$$

Simons (1987) lists seventy theorems of Classical Extensional Mereology, preceded by a detailed analysis of the significance of the more important ones and various subsystems that they generate.

The significance of the General Sum Principle is as follows. It tells us that for any identifiable class of individuals (identified as the individuals which satisfy the predicate Φ), there is an individual which overlaps all and only those individuals which overlap some member of the class. This new individual is the 'mereological sum', or 'fusion' of the individuals in the class. Because of the '$\leftrightarrow$' connective appearing in **GSP**, this fusion is unique for a given Φ, and hence we may denote it $\bigsqcup_x \Phi(x)$. With this notation, **GSP** can be written as

$$y \circ \bigsqcup_x \Phi(x) \leftrightarrow \exists z(\Phi(z) \wedge y \circ z),$$

with the proviso that this is only well formed if $\exists x \Phi(x)$. (Thus the null class has no fusion, so there is no 'null' individual.)

Some important applications of **GSP** are as follows:

- Given individuals y, z, define $\Phi(\xi)$ to be the predicate $\xi = y \vee \xi = z$. Then $\exists x \Phi(x)$ (since $\Phi(y)$ and $\Phi(z)$), so we can form the fusion $\bigsqcup_x \Phi x$. This is the 'binary fusion' of y and z, denoted $y \sqcup z$. Its defining property can be written as

  $$x \circ (y \sqcup z) \leftrightarrow x \circ y \vee x \circ z.$$

  Thus, to overlap $y \sqcup z$ is tantamount to overlapping at least one of y and z; so $y \sqcup z$ is an individual composed of y and z and nothing else.

- Given individuals y, z, define $\Phi(\xi)$ to be the predicate $\xi \sqsubseteq y \wedge \neg(\xi \circ z)$. Suppose $\exists x \Phi(x)$, so that $\exists x(x \sqsubseteq y \wedge \neg(x \circ z))$. By **WSP** this will certainly be true if $z \sqsubset y$. (In fact it can be proved more generally that it will be true under the weaker condition that $\neg(y \sqsubseteq z)$: if y is not part of z then it has a part which does not overlap z.) Then we can form the fusion $\bigsqcup_x \Phi(x)$. This is the mereological difference, denoted $y - z$, with the defining property

$$x \circ (y - z) \leftrightarrow \exists u(u \sqsubseteq y \wedge \neg(u \circ z) \wedge x \circ u).$$

Thus to overlap $y - z$ is to overlap part of y which does not overlap z; so $y - z$ is the individual obtained by cutting away from y any part which it shares with z.

- Now let $\Phi(\xi)$ be the predicate $\forall x(\Psi(x) \rightarrow \xi \sqsubseteq x)$. Thus $\Phi(y)$ means that y is part of every individual satisfying Ψ. So $\exists y \Phi(y)$ means that all the individuals satisfying Ψ have a common part. The fusion of all such common parts is $\bigsqcup_y \Phi(y)$. This is called the mereological product of the class of Ψ individuals, and is denoted $\bigsqcap_x \Psi(x)$. Its defining property is

$$u \circ \bigsqcap\nolimits_x \Psi(x) \leftrightarrow \exists y(u \circ y \wedge \forall z(\Psi z \rightarrow y \sqsubseteq z)),$$

in other words, u overlaps the product of the Ψ individuals just so long as it overlaps some common part of all the Ψ individuals. In fact one can prove that the mereological product also has the following property:

$$y \sqsubseteq \bigsqcap\nolimits_x \Psi(x) \leftrightarrow \forall z(\Psi(z) \rightarrow y \sqsubseteq z),$$

in other words, y is part of the product of the Ψ individuals if and only if it is part of every Ψ individual.

A special case of the mereological product is the binary product, denoted $y \sqcap z$, defined as $\bigsqcap_x (x = y \vee x = z)$, which satisfies the rule

$$x \sqsubseteq (y \sqcap z) \leftrightarrow x \sqsubseteq y \wedge x \sqsubseteq z.$$

The binary fusion and binary product are strongly analogous to the join and meet operators of Boolean algebra. There are important differences, however, owing to the fact that the product is not defined if the two regions do not overlap. This is a reflection of the fact that there is no null individual in the mereology. An unfortunate result of this is that many of the theorems of Classical Extensional Mereology, which are strongly analogous to those of Boolean algebra, are considerably more complicated to state than their analogues, simply because they have to be broken down into special cases according as this or that binary product does or does not exist. For example, a theorem of Boolean algebra states that

$$X - (Y - Z) = (X - Y) + (X \cdot Z).$$

The corresponding result in Mereology is:

If $\neg(x \sqsubseteq y)$ and $\neg(y \sqsubseteq z)$ then, if $x \circ z$,

$$x - (y - z) = (x - y) \sqcup (x \sqcap z)$$

but otherwise,

$$x - (y - z) = x - y.$$

If $x \sqsubseteq y$ and $\neg(y \sqsubseteq z)$ and $x \circ z$ then

$$x - (y - z) = x \sqcap z.$$

In all other cases, $x - (y - z)$ is undefined.

Philosophical qualms about admitting null individuals clearly come with a heavy price in terms of technical complications.

I have referred to the elements of mereology as 'individuals', but in the present context it is natural to consider these as regions. One of the objections that has been levelled against extensional mereology as a general calculus of individuals is precisely that it is extensional, so that individuals with exactly the same parts are the same individual. It is sometimes felt that this is not realistic for some kinds of individuals. For spatial regions, however, this objection does not seem to carry much weight, and for that reason extensional mereology seems to be an appropriate tool for handling regions if one does not wish to take the standard mathematical route of treating regions as sets of points, and is unwilling to admit a null spatial region into the ontology.[10]

2.5.2. Connection calculi

The part–whole relation, whether modelled formally by means of a Boolean algebra, mereology, or some other structure, is in itself insufficient to handle topological notions such as connection and adjacency which form an important part in everyday spatial reasoning.

[10] See §3.2.2 for another application of extensional mereology.

Regions which overlap must certainly be regarded as connected; the problematic case is that of regions which are connected but do not overlap. In point-set topology, this may or may not occur, depending on how we define overlap and connection. Overlap between X and Y could be understood to mean non-null intersection ($X \cap Y \neq \emptyset$) or it could mean intersection with an interior ($int(X \cap Y) \neq \emptyset$). These definitions coincide if X and Y are open. Connection between X and Y could mean (as in the standard topological definition) that any open supersets of X and Y have non-empty intersection; or it could mean that the closures of X and Y have non-empty intersection. If all regions are open, only the second of these two definitions allows there to be connected but non-overlapping regions.

If we adopt a 'pointless' approach, we cannot define connection in terms of point-sets. Indeed, the most usual procedure is not to *define* connection at all, but rather to take it as a primitive notion of the theory. This approach was taken by Clarke (1981), who proposed connection as an alternative to parthood to act as a primitive basis for a theory of individuals. Theories of this kind are known as 'connection calculi'. Clarke's original theory has been modified and further developed in a series of papers originating from the Artificial Intelligence research group at Leeds University, presided over by A. G. Cohn (Randell *et al.* 1992; Randell and Cohn 1992; Cui *et al.* 1992; Cohn 1997; Cohn *et al.* 1998); a rather different theory, also deriving from the work of Clarke, is that of Asher and Vieu (1995). I shall consider these two approaches in turn.

The RCC theory

The form of connection calculus developed by the Leeds group is known as the RCC theory, where 'RCC' can be taken as standing for either Randell, Cui, and Cohn, the authors of the original paper in which it was presented, or Regional Connection Calculus.[11]

In RCC, we write $C(x, y)$ to mean that region x is connected to region y. The relation C is stipulated to be both reflexive and symmetric. We can then define a part relation P in terms of C as follows:

$$P(x, y) \triangleq \forall z(C(z, x) \rightarrow C(z, y)).$$

In other words, x is a part of y just so long as anything connected to x is connected to y. The intended intuitive meaning of $P(x, y)$ is the same as

[11] The first use of the designation 'RCC' is, I think, in Galton 1993.

that of our earlier notation $x \sqsubseteq y$; we use the former notation in pure mereological contexts where this relation is understood as defined in terms of the primitive $\sqsubset$, and the RCC notation $P(x, y)$ in contexts such as RCC where it is understood as defined in terms of the primitive C. As Cohn himself has pointed out, this definition of P leads to a problem if space is discrete, since according to RCC any region connected to an atomic region is connected to the complement of that region, which would make an atomic region part of its complement. Thus there is a presumption here that space is not discrete.

In terms of P we can define RCC analogues of the mereological relations $x \sqsubset y$ and $x \circ y$ as follows. For proper part, we put

$$PP(x,y) \quad \triangleq \quad P(x,y) \wedge \neg P(y,x),$$

and for overlap we put

$$O(x,y) \quad \triangleq \quad \exists z(P(z,x) \wedge P(z,y)).$$

Using connection as a primitive, the RCC theory defines a set of eight relations such that any two regions of the same dimension and co-dimension must stand in exactly one of the eight relations. Six of these are

- x is disconnected from y:

$$DC(x,y) \quad \triangleq \quad \neg C(x,y).$$

- x is externally connected to y:

$$EC(x,y) \quad \triangleq \quad C(x,y) \wedge \neg O(x,y).$$

- x partially overlaps y:

$$PO(x,y) \quad \triangleq \quad O(x,y) \wedge \neg P(x,y) \wedge \neg P(y,x).$$

- x coincides with (or equals) y:

$$EQ(x,y) \quad \triangleq \quad P(x,y) \wedge P(y,x).$$

- x is a tangential proper part of y:

$$TPP(x,y) \quad \triangleq \quad PP(x,y) \wedge \exists z(EC(z,x) \wedge EC(z,y)).$$

- x is a non-tangential proper part of y:

$$NTPP(x,y) \triangleq PP(x,y) \wedge \neg TPP(x,y).$$

The remaining two relations are the inverses of TPP and $NTPP$. All eight relations are illustrated using circular regions in Figure 2.9.

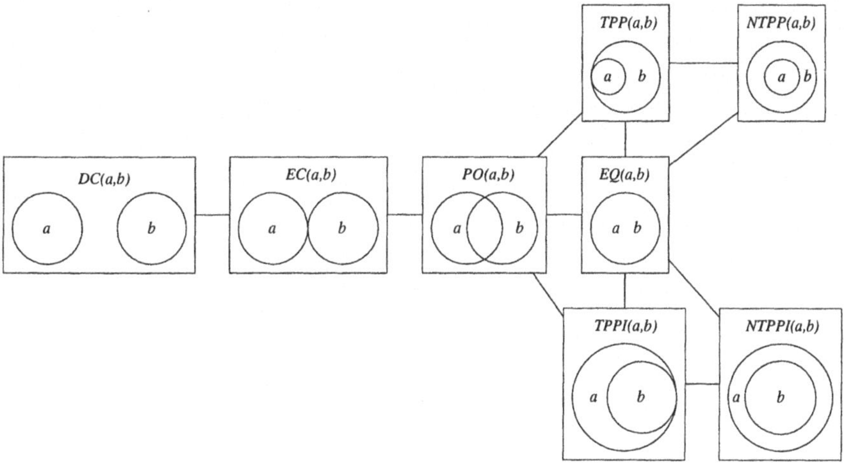

Figure 2.9. The eight irreducible region–region relations

An important feature of the RCC approach is the use of a composition table for computing logical consequences of pairs of relations. Given, for example, that x is externally connected to y, and y contains z as a non-tangential proper part, what is the relation between x and z? The answer is that they are disconnected, and this fact is duly noted as one entry in the table. In many cases things are rather less simple than this: for example, if x partially overlaps y, which itself is a non-tangential proper part of z, then the relation in which x stands to z may be any of PO, TPP, or $NTPP$. The corresponding entry in the table is therefore a disjunction of these three relations. In some cases things are even worse, an extreme example being that if x is disconnected from y, which is in turn disconnected from z, then the relation between x and z can be *any* of the eight RCC relations.

The idea of such a composition table derives from Allen's work on temporal intervals, which I shall discuss below in §5.3.3. Similarly, the important idea of conceptual neighbourhood, which I discuss next, was introduced by Freksa (1992*a*) in connection with Allen's work on intervals rather than the RCC work on regions. The idea is to consider the way in which the RCC relation between two regions varies as the regions themselves are subject to continuous change. Figure 2.10 shows a small circle approaching a larger circle from outside and entering it. The sequence of RCC relations realized by this chain of events is *EC-DC-PO-TPP-NTPP*. We say that *EC* is a 'conceptual neighbour' of *DC*, *DC* of *PO*, and so on. On the other hand, *DC* is *not* a conceptual neighbour of *PO* since it is not possible under continuous motion or deformation of the regions for a situation in which two regions are *DC* to be transformed directly into one in which they are *PO*—the only way to effect the transformation is by way of *EC*.[12] The organization of Figure 2.9 can now be explained: two RCC relations are joined by a line just in case they are conceptual neighbours of each other. The figure may thus be described as a *conceptual neighbourhood diagram*; the abstract space of RCC relations which it portrays is a mode space in the sense introduced in §1.4. Here we may simply note the conceptual neighbourhood links: later I shall derive them from first principles in the context of a theory of qualitative continuity (see §8.6.1).

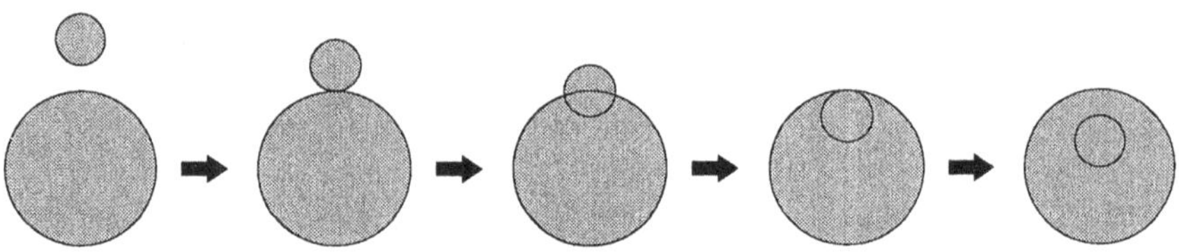

Figure 2.10. Continuous transformation of the RCC relation between two regions

Other relations between regions can be expressed as disjunctions of the eight RCC relations as follows:

[12] A possible counter-example which has been much discussed by the RCC group is when a two-part scattered region x has one part inside y, and the other part outside y. If the inside part dwindles continuously to a point and then disappears, we have *PO* transformed into *DC* with no intervening instance of *EC*. The question is whether this kind of change, in which a component of a region disappears, is to count as continuous. This topic is discussed in detail in Chapter 7.

- x is a proper part of y:

$$PP(x, y) \equiv TPP(x, y) \vee NTPP(x, y).$$

The inverse $PPI(x, y)$ is equivalent to $PP(y, x)$.

- x is a part of y:

$$P(x, y) \equiv PP(x, y) \vee EQ(x, y).$$

The inverse $PI(x, y)$ is equivalent to $P(y, x)$.

- x overlaps y:

$$O(x, y) \equiv PO(x, y) \vee P(x, y) \vee PI(x, y).$$

- x is connected to y:

$$C(x, y) \equiv O(x, y) \vee EC(x, y).$$

- x is disjoint from y:

$$DR(x, y) \equiv EC(x, y) \vee DC(x, y).$$

This is equivalent to $\neg O(x, y)$.

These relations are summarized in Figure 2.11. The same information is portrayed in the form of a lattice in Randell *et al.* 1992.

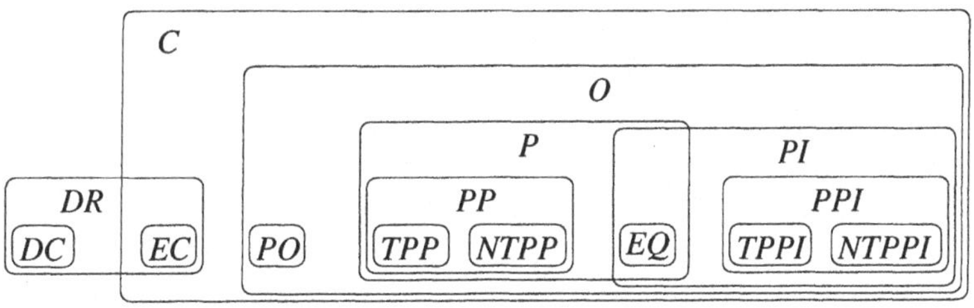

Figure 2.11. RCC relations and their disjunctions

Operations of sum, product, difference, and complementation have been introduced into the RCC theory and defined indirectly by means of axioms such as

$$C(z, sum(x, y)) \leftrightarrow C(z, x) \vee C(z, y),$$

which says that the regions connected to $sum(x, y)$ are precisely those regions which are connected to at least one of x and y. In general, the RCC way of defining an operation on regions is to give necessary and sufficient conditions for an arbitrary region to be connected to the region resulting from the operation. As Bennett (1998) points out, this style of definition will not characterize the defined operation uniquely unless connection is extensional, so that if the regions connected to x are exactly the regions connected to y then x and y must be the same region.

Of the other axioms of this type the most problematic is that for complementation,

$$[C(y, compl(x)) \leftrightarrow \neg NTPP(y, x)] \wedge [O(y, compl(x)) \leftrightarrow \neg P(y, x)].$$

As Bennett remarks, each conjunct of this axiom looks as though it alone ought to be sufficient to characterize complementation exactly, yet all his efforts to prove their equivalence (in the context of the other definitions and axioms), even when aided by a sophisticated automated theorem prover, resulted in failure. He also failed to prove, despite considerable effort, that an arbitrary region is the complement of its own complement! In the light of these observations it appears that the addition of 'Boolean' type operations to RCC theory cannot be assumed to be unproblematic. A consequence of the RCC definition of complementation is that every region is connected to its own complement, which is why an atomic region, if such were to exist, would have to be *part* of its complement. As noted earlier, this means that RCC cannot be modelled using a discrete space.

Although the standard presentation of RCC, as here, is framed in a first-order logical language, it has been shown by Bennett (1998) that essentially equivalent theories can be presented in zero-order form, that is, using a logical language lacking individual variables or constants. Regions are now represented as proposition-like entities, and the logic of regions is essentially a form of propositional logic. Bennett considers in turn Classical Propositional Logic, which is only adequate for representing the pure mereological part of the theory (exploiting the correspondence between elementary set theory and propositional logic arising from the fact that both can be viewed as models for Boolean algebra); Modal Propositional Logic, which is able additionally to express topological properties as a result of a formal correspondence between the modal necessity operator and the interior operator of topology; and finally Intuitionistic Propositional Logic, which exploits an interpretation originally due to Tarski which makes use of a topological interior operator.

It will not have escaped the reader's attention that we now have *three* different sets of notations for a number of spatial relations and operations, presented above in the respective contexts of Boolean algebra, Mereology, and the Connection Calculus. For convenience the correspondences between these notations are tabulated in Table 2.2. It must be emphasized that the different notations are not straightforwardly equivalent to each other, since they are defined using different primitive bases. They represent alternative, if closely similar, approaches to the problem of systematizing certain qualitative aspects of relationships amongst spatial regions.

Table 2.2. *Notations for spatial relations and operations on regions*

Relation	Boolean algebra	Mereology	RCC
Part	$\leq$	$\sqsubseteq$	P
Proper part	$<$	$\sqsubset$	PP
Overlap		$\circ$	O
Sum	$+$	$\sqcup$	sum
Difference	$-$	$-$	diff
Product	$\cdot$	$\sqcap$	prod

The theory of Asher and Vieu

Another mereotopological system owing its origins to Clarke (1981) is due to Asher and Vieu (1995), who state that their aim is to 'develop the foundations of a common-sense geometry'. A particular feature of the latter which they wish to capture in their system is a distinction between connection proper and a relation they call 'weak contact'. In their own words,

> Examples of weak contact are the relation between a glass and the table on which it is standing and the relation between the glass and the wine it contains, while an example of external connection or 'strong contact' is the relation between the stem of the glass and the cup of the glass.
>
> (Asher and Vieu 1995)

In order to capture this distinction they found it necessary to import into their mereotopology the notions of open and closed sets from point-set topology, in contradistinction to the RCC theory which has no place for these notions. Closed regions are used to model the spatial extent of material objects, open regions for their complements (the space outside the bodies). Disconnected regions X and Y can then be said to be in weak contact so long as Y is connected to the closure of any open region containing X.

Amongst the axioms proposed by Asher and Vieu is one which states that for any region there is a *minimal* open region containing it, a 'smallest neighbourhood' of that region. This means that in contrast to RCC, space is not allowed to be dense—more exactly, it cannot in general be asserted that if the open region X is part of the open region Y then there is an open region Z which contains X and is contained in Y and is distinct from both. Despite this, they use the same definition of parthood (P) in terms of connection (C), which in the case of RCC gave rise to the requirement that space be dense. The explanation for the apparent discrepancy is that for Asher and Vieu a region is not connected to its complement, whereas for RCC it is.

The theory of Asher and Vieu has been developed further by Philippe Muller, who integrates the time dimension into the theory in order to provide a suitable basis for a theory of movement (Muller 1998*a*; Muller 1998*b*).

2.5.3. Models for connection calculi

The obvious framework within which to interpret the RCC theory is point-set topology. Within this framework there are, as already indicated, various ways in which the connection relation C might be interpreted. Here I discuss two possibilities, which are appropriate for the RCC theory and the theory of Asher and Vieu respectively.

For the first interpretation, which is appropriate for RCC, I shall assume that regions are represented by regular open sets (without committing myself to regarding every regular open set as representing a region), and say that regions are to count as connected so long as their closures have at least one point in common:

$$C(x,y) \triangleq cl(X) \cap cl(Y) \neq \emptyset,$$

where X and Y are the regular open sets representing the regions x and y. Since all the other RCC relations are defined in terms of connection, the interpretations of these are now fixed. By reasoning within topology it can be shown that the interpretations of region–region relations in this model are as shown in Table 2.3. Similar interpretations are given by Gotts (1996), with the difference that he imposes certain restrictions on the topology in order to ensure that the axioms for sum, diff, and prod, which are regarded as an integral part of the RCC system, are satisfied.

Table 2.3. *Topological interpretation of the RCC relations*

Region–region relation	Topological interpretation
$P(x,y)$	$X \subseteq Y$
$PP(x,y)$	$X \subset Y$
$O(x,y)$	$X \cap Y \neq \emptyset$
$DC(x,y)$	$cl(X) \cap cl(Y) = \emptyset$
$EC(x,y)$	$\partial X \cap \partial Y \neq \emptyset \wedge X \cap Y = \emptyset$
$PO(x,y)$	$X \cap Y \neq \emptyset \wedge X \not\subseteq Y \wedge Y \not\subseteq X$
$EQ(x,y)$	$X = Y$
$TPP(x,y)$	$X \subset Y \wedge \partial X \cap \partial Y \neq \emptyset$
$NTPP(x,y)$	$X \subset Y \wedge \partial X \cap \partial Y = \emptyset$

In the second interpretation, used by Asher and Vieu (1995), regions are interpreted as regular open or closed sets, and the connection relation $C(x,y)$ is interpreted as non-empty intersection, $X \cap Y \neq \emptyset$, where X and Y are the point-sets corresponding to regions x and y. The relation $EC(x,y)$ will hold when $X \cap Y$ is non-empty but has no regular open subset: for this to be the case, X and Y must be closed. This is to be distinguished from 'weak contact' (see above), $WCont(x,y)$, which holds when $X \cap Y$ is empty but $cl(Z) \cap Y$ is non-empty for any open region Z containing X. Asher and Vieu's system has the advantage over RCC that their axiomatization is provably both sound and complete with respect to the given interpretation.

An alternative form of connection calculus is considered in the work of Pratt and Lemon (1997). Here, in keeping with the normal practice in mathematics, the predicate C is unary rather than binary, a formula such as $C(x)$ being understood to say that the region x is self-connected, that is, any two regions of which it is the sum must be connected in the binary sense.[13] Pratt and Lemon aim to show that in a certain sense point-set models of connection are the simplest that can be obtained, and that the drive to distance oneself from point-set models of space is therefore misconceived.

Pratt and Lemon use the techniques of mathematical logic to investigate the set of possible models for a theory of connection. They consider a first-order language $\mathcal{L}$ containing a one-place predicate 'C' (for 'is connected'), constant symbols '0' and '1' to represent the null and universal regions respectively, a one-place function '$-$' (for complementation), and two-place

[13] This equivalence must be handled with care: in particular, if we use the first topological interpretation of binary connection above then the corresponding notion of unary connection is *not* the standard topological one, but is strictly weaker.

functions '+' and '·' (for join and meet respectively). They then propose a particular interpretation $\mathcal{Q}$ of this language over the domain of polygons in $\mathbb{Q}^2$ (as described briefly earlier in §2.3.2). The set of formulae of $\mathcal{L}$ which come out true under this interpretation constitutes a formal theory which Pratt and Lemon call 'computer mereotopology', claiming it to be the real intended target of connection calculi and related enterprises. What they prove is that the interpretation $\mathcal{Q}$ is minimal in the sense that any other model for computer mereotopology must contain $\mathcal{Q}$ as an elementary submodel.

Pratt and Schoop (1997) show that computer mereotopology can be completely axiomatized so long as one is prepared to allow an infinitary rule of inference which permits one to infer that every region has a certain property if it can be shown, for every positive integer n, that any sum of n connected regions has that property. The necessity for including such rules is a reflection of the fact that computer mereotopology is in fact undecidable, as shown for a closely related system by Dornheim (1998).

The exact relationships amongst the various forms of connection calculi considered here have yet to be elucidated, although it would appear that apparently small differences in the detailed formulation of such calculi can lead to significant differences in their characters as logical theories. A useful survey of the range of different varieties of connection calculus is given by Cohn and Varzi (1998), and further discussion can be found in Casati and Varzi (1999: ch. 4).

Although we can model these connection calculi in point-set topology, some researchers have questioned the desirability of doing so, given that the spirit of mereotopology in general, and connection calculi in particular, is precisely to break away from a dependence on point-set models of space. For that reason it might be worthwhile looking for alternative mathematical models which do not represent regions as sets of points. Very little work has been done on this, although we can refer the reader to Stell and Worboys 1997, which does provide an algebraic model for RCC in terms of Heyting algebras.

2.5.4. Connection beyond co-dimension zero

The connection calculus makes the tacit assumption that the regions of interest are all of co-dimension zero. It can be applied equally well to volumes in space, areas in a surface, or line segments along a line. When we turn to regions of positive co-dimension, we find that there are more than just the eight RCC relations described above.

The topological interpretations for the RCC relations can be extended in various ways to provide a more comprehensive taxonomy for handling the more general case of regions of positive co-dimension. A useful matrix-based formalism has been devised by Max Egenhofer, which is well adapted to serve this purpose (Egenhofer 1989; Egenhofer 1991).

Egenhofer's '4-intersection' matrix

Egenhofer defines the 4-intersection of two regions X and Y to be the matrix

$$\begin{pmatrix} \partial X \cap \partial Y & \partial X \cap int(Y) \\ int(X) \cap \partial Y & int(X) \cap int(Y) \end{pmatrix}$$

The entries here are sets of points, but from the simplest qualitative point of view what matters is whether each set is empty or non-empty. Putting '0' for empty and '1' for non-empty, there are sixteen different ways of filling the matrix. Of these sixteen, seven correspond precisely to the RCC relations apart from PO; an eighth corresponds to PO so long as both the regions have a single connected component:

DC	EC	PO	EQ
$\begin{pmatrix} 0 & 0 \\ 0 & 0 \end{pmatrix}$	$\begin{pmatrix} 1 & 0 \\ 0 & 0 \end{pmatrix}$	$\begin{pmatrix} 1 & 1 \\ 1 & 1 \end{pmatrix}$	$\begin{pmatrix} 1 & 0 \\ 0 & 1 \end{pmatrix}$

TPP	$NTPP$	$TPPI$	$NTPPI$
$\begin{pmatrix} 1 & 1 \\ 0 & 1 \end{pmatrix}$	$\begin{pmatrix} 0 & 1 \\ 0 & 1 \end{pmatrix}$	$\begin{pmatrix} 1 & 0 \\ 1 & 1 \end{pmatrix}$	$\begin{pmatrix} 0 & 0 \\ 1 & 1 \end{pmatrix}$

The other eight possibilities can only be realized if at least one of the regions has positive co-dimension. All the possibilities in two-dimensional space are illustrated in Figure 2.12. Reflection of a 4-intersection matrix about its leading diagonal corresponds to taking the inverse of the relation it represents; the diagram shows only one of each pair of inverses. It will be seen that the greatest freedom is exhibited by regions of positive co-dimension: all the 4-intersection relations can be realized for regions of co-dimension 1, whereas if one or both of the regions has co-dimension zero then the options are restricted.

It is important to notice here that for the one-dimensional regions we are interpreting 'boundary' to mean the end-points, and 'interior' to mean the

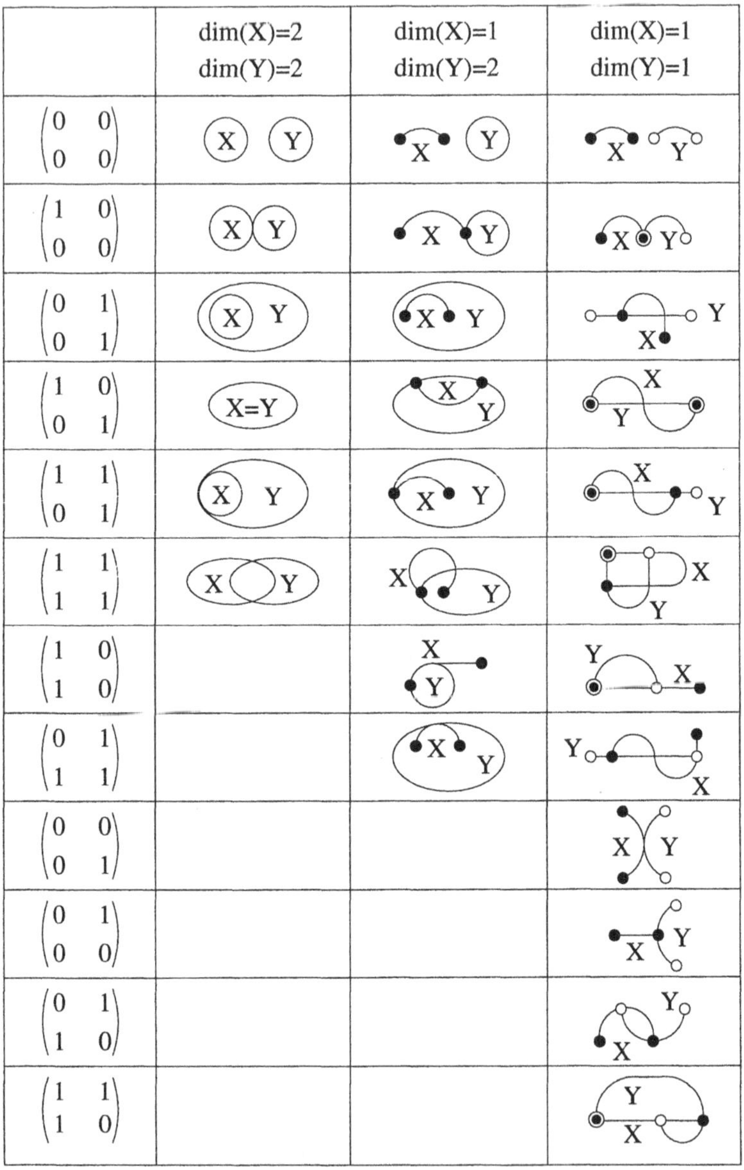

Figure 2.12. Egenhofer's 4-intersection relation for regions of co-dimension 0 or 1 in two dimensions (end-points of X are represented by black circles, of Y by white circles)

region apart from its end-points. We are thus not using, for these regions, the topology of the two-dimensional space within which they are embedded, but rather an implicit subspace topology on some line of which the region is regarded as a segment. There are infinitely many suitable choices of such lines, any of which would yield the desired interpretation.

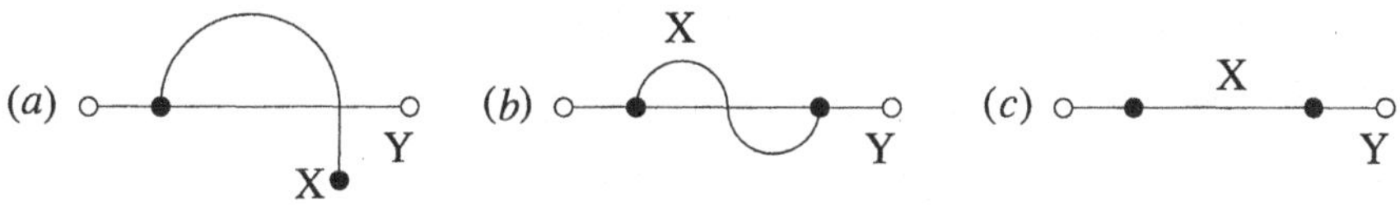

Figure 2.13. Three pairs of regions with the same 4-intersection but different 9-intersections

Egenhofer's '9-intersection' matrix

There are some significant distinctions which the 4-intersection matrix is incapable of making. An example is shown in Figure 2.13, where three different realizations of the relation denoted by the matrix

$$\begin{pmatrix} 0 & 1 \\ 0 & 1 \end{pmatrix}$$

are shown for linear regions of co-dimension 1. All three figures agree in having both the interior and the boundary of X intersecting the interior but not the boundary of Y. They differ, however, in how much of the interior and boundary of X falls within the interior of Y. In (*a*), both the interior and boundary of X fall partly within the interior of Y and partly outside it; in (*b*), the boundary of X falls wholly within the interior of Y, but the interior still falls partly within and partly outside it; in (*c*), both the boundary and the interior of X fall wholly within the interior of Y. The case of (*a*) is in some ways the most generalized of the the three possibilities, which is why it was chosen as their representative in Figure 2.12. In order to discriminate between examples like these, Egenhofer introduced a more complex index, the 9-intersection, which is the matrix

$$\begin{pmatrix} \partial X \cap \partial Y & \partial X \cap int(Y) & \partial X \cap comp(Y) \\ int(X) \cap \partial Y & int(X) \cap int(Y) & int(X) \cap comp(Y) \\ comp(X) \cap \partial Y & comp(X) \cap int(Y) & comp(X) \cap comp(Y) \end{pmatrix}$$

where $comp(X)$ denotes the complement of $X \cup \partial(X)$. By taking account of all the intersections between the boundary, interior, and complement of X and the boundary, interior, and complement of Y, the 9-intersection gives a much more comprehensive account of the spatial relation between two regions. The three cases of Figure 2.13 can now be distinguished as follows:

$$(a)\begin{pmatrix} 0 & 1 & 1 \\ 0 & 1 & 1 \\ 1 & 1 & 1 \end{pmatrix} \quad (b)\begin{pmatrix} 0 & 1 & 0 \\ 0 & 1 & 1 \\ 1 & 1 & 1 \end{pmatrix} \quad (c)\begin{pmatrix} 0 & 1 & 0 \\ 0 & 1 & 0 \\ 1 & 1 & 1 \end{pmatrix}$$

Note that, tempting though it may be, we cannot identify the complement of a region with its topological exterior. Although it *is* its exterior in the topology of the full space in which our regions are embedded, it is not equal to the exterior with respect to any of the (infinitely many) subspace topologies with respect to which what we have called the boundary and interior are the topological boundary and interior.

Further possibilities

Both 4-intersection and 9-intersection can be elaborated by being more specific about the nature of the various intersections. One might, for example, record the dimension of the intersection (for example, two linear features might intersect in a point or a line), or the number of connected components (Galton 1998), or, more comprehensively still, the number of connected components of each dimension. Thus an entry in one of the matrices might read something like '$1^2 0^3$', indicating an intersection consisting of two components of dimension 1 together with three components of dimension 0. Of course, as soon as we begin counting things in this way, we are being drawn away from the qualitative and towards the quantitative, or at least that half-way house between qualitative and quantitative which follows the introduction of counting prior to the introduction of measurement.

2.6. Discrete space

Up to this point I have had little to say on the issue of discreteness versus density. My standard examples of spaces have been based on the real numbers and therefore dense. Density manifests itself in various ways. Focusing on points, for example, we note that in dense space a point has no nearest neighbours: given any two points we can find a third point that is nearer to each of the two than either is to the other. By iteration, we can find infinitely many such points. The presence of infinitely many points within any bounded region is characteristic of dense space. If instead we focus on regions, we note that any region in dense space can be divided up into disjoint proper subregions, and again this infinite divisibility is characteristic of dense space.

For some purposes, however, it is more convenient to model space as discrete. In a discrete space, a point does have nearest neighbours, and some regions are *atomic*, being incapable of subdivision into proper subregions. These properties give a very different flavour to both the intuitive aspects and the mathematical and logical aspects of discrete space. In this section I shall examine informally a number of different discrete models for space, and indicate briefly the lines along which a formal treatment of such models might proceed.

2.6.1. Varieties of discrete space

In the first place we may distinguish models for discrete space in which points and atomic regions are distinct types of entity, from models in which there is no distinction between points and atomic regions. The latter type of model is surely the simpler. In such a model, we have a single type of primitive element, called a *cell*, together with a specification of which cells are neighbours of each other—a symmetric *adjacency* relation. A model of this kind is essentially a *graph*, which is a set of *nodes* (the cells) connected by a network of *edges* (representing adjacency). There are many different ways in which such a structure can be set up; some examples are portrayed in Figure 2.14.

I shall call spaces of this kind *adjacency spaces*. Somewhat more richly structured are the *multiple-adjacency spaces*, in which we again have only a single type of primitive element—cells—but now distinguish between different types of adjacency. An example to motivate this is the chessboard with its sixty-four cells. We could regard this as an adjacency space in more than one way; for example, we could take the point of view of the rook, and regard two

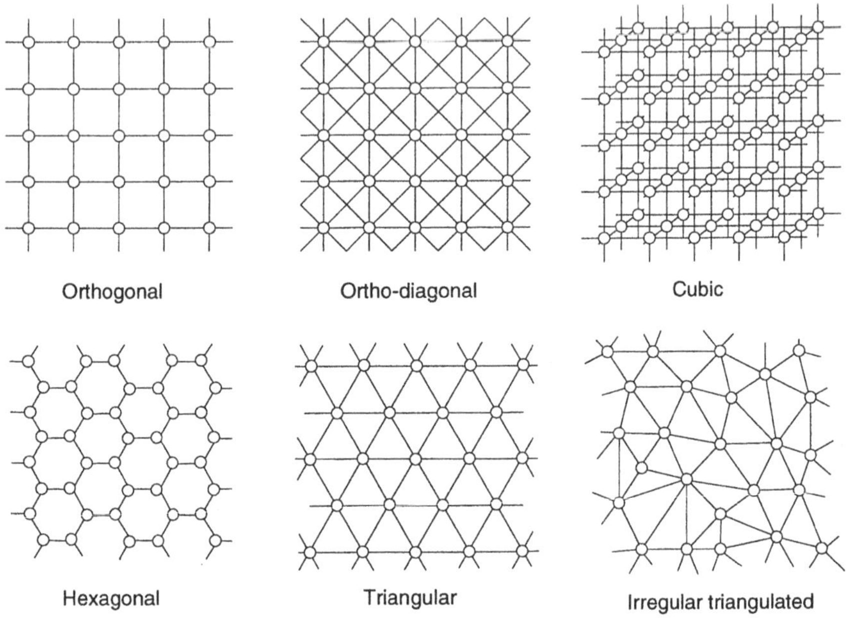

Figure 2.14. Examples of discrete spaces

cells as adjacent only if they meet along a common edge. This gives us an adjacency space like the orthogonally connected rectangular lattice at the top left of Figure 2.14. Or we could take the king's point of view, and regard two cells as adjacent if they meet either along a common edge or at a common vertex. This gives us the ortho-diagonally connected rectangular lattice (middle of top row in the figure). Each of the other chess-pieces has its own view of what cells count as adjacent.[14] In a multiple-adjacency space this multiplicity of viewpoints is explicitly represented by means of more than one adjacency relation. A simple example of this is a rectangular array of cells on which are defined relations corresponding to both orthogonal and diagonal adjacency.

An alternative way of encoding this additional richness of structure is to cross over to the other type of discrete space, in which we have more than one type of primitive element. The rectangular lattice with both orthogonal and diagonal adjacency could be represented by a space consisting of three sets

[14] The case of the knight is particularly interesting: its adjacency space is a quotient of the four-dimensional hypercube lattice under a certain simple equivalence relation.

of primitives, called cells, edges, and vertices, together with a single relation of *bounding*. In this model a cell is a two-dimensional atom; it is bounded by four edges, which are one-dimensional atoms, and also by four vertices, which are zero-dimensional atoms (points). Each edge is itself bounded by two points. The union of the bounding relation with its converse is called *incidence*: an element is incident with all those elements which either bound or are bounded by it. Incidence resembles adjacency in being symmetric, and I shall call spaces of this kind *incidence spaces.*

2.6.2. Formal description of discrete space

I shall indicate briefly the lines along which a formal treatment of the different varieties of discrete space might proceed, together with some pointers to the literature.

Adjacency spaces

Adjacency spaces are described in Galton 2000*a*; see also Galton 1999. An adjacency space is specified by a pair (U, A), where U is a set of primitive elements called 'cells' and A is an irreflexive, symmetric relation on U called 'adjacency'. Given such a space, we are able to define various mereotopological notions which, however, differ in various ways from their standard topological counterparts:

- A region is a (possibly empty) set of cells.
- Region X is connected to region Y if some cell of X is adjacent or equal to some cell of Y:

$$C(X,Y) \triangleq \exists x, y(x \in X \wedge y \in Y \wedge (x = y \vee A(x,y))).$$

- The immediate neighbourhood of a cell is the region consisting of the cell itself and all cells adjacent to it:

$$N_x \triangleq \{x\} \cup \{y \in U \mid A(x,y)\}.$$

- The interior of a region consists of those cells whose immediate neighbourhoods lie wholly within that region:

$$int(X) = \{x \in U \mid N_x \subseteq X\}.$$

- The closure of a region consists of those cells whose immediate neighbourhoods overlap that region:

$$cl(X) = \{x \in U \mid N_x \cap X \neq \emptyset\}.$$

- The boundary of a region consists of those cells in its closure but not in its interior:

$$\partial X \triangleq cl(X) \setminus int(X).$$

  Note that in general the boundary of a region lies partly inside and partly outside that region. We shall call that part of the boundary of a region which forms part of the region itself the *margin* of the region. Then a region's boundary is the union of its margin with the margin of its complement.

Figure 2.15 illustrates these concepts in the case of a region defined on an ortho-diagonally connected rectangular grid. The region itself is the shaded

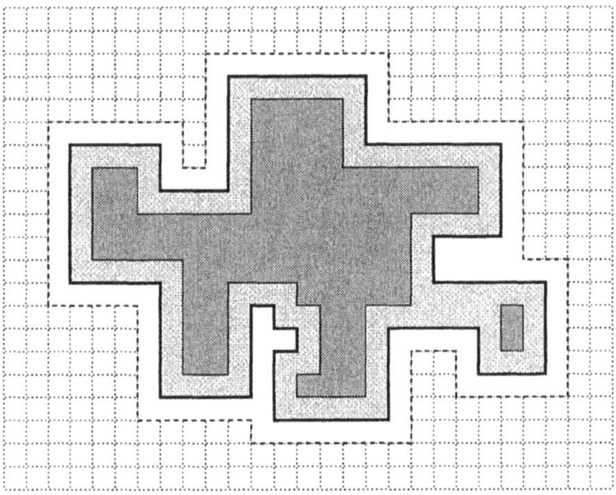

Figure 2.15. A region in an ortho-diagonally connected rectangular grid

area. Its interior is distinguished by being shaded more heavily. Thus the lighter shading picks out the margin of the region. The closure is the region delineated by the dashed line. The unshaded part of the closure is the margin of the complement of the region; this, together with the lightly shaded margin of the region itself, constitutes the boundary of the region.

As defined here for adjacency spaces, these concepts share many, but not all, of the properties enjoyed by their standard topological versions. In particular, the duality between interior and closure is preserved: the closure of the complement is the complement of the interior, and the interior of the complement is the complement of the closure. Again, the closure of the union of two regions is the union of their closures, and the interior of their intersection is the intersection of their interiors. To set against these shared properties, there are also some significant disanalogies. Thus whereas in standard topology closure and interior are both idempotent operations, that is, we have $cl(cl(X)) = cl(X)$ and $int(int(X)) = int(X)$, in adjacency space these do not hold in general. Contrariwise, in adjacency space we have such results as:

$$\begin{array}{l} cl(int(X)) \subseteq X \\ X \subseteq int(cl(X)) \\ cl(X) \subseteq Y \leftrightarrow X \subseteq int(Y) \end{array}$$

which do not hold universally in standard topology.

In adjacency space we have little use for the notions of open and closed sets. If we follow the standard definitions and call a set open if it is equal to its own interior and closed if it is equal to its own closure, then we have the curious consequence that in adjacency space a set is open if and only if it is closed! Such a set is *isolated* since it is not connected to its complement. If the universe U is connected then the only open or closed sets are U itself and the empty set.

Adjacency spaces are, in fact, a special case of what have been called in the mathematical literature 'closure spaces' (Čech 1966; Smyth 1995), of which topological spaces are a different special case. Thus the properties shared between the two types of space include all properties of closure spaces generally. Closure spaces and topological spaces, in turn, are just two of a range of different types of space that may be characterized in terms of the properties assigned to a closure-like operator (a mapping from sets to sets); these very general spaces were investigated in some detail in the 1950s and 1960s by Hammer (Hammer 1955; Hammer 1960; Hammer 1964; Hammer 1969).

Because the closure operation is not idempotent in an adjacency space, each region gives rise to a sequence of closures:

$$X, cl(X), cl(cl(X)), cl(cl(cl(X))), \ldots, cl^n(X), \ldots$$

This in turn gives us a natural measure of the distance of a cell from a region: the distance of cell x from region X is the smallest n for which $x \in cl^n(X)$. This is zero if and only if $x \in cl^0(X) = X$. For more details, see Galton 1999.

Likewise the sequence of interiors gives us a measure of the 'bulkiness' of a region. It is the smallest n such that $int^n(X) = \emptyset$. Bulky regions are ones which contain cells deep within their interiors. They are to be contrasted with 'thin' regions all of whose cells are in or near the boundary. The thinnest non-null regions, those with bulkiness 1, have empty interior and are thus equal to their own margin. Every cell in such a region is adjacent to a cell outside the region.

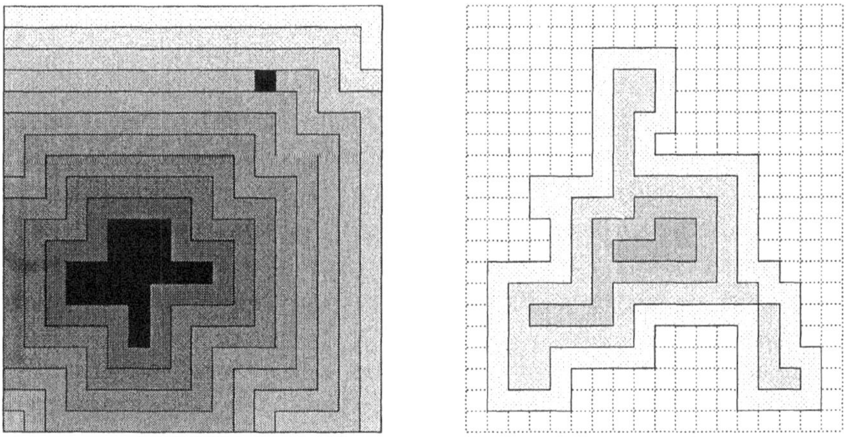

Figure 2.16. Distance and bulkiness in an adjacency space

Distance and bulkiness are illustrated in Figure 2.16. On the left, we see a black region and a single black cell outside the region. The region is embedded in its nested sequence of closures, indicated by progressively lighter shading. The cell is found in the seventh closure of the sequence, hence the distance from the cell to the region is seven units. This is the smallest number of steps in which it is possible to travel from the cell to the region, assuming that each step is between adjacent cells. On the right of the figure we see a region together with its nested sequence of interiors. The fourth interior in the sequence is empty, so the region has bulkiness 4. This is the largest number of steps that are needed in order to escape from the region from a cell inside it. These measures arise naturally from the discrete character of the adjacency space. There are no direct analogues of these for dense spaces.

Multiple-adjacency spaces

There is much of interest to be gained from adopting the adjacency approach to discrete space, but when it was investigated in the 1970s and 1980s in the context of analysing digital images on computer screens, it was found to fall foul of some debilitating flaws, the so-called 'connectivity paradoxes' (Kong and Rosenfeld 1989). The paradoxes arose in the context of rectangular two-dimensional pixel-arrays. Modelling an infinite computer screen as $\mathbb{Z}^2$, there are two natural choices for an adjacency relation on the pixels: 4-adjacency (orthogonal adjacency), by which each pixel (x, y) is considered to be adjacent to the four pixels $(x, y \pm 1)$, $(x \pm 1, y)$ and to no others, and 8-adjacency (orthodiagonal adjacency), by which in addition each pixel is considered to be adjacent to the four pixels $(x \pm 1, y \pm 1)$. I shall use A_4 and A_8 to denote these adjacency relations. The paradoxes arise when we look for an analogue to the Jordan Curve Theorem.

In the continuous space $\mathbb{R}^2$, a Jordan Curve is the image of a continuous injective mapping $f : [0, 1) \to \mathbb{R}^2$ such that $\lim_{x \to 1} f(x) = f(0)$. The theorem states that any Jordan Curve C divides $\mathbb{R}^2 \setminus C$ into exactly two connected components (the 'inside' and the 'outside'). In $\mathbb{Z}^2$, the analogue of a Jordan Curve is a sequence $p_1, p_2, \ldots, p_n$ of distinct primitive elements such that for $i = 1, \ldots, n-1$, $p_i A_j p_k$ if and only if $|i - k| \equiv 1 \bmod n$ (where $j = 4$ or 8). According as $j = 4$ or 8 we may speak of a 4-curve or an 8-curve. Likewise, I shall speak of 4-components or 8-components, according as the connectedness is defined in terms of 4-adjacency or 8-adjacency. In Figure 2.17, adapted from Kong *et al.* 1991, the left-hand illustration shows a set of black pixels which (i) forms a Jordan 8-curve but does not divide its complement into two 8-components, and (ii) is 4-disconnected but manages to divide its complement into two 4-components. The right-hand illustration shows a set of pixels which (i) forms a Jordan 4-curve which divides its complement into *three* 4-components and (ii) divides its complement into two 8-components but is not a Jordan 8-curve.

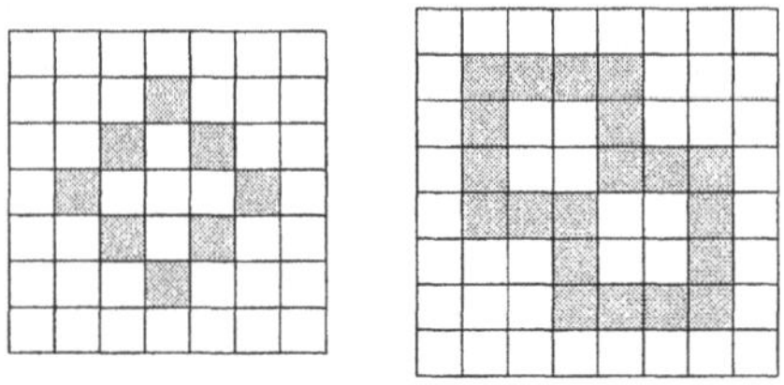

Figure 2.17. The connectivity paradoxes

Since the Jordan Curve Theorem is regarded as fundamental for reasoning about figures in the plane, approaches which model the digital plane in terms of a single adjacency relation were regarded as deeply flawed. To overcome this problem, Rosenfeld proposed using different adjacencies for a set of pixels and its complement. This makes sense if one can systematically distinguish between 'foreground' and 'background': the former can be represented as sets of black pixels, and the latter, their complements, as sets of white pixels. We can either use 8-adjacency for black pixels and 4-adjacency for white, or the other way round; in either case, adjacency between a black pixel and a white pixel follows the rule for white pixels. The left-hand illustration of Figure 2.17 can now be regarded as a Jordan 8-curve dividing its complement into two 4-components, or as a non-Jordan 4-curve leaving its complement 8-connected, both viewpoints being in accordance with the Jordan Curve Theorem; and similar remarks apply to the right-hand illustration. This approach to digital space was extensively developed by Rosenfeld during the 1970s. It was used for defining image-processing operations such as thinning and segmentation, with particular applications to pattern recognition. See Kong and Rosenfeld 1989 for a survey.

More generally, we may define a multiple-adjacency space by means of a tuple $(U, A_1, \ldots, A_n)$, where U is a set of cells and the A_i are distinct adjacency relations. The two-dimensional 'digital topology' of Rosenfeld (Rosenfeld 1979*b*; Rosenfeld 1979*a*; Kong and Rosenfeld 1989) can be regarded as a multiple-adjacency space $(\mathbb{Z}^2, A_4, A_8)$, where the adjacency relations A_4 and A_8 are as defined above. If a set of cells F is selected as the 'foreground', then Rosenfeld's method is tantamount to using a single adjacency relation A defined in terms of A_4 and A_8 as follows:

$$xAy \triangleq \begin{cases} xA_8y & \text{(if } x, y \in F) \\ xA_4y & \text{(otherwise)} \end{cases}$$

(Alternatively, one could reverse the roles of A_4 and A_8 in this definition.) The choice of foreground and background is dependent on the figure under consideration. The relations A_4 and A_8 also play an important part in Egenhofer and Sharma 1993.

This approach is difficult to apply if the distinction between foreground and background is not rigorously enforceable. In Figure 2.18 (adapted from Kovalevsky 1989), for example, we see a white 'X' on a black background and a black 'X' on a white background. Whether we regard white or black as the foreground colour, and whether we choose 4-adjacency for the background and 8-adjacency for the foreground, or vice versa, it is impossible simultaneously to assign the same connectivity to both figures. And yet on the computer screen it is not uncommon to find both white text on a black background and black text on a white background at the same time. Problems of this kind led Kovalevsky to adopt a completely different approach to the formal modelling of discrete space, the topological approach based on the notion of incidence. This approach is discussed in the next section.

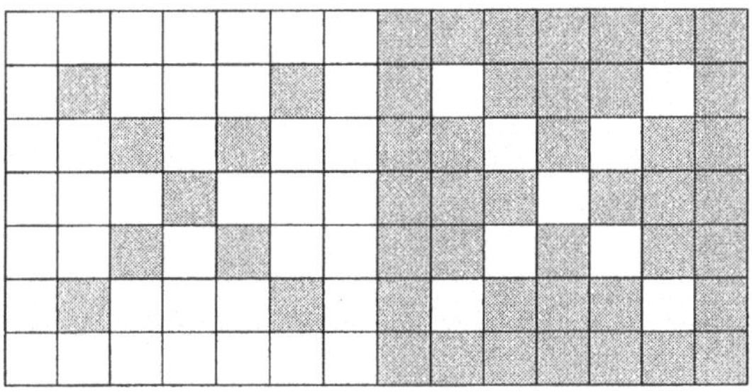

Figure 2.18. Interchange of foreground and background

Incidence spaces

The key notion here is that of 'bounding'. The idea is that in modelling the digital plane, for example, we employ a formal ontology consisting not only of the pixels which overtly appear in an image, but also lower-dimensional entities which form the boundaries between pixels: edges and vertices (sometimes called 'linels' and 'pointels'). In a rectangular array, a pixel is considered to be bounded by four primitive elements called edges, each of which is in turn bounded by two primitive elements called vertices. The bounding relation is assumed to be irreflexive, asymmetric, and transitive, and therefore a strict partial order on the set of primitive elements. Thus a pixel is also bounded by the vertices bounding the edges which bound it. On this basis it is possible to specify a topology on the set of primitive elements. Each element has a *minimal neighbourhood* consisting of the element itself together with any elements which it bounds. Thus, in the plane, there are three distinct types of minimal neighbourhood: a cell on its own, an edge with the two cells it bounds, and a vertex with the four edges and four cells which it bounds. A set of primitive elements is regarded as open so long as it contains the minimal neighbourhood of each of its members. This topology has been called the 'star' topology (Ahronovitz *et al.* 1995); it is also known as a Khalimsky space (Khalimsky *et al.* 1990).

More formally, the Khalimsky space of the plane consists of a rectangular grid, within which we distinguish three types of elements: cells, edges, and vertices. These can be labelled using integers as follows:

1. A vertex is labelled by a pair of integers: $V_{mn} \triangleq \langle m, n \rangle$.
2. An edge is labelled by a pair consisting of an integer and a pair of consecutive integers:

$$\begin{array}{lll} E^v_{mn} & \triangleq & \langle m, (n, n+1) \rangle \quad (\text{`vertical' edges}) \\ E^h_{mn} & \triangleq & \langle (m, m+1), n \rangle \quad (\text{`horizontal' edges}) \end{array}$$

3. A cell is labelled by a pair consisting of two pairs of consecutive integers:

$$C_{mn} \triangleq \langle (m, m+1), (n, n+1) \rangle.$$

Khalimsky space is a true topological space, with open sets as follows:

1. A cell on its own makes an open set: $\{C_{mn}\}$.
2. An edge together with two ('neighbouring') cells forms an open set: $\{C_{(m-1)n}, E^v_{mn}, C_{mn}\}$ and $\{C_{m(n-1)}, E^h_{mn}, C_{mn}\}$.
3. A vertex together with four edges and four cells forms an open set:

$$\{V_{mn}, E^v_{m(n-1)}, E^v_{mn}, E^h_{(m-1)n}, E^h_{mn}, \\ C_{(m-1)(n-1)}, C_{(m-1)n}, C_{m(n-1)}, C_{mn}\}.$$

4. Any union of open sets is an open set.

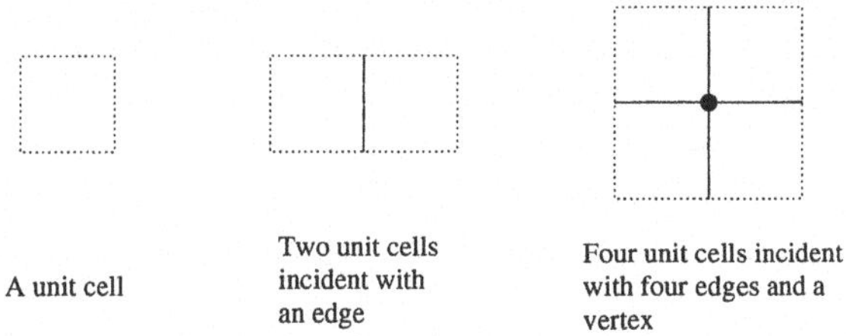

Figure 2.19. Three kinds of open set forming a basis for two-dimensional Khalimsky space

The basic open sets are illustrated in Figure 2.19. The idea is that an open set cannot contain any 'naked' edges or vertices: an edge in an open set must be accompanied by the cells which it bounds, and a vertex must be accompanied by the edges and cells which it bounds. This bounding relation can be defined as the irreflexive part of the 'specialization order' of the topology (Johnstone 1982). To be precise: an element x bounds an element y distinct from x if and only if every open set containing x also contains y. Notice that this definition would lapse into vacuity if the topology were a T_1 topology, that is, a topology in which, for any two elements, each is contained in an open set not containing the other; so it is an important feature of Khalimsky space that it is non-T_1 (and *a fortiori* non-Hausdorff—see p. 66). It is however T_0: given any two distinct elements, *at least one* is in an open set not containing the other.

Since Khalimsky space is a true topological space, notions such as closure, interior, boundary, connectedness, and continuity can all be defined in the standard way. Khalimsky space can be derived from the standard topology on $\mathbb{R}^n$ by partitioning the latter set into equivalence classes and forming the quotient topology on the set of equivalence classes. Points in $\mathbb{R}^n$ are deemed equivalent so long as the members of each pair of corresponding coordinates either are equal integers, or both fall strictly between the same pair of integers. This partitions the space into a rectangular grid structure of n dimensions. In general, given a set T and an equivalence relation E on T, the set-theoretic quotient T/E is the set of equivalence classes of T under E. We write $[\![x]\!]$ to denote the equivalence class to which x belongs. Thus

$$
\begin{aligned}
V_{mn} &= [\![(m, n)]\!] \\
E^h_{mn} &= [\![(m + \tfrac{1}{2}, n)]\!] \\
E^v_{mn} &= [\![(m, n + \tfrac{1}{2})]\!] \\
C_{mn} &= [\![(m + \tfrac{1}{2}, n + \tfrac{1}{2})]\!]
\end{aligned}
$$

If T is a topological space, then the quotient topology on T/E is defined by stipulating that a set of equivalence classes is to count as open in T/E so long as the union of those equivalence classes is open in T:

> X is open in the quotient topology on T/E if and only if $\{x \in T \mid [\![x]\!] \in X\}$ is open in the original topology on T.

More generally, we may similarly define Khalimsky spaces for rectangular grids in any number of dimensions. These in turn are special cases of the more general notion of *incidence space* which we define as follows:

> An incidence space consists of a set S together with an irreflexive, asymmetric, transitive relation $B \subseteq S \times S$.

B is the bounding relation. We define a topology on an incidence space (S, B) by defining the open sets. The rule is that a subset $X \subseteq S$ is open if and only if the following condition holds:

$$\forall x \in X \, \forall y \in S \, (B(x, y) \rightarrow y \in X).$$

This is the condition which says that an element in an open set must be accompanied by all the elements which it bounds. Khalimsky space is just one example of an incidence space. Later, in Chapter 8, I shall make use of a wider variety of incidence spaces to support an account of continuous qualitative change.

A *rapprochement* between our adjacency spaces and incidence spaces is provided by some work reported by Stell and Worboys (1997). As remarked above, adjacency spaces are essentially *graphs*. Thinking of them in this way enables us to draw a connection with the theory of bi-Heyting algebras as expounded by Stell and Worboys. A bi-Heyting algebra is like a Boolean algebra in which the complementation operation has been 'split' into two operations $\neg$ and $\sim$, called negation and supplementation respectively. The negation $\neg X$ is the 'largest' element for which $X \cdot \neg X = 0$ holds, while the supplement $\sim X$ is the 'least' element for which $X + \sim X = 1$ holds. In a Boolean algebra these coincide. Stell and Worboys give as one example of a bi-Heyting algebra the set of regions defined on networks (graphs). The regions they consider, however, are not just subsets of the nodes of the graph, but rather subgraphs of the graph. A subgraph is specified by a subset of the nodes of the graph together with a subset of its set of edges, where the edges included in the subgraph must link nodes which are also in the subgraph. If we think of the nodes as having dimension 1 and the edges as having dimension 0, and say that an edge bounds the two nodes that it links, then what we have is an incidence space in which the open sets are precisely the subgraphs as defined above. It is not, however, the topological aspect of the graph that interests Stell and Worboys so much as the fact that the set of subgraphs can be viewed as a bi-Heyting algebra. For this purpose the operations of negation and supplementation are defined as follows:

- The negation of a subgraph consists of all the nodes not in the subgraph, together with all edges with end-points from amongst those nodes.
- The supplement of a subgraph consists of all the edges not in the subgraph, together with all their end-points.

In terms of these operations, Stell and Worboys define further operations that are strongly analogous to the closure and interior operations on adjacency spaces, namely the *supplement of the negation* (analogous to closure) and the *negation of the supplement* (analogous to interior). These are illustrated for the case of an orthogonally connected rectangular grid in Figure 2.20. For convenience, the original subgraph on the left is highlighted in all three diagrams by shading. If we attend to the nodes only, and not the edges, thereby in effect regarding the space as an adjacency space, we see that the operations shown here coincide exactly with the closure and interior operations on such spaces; but the edge information is lost when this viewpoint is adopted.

We can also think of a graph as an incidence space, and thus as a topological space; the topological closure and interior operations are different from the adjacency-based operations, and in this case are of considerably less interest. There are two possibilities: we either stipulate that nodes bound edges or the other way round. If nodes bound edges, then a subgraph is a closed

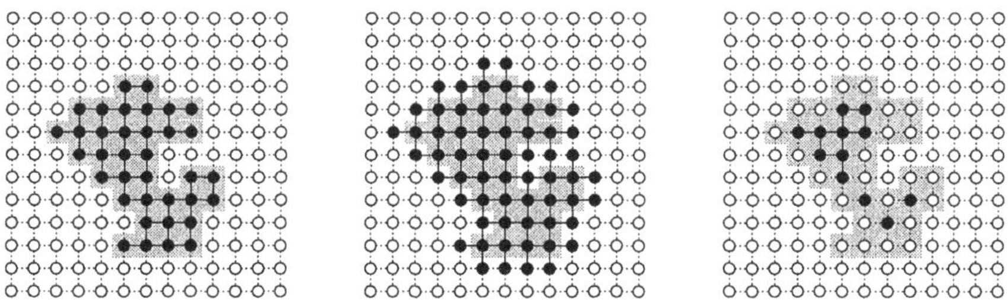

Figure 2.20. A subgraph of an orthogonally connected rectangular grid, together with the supplement of its negation and the negation of its supplement

subset of the full graph; hence topological closure is the identity operation on subgraphs, whereas the topological interior of a subgraph, obtained by removing those nodes which bound edges not in the subgraph, has 'naked' edges and hence is not itself a subgraph. Conversely, if edges bound nodes, a subgraph is an open subset, so it is now the interior operation that reduces to identity; closure consists in adding to a subgraph all edges joining a node in the subgraph to a node outside it, the resulting object again having 'naked' edges and thus not itself a subgraph.

2.7. Summary

In this chapter I have surveyed a range of different theoretical approaches to the description of space, from well-established mathematical theories such as Euclidean geometry, co-ordinate geometry, and point-set topology, to more recent approaches such as the mereotopologies developed in the context of philosophy and artificial intelligence. There is no question here of having to decide definitively in favour of one approach at the expense of the others; all the approaches usefully capture certain aspects of space as we understand it both from our everyday experience and our more technical or scientific enterprises, and the many interconnections between the different theories further testify to a deep underlying unity. In the remainder of this book I shall freely draw from whatever spatial theory most suits my immediate purpose.

3

Objects

Since we are interested in spatial change, we are interested in objects in so far as they can participate in spatial change. That means we are interested in objects which possess spatial attributes susceptible to change. These objects form a very varied class; the purpose of this chapter is to survey the variety of different kinds of spatial objects and attempt to bring some system into their classification. In the next chapter I shall look more closely at the spatial attributes that these objects possess.

3.1. Matter and material objects

It is surely fair to say that for an object to have spatial attributes it must either be a material object or must depend for its existence in some way on the existence of one or more material objects. This claim will be amplified in the discussion that follows. I shall organize the survey of spatial objects on the basis that matter is fundamental to their existence, and shall move outward, as it were, from the central core of 'bare matter' to successively more elaborate kinds of dependence on bare matter.

Bare matter, then, is simply the stuff of which the material world is made. The matter we encounter mostly comes in the form of substances composed of molecules which are in turn composed of atoms, although this type of structure is not manifest in the way we experience it, but has to be teased out of it by careful scientific theory and experiment. From the point of view of our everyday common-sense understanding of the world, matter can mostly be regarded as having a continuous fine structure, so that, for example, a lump of iron is iron all the way through.[1] We have a compulsive tendency to regard much of the matter we see about us as differentiated into large numbers of discrete *objects*; it is a point of some dissension amongst philosophers to what

[1] It is harder than we suppose to imagine everyday substances as made of vast numbers of exceedingly tiny atoms. It is easy to make the mistake of imagining the atoms to be far too large, conferring on matter a graininess that is only a little way below naked-eye visibility, whereas in reality the scale of the atomic structure is vastly smaller than this.

extent this differentiation reflects inherent organization in the matter itself and to what extent it rather stems purely from the systematizing propensities of the human mind.

The idea of an object is subtler than one might think. A typical object is not just a chunk of matter. To see clearly why this is, I shall examine in some detail a number of individual cases, ranging from one in which the equation of an object with a chunk of matter is pretty nearly right to ones in which it is disastrously wrong.

My first example is a roughly hemispherical pebble of sandstone which I picked up on a beach some years ago and have used as a paperweight on my desk since then. During that time, the material constitution of the pebble has remained pretty nearly constant. To be sure, each time it is picked up, some of the outermost atoms get knocked off, and a small amount of slightly dirty grease gets deposited on its surface; but these changes are to all intents imperceptible. During the time I have owned it, the history of the pebble and the history of the matter composing it are almost exactly the same. If we look outside that time, both into the past and the future, however, we find things are very different. For example, before I picked the pebble up, it must have spent a considerable time on the beach being buffeted and rolled over and over by the twice-daily tides, a process that must have resulted in a steady attrition of material from its surface, so that what we now see as its surface must not so long ago have been entirely hidden within the interior of the pebble. The approximately hemispherical form suggests that at some time in the past my pebble formed half of a more nearly spherical pebble, which in some comparatively violent event was smashed in two. Perhaps the other half even now adorns some other desk. Was this the moment at which my pebble came into existence, or was the large pebble of which it is a part already my pebble, which merely lost weight at that time? But if the other half also remained intact as a single hemispherical piece, then it has equal claim to historical identity with the larger pebble. If identity is to remain transitive, then something has to give. Probably the best solution is to say that the larger pebble ceased to exist as such at the moment of splitting, and that two new smaller pebbles came into existence; but the important point to realize is that the history of the matter itself does not provide any help here: the decision rather lies in our hands as language users to determine the 'correct' usage of the phrase 'this pebble'.

Assuming, then, that we cannot trace the history of the pebble back beyond the fission event that gave it birth, we can now turn to consider the history of the matter that constitutes it. If Q is the quantity of matter[2] of which the pebble is now composed, we see that as we go back in time, although Q was a quantity of stone in a pebble, it was only part of *the* matter of which the pebble was composed. If we go back far enough we shall find Q embedded within a much larger chunk of rock, with nothing about it to give anyone any reason to single out that particular quantity of matter from all the infinitely many different quantities one might have chosen instead; only its later history confers any kind of individuality on Q, but, for all that, it existed unrecognized, indeed unrecognizable, even then. The rock containing Q was at some earlier time part of a section of cliff that tumbled down during a storm one winter. The cliff only became a cliff because of the slow coastal erosion eating into the land; earlier still, our quantity Q was buried under the landscape. We are now entering a geological time-scale; over many millions of years, not much happened to Q, but at some point the sand grains which compose it must have come together in a process of deposition. Tracing further back, each grain has its own separate history, including erosion from a much earlier rock. It is unlikely that at this distant time many of the sand-grains were in each other's company as they are now.

My pebble can reasonably be regarded as a bare chunk of matter so long as we confine our attention to the period during which the history of the pebble and the history of the matter coincide, at least very nearly. Since this period spans several years, and in particular covers the whole period of my interaction with it, this identification is quite workable. Our first kind of material object, then, is simply a *piece of matter* which persists in the same form over a period of time. During that time, the object can be identified with the piece of matter, but it must be understood that the object, as opposed to the matter, came into being when the matter first assumed that form whose stability allows us to identify it *as* an object, and ceases to exist once it assumes a sufficiently different form. Since some leeway is possible in the identification of forms, it may not be possible to pinpoint precise moments at which the coming into being and the ceasing to exist occur.

For my second example I turn to my bicycle, which I acquired in 1979 and have been using ever since then. For the most part, its material constitution remains much the same from one day to the next. Of course, imperceptibly little bits are constantly getting knocked off it, just as with the pebble, and sometimes more perceptible changes occur: here a flake of paint peels off the frame, there a bit of exposed metal starts going rusty (so that oxygen from

[2] The sense of 'quantity' here is due to Bertrand Russell; see Cartwright 1970.

the air is being *added* to the bicycle). Once in a while, a screw comes loose and drops off, and I have to replace it; sometimes the tyres or the brake-blocks need changing. On a couple of occasions there have been more major changes, in which the saddle-pin, the front wheel, and the front forks were replaced. The chain and cogs have all been replaced. In fact the only substantial parts of the bicycle which remain the same now as when I bought it are the frame itself, the handlebars and brake-levers, the back wheel, and the saddle. Still, through all this time it has been in a fairly clear sense the same object: my bicycle, with a history that is significantly different from the history of the various bits of matter which make it up at any one time. Perhaps if I had to get a new frame, to which all the other parts were transferred, one might be less willing to regard it as the same bicycle as before. Are bicycles individuated by their frames? Maybe; but again the important thing to realize is that there is nothing about the comings and goings of bare matter that determines the answer in any way. It is up to us to sort out the principles of individuation for bicycles and other objects. One can approach this issue in various spirits. In the spirit of 'ordinary language' philosophy one might suppose that the answer is built into the way we use words, and can be revealed by a careful examination of linguistic usage; but what if usage is inconsistent? Then it seems that if one wants clear-cut criteria for individuation then one is going to have to stipulate something, thereby introducing new usage.[3]

My third example is the apple-tree in my garden. The history of the tree involves a flux of matter at several different time-scales. On the shortest scale there is, during the warmer months, a continuous flow of water sucked in by the roots, carrying dissolved nutrients which are incorporated into the living tissue, the water itself then being transpired from the leaves. There is the intake of carbon dioxide from the air, and the expulsion of oxygen. Then, on a larger time-scale there is the yearly growth and shedding of leaves and fruit, not to mention the occasional pruning, sometimes quite drastic, involving the severance of substantial branches. If I were to chop the tree right down it would very probably grow again, though all of the matter in the above-ground part would be different. What we have said here about the tree could be said, suitably modified, of any living thing, including me, the writer, and you, the

[3] Compare Sharvy (1971): 'I might say that I have had this beard for five years, but this hair on my face for less than a week. But in a fit of materialism I might say that I have had this beard for less than a week; and in the opposite frame of mind, I might say that I have had this hair on my face for five years. There seems to be some truth in all of these, even the last two. The task of separating the strict talk from the loose is not merely automatic here.'

reader, of this book. A living thing is very far removed from being a mere piece of matter, even though at any particular time there is a certain quantity of matter of which it is composed. The importance of form and continuity is manifest here.

The fourth example, a lake, illustrates a more extreme dissociation between objects and the matter they are made of (Hayes 1985). A lake is a body of water, and as such it is a material object: an object made of water, so to speak. It is not, however, a bare quantity of water: water enters the lake by way of streams flowing into it, and by rainfall. It leaves the lake by way of streams flowing out of it, and by evaporation. The precise quantity of water making up the lake is never the same from one day to the next, and over a long enough time we can expect there to be a total replacement of the water in the lake. The identity of the lake is largely a matter of its being located in a particular place; it can expand or contract depending on the shifting balance between inflow and outflow, but this does not prevent it from retaining a clear identity over a long period of time. Of course, if it contracts sufficiently it may split into two lakes, and if it expands sufficiently it may merge with a previously separate lake to become a new, larger lake. But these cases differ from the case of the pebble because these objects which are merging and splitting are not, even approximately, bare quantities of matter.

So far, then, I have considered *quantities of matter* and *material objects*. A material object is made of matter in the sense that at any time it can be identified with a quantity of matter, but because this identification does not remain fixed over time, the material object cannot be identified with a quantity of matter without further qualification. More plausibly, a material object can be identified with a (partial) function from times to quantities of matter, but this still leaves a good deal unsaid, since an arbitrary function from times to quantities of matter will not, in general, pick out anything that we would want to call an object, notwithstanding Quine's assertion that 'any arbitrary congeries of particle-stages, however spatiotemporally gerrymandered or disperse, can count as a physical object' (Quine 1974: §14). Thus the issues of form and continuity, mentioned earlier, are paramount in the definition of material objects. It does not, however, seem possible to lay down general conditions here since different types of object have different requirements.

The simplest case is a *piece of matter*, like the pebble I began with. A piece of matter is a quantity of matter forming a connected whole over a period of time. So long as that quantity of matter continues to remain in the form of a connected whole, the piece of matter that it constitutes continues to exist. It ceases to exist when the quantity of matter becomes scattered into disconnected parts. Of course, even the pebble was only a very good approximation to this, since its material constitution does not remain totally identical.

The next example, my bicycle, is an example of an *assembly*. An assembly consists of various parts put together in a particular way. Amongst the parts may be some that are themselves assemblies, but this recursive decomposition cannot proceed indefinitely. Ultimately the assembly can be analysed into constituent parts that are themselves pieces of matter (the individual spokes of a bicycle wheel, for example). In the context of the larger assembly, all or at any rate most of the constituent parts are replaceable in the sense that they can be replaced by other more or less identical parts without compromising the identity of the assembly itself. Typically, however, such replacement occurs only sporadically, so that the normal state of existence of an assembly is to retain the same component pieces of matter over substantial portions of its lifetime, and because of this the history of an assembly often remains coupled, albeit sometimes rather loosely, to the history of a quantity of matter.

Assemblies are typically artefacts, not necessarily of human construction (a bird's nest, for example, is a kind of assembly); and they can be more or less precisely specified. They grade into *heaps*, which are mere aggregations of pieces of matter with little in the way of significant form. In a brick wall each brick is fairly precisely positioned within an overall pattern; individual bricks are freely interchangeable, but the pattern is important, so it is appropriately regarded as an assembly. In a heap of bricks, while it is also true that each brick has a precise position within the rather complex pattern formed by the whole, this pattern is regarded as inessential to the heap's existence; the bricks can be rearranged into a completely different pattern and still be accepted as the same heap—which might be identified solely by its position, as for example the heap of bricks at the bottom of my garden.

The tree is an example of an *organism*. Organisms differ from assemblies in a number of important ways. They are not artefacts, and because of this they are not composed of discrete pieces of matter. As a result the history of an organism is very much less tightly bound to the history of any quantity of matter. There is a continuous flow of matter through the organism, and this is its normal state of existence. As Vogel (1999) puts it, 'organisms continuously rebuild nearly everything within themselves', attributing this insight to Schoenheimer (1942). Vogel goes on to say

> This dynamic state represents a profound difference between living and non-living systems. The same molecules the ancient Egyptians hauled into place more than four thousand years ago still make up the pyramids, but you're not the same person you were a year ago. Organizationally, yes, but materially, no. The organization of the individual organism persists far beyond the material of which it's made.
>
> (Vogel 1999: ch. 11)

Thus although the matter that constitutes the organism is in a continuous state of flux, it would be misleading to describe an organism as merely a flowing of matter, as a river or waterfall *is* a flowing of water. The complex internal structures of organisms are essential to their being the particular individuals that they are.

A *flow* can be characterized as an object, usually rather homogeneous and unstructured, whose constituent matter is being constantly and continuously replaced. Rivers, waterfalls, and fountains are flows of water. Flows in which the replacement of matter is sufficiently slow, or sufficiently localized within the body of matter, seem to grade into mere pieces of matter: a lake with no inflow or outflow is simply a piece of water, whereas a lake with strong in- and outflows can be viewed as a flow. A cloud is another kind of object which may occupy various positions on the continuum between pieces of matter and flows. Flows do not have to be liquid. A sandbank may be being continuously eroded at one end and continuously built up at the other, thus constituting a slow flow of sand. Although the flow itself is generally rather unstructured, the constituents may themselves be highly structured objects: a queue is a flow of people, traffic is a flow (sometimes all too slow) of vehicles.

Assemblies, heaps, and flows are related as follows. A component of an assembly may be said to *participate* in the assembly; and its participation is associated with a specific role that it plays in the functioning of the assembly as a whole. A member of a heap can also be said to participate in the heap, but here the notion of functional role is absent. Functionality is not an all-or-nothing affair, though: for example, one might say that in a pile of stones, the stones at the bottom function as supports for those above. This is a very weak notion of function, but it is not entirely absent. In general we may suppose that assemblies grade into heaps as the functional roles are attenuated or removed altogether. We can speak of participation in a flow as well, by the quantities of matter which pass through it. It is in the nature of a flow that such participation is systematically transient: as for example the quantities of water that flow along a river enter the river at various places along its length and leave it at the mouth. Participation in an assembly or heap can also be transient, of course, but this transience is not systematic, not part of the essential nature of the assembly or heap. We might therefore suppose that heaps and assemblies grade into flows as the transience of participation becomes more systematic. Although we speak here of participation, or taking part, it should be noted that by a 'part' of a flow we do not normally mean one of the participating quantities of matter, but rather a position within the flow through which such quantities pass, for example a particular stretch of river, which does not move downstream as the water flows.

3.2. Formalizing the ontology of material objects

The foregoing considerations all lead us to the idea of a *two-level* ontology consisting of, on the lower level, quantities of matter, and, on the upper level, material objects. Formalizations based on this idea have been proposed by a number of authors, including Burge (1979) and Hayes (1985). The idea itself is very ancient, and corresponds to the distinction between matter (*hylē*) and form (*morphē*) that played such an important part in the philosophy of Aristotle—our two-level ontology corresponding to Aristotle's 'hylomorphic' dualism.

3.2.1. Mass terms and count terms

Before embarking on the formalization itself I must introduce here a crucial distinction between *mass terms* (such as 'milk', 'water', 'stone', 'chocolate', 'wood', 'metal') and *count terms* (such as 'bottle', 'lake', 'pinnacle', 'cake', 'table', 'ring'). The 'terms' here are typically nouns, as in the examples just given, but they can also be noun-phrases such as 'milk chocolate' or 'bottle of whisky'.

In English grammar, the two classes can be differentiated by a number of purely syntactical criteria. In the following list, 'M' stands in place of an arbitrary mass term, and 'C', a count term.

- We say 'a C', 'each C', 'every C', but not 'an M', 'each M', 'every M'.
- We say 'Cs', 'many Cs', 'n Cs' (where n is a natural number), but not 'Ms', 'many Ms', 'n Ms'.
- We say 'much M', 'less M', but not 'much C', 'less C'.
- We say 'all M', 'more M', 'a lot of M', but 'all Cs', 'more Cs', 'a lot of Cs'.

A glance at our examples above will show that not all of them come out as unequivocally mass or count on these criteria. 'Water', 'chocolate', and 'metal' behave as mass terms most of the time, but we can still speak of 'the waters of Babylon', 'a box of chocolates', and 'precious metals'. Similarly, 'cake', which usually behaves as a count noun, behaves as a mass term in 'Let them eat cake'.

We can salvage the distinction by noting that many terms are systematically ambiguous between a 'mass reading' and a 'count reading'. So *stone* is a material; a piece of that material is *a stone*. *Chocolate* is a substance; *a chocolate* is a piece of confectionery made (at least in part) of that substance. *A water* is a body of *water* delineated by a shore. *A metal* is a kind of *metal*. If you say 'an M' then I must understand you as meaning to refer to a piece of M, or something made out of M, or a kind of M; and if you use C as a mass noun then I will probably understand you as referring to a substance of which Cs are typically made. Quine (1960: §19) draws attention to the ambiguity of 'Mary had a little lamb', which is compounded of an ambiguity between count and mass uses of 'lamb', an ambiguity between two senses of 'little', and an ambiguity between two senses of 'had'.

Bunt aligns the mass/count distinction with the continuous/discrete distinction. He points out how pervasive continuous (and 'quasi-continuous') concepts are in everyday life:

> Your *tea* is cold, and there's no more *hot water*, and your computer program still doesn't work as it should; you had better take some *time* to figure out what the problem is. Perhaps you can get some *help* from Jeff. And all that *traffic noise*. Maybe you should switch on the radio and find some pleasant *music*. Or go out and get some *fresh air*.
>
> (Bunt 1985, my italics)

This quotation shows clearly that the mass/count distinction does not only apply to terms denoting material objects, although for us it is the latter class of referents that is of particular interest.

Mass terms are problematic because almost all our analytical tools are *object-oriented*; that is, they have built into them the assumption that the world we want to talk about consists of objects, properties of objects, and relationships between objects. This is particularly true of the predicate calculus. Many attempts have been made to accommodate the logic of mass terms within the predicate calculus.

Quine, for example, treats mass terms as general terms—in other words, predicates—when they occur after 'is' (in *predicative* position), as in 'That puddle is water', and as singular terms—individual constants—when they occur before 'is', as in 'Water is a fluid'. He says:

> In general a mass term in predicative position may be viewed as a general term which is true of each portion of the stuff in question, excluding only the parts too small to count. Thus 'water' and 'sugar', in the role of general terms, are true of each part of the world's water or sugar, down to single molecules but not to atoms; and 'furniture', in the role of general term, is true of each part of the world's furniture down to single chairs but not to legs and spindles.
>
> In 'Water is a fluid', on the other hand, ... the mass term is much on a par with the singular term of 'Mama is big' or 'Agnes is a lamb'. A mass term used thus in subject position differs none from such singular terms as 'mama' and 'Agnes', unless the scattered stuff that it names be denied the status of a single sprawling object.
>
> (Quine 1960: §20)

In this way mass terms are accommodated into the ontological framework of first-order logic, but at the expense of countenancing some pretty weird individual objects: for to consider the totality of the world's water as a 'single scattered object' is surely to depart radically from our everyday conceptualization of water. We can, with a little effort, get into the frame of mind of thinking of water in this way, but it is very far removed from how we normally think of it.

Hayes (1985) agrees with Quine in seeking to use first-order logic as the basis for analysing mass terms, asserting that 'Every representational formalism I know which has even the glimmerings of a clear semantics, is based on the idea of individual entities and relationships between them.' Hayes then goes on to attempt with considerable ingenuity to formalize the ontology of liquids on this basis, making use of a two-level ontology similar to what we suggested above:

> Liquid objects endure through time, as do pieces of liquid. It always makes sense to equate two liquid-object expressions, or two piece-of-liquid expressions, independently of time. But since the criteria for individuating such objects are different, it makes sense to equate a liquid-object expression and a piece-of-liquid expression only during a time-period when these criteria agree. Such an equality *must* be understood as temporary, as liable to change in truth-value.
>
> (Hayes 1985: §10)

Here 'liquid objects' are material objects whose constituent material happens to be liquid; and pieces of liquid are quantities of liquid matter. Thus I am essentially in agreement with Hayes as to the broad ontological basis for the treatment of mass terms, although I differ from him as to the details of the formalism, to which I now turn.

3.2.2. Formalizing the two-level ontology

An appropriate calculus for the lower-level ontology is extensional mereology (see §2.5.1). The domain of this theory will be a set $\mathcal{Q}$ whose elements are *quantities of matter*. In addition I shall use a set of *mass-term predicates*, corresponding to everyday mass terms such as 'water', 'iron', 'paper', and 'sand'. A mass term M can be predicated of a quantity of matter q just when all the material of q is of type M. Thus mass-term predication is *homogeneous* in the sense that if q is M then any part of q must be M as well. We must bear in mind, though, that the nature of a quantity of matter can change. The water in the bird-bath may be ice tomorrow morning: it is the very same quantity of matter but now falls under a different description. Again, a quantity of matter which is now pastry is the fusion of quantities which were formerly flour, fat, and water. Hence the mass-term predicates must be relations between quantities and times: I shall write $M(q,t)$ to mean that quantity q was of type M at time t. The homogeneity rule comes out as:

$$M(q,t) \wedge q' \sqsubset q \rightarrow M(q',t).$$

Note that this rule embodies a definite decision on my part as to how I am using these mass terms, to regard arbitrarily small parts of any quantity of some material M as M. This is a decision about granularity, which may not always seem realistic. Taylor (1985), for example, says that

> Division of a lump of fruit-cake will produce a lump of fruit-cake only until a sample of some minimal size is reached; a mere sultana does not in itself constitute a lump of fruit-cake, and can at best be part of such a lump.
>
> (Taylor 1985: 70)

It may seem hard to quarrel with this, yet there is a case to be made for saying that even an individual sultana can be considered to be a lump of fruit-cake so long as it actually forms part of a larger lump with the requisite heterogeneity of ingredients. If one isolates the sultana, physically removing it from the cake, then it ceases to be fruit-cake. It is sultana all the time, of course, both when it forms part of the fruit-cake and when it does not. This shows that there is no unique species of material to which a given quantity of matter must belong at any one time; but some mass terms belong to a higher level of analysis than others. What makes something a bit of fruit-cake is the fact of its being or forming part of a particular type of heterogeneous ensemble of ingredients, whereas for a quantity to be water, say, does not in any way depend on what larger quantities the quantity is part of. So while Taylor is certainly correct to draw attention to the phenomenon of granularity, this does not invalidate the homogeneity postulate for mass-term predicates.

At the higher level we have an ontology of *material objects*. The domain of the theory of objects is a set $\mathcal{O}$ whose elements are individual objects. Objects are described by means of count-term predicates corresponding to everyday terms such as 'ball', 'table', 'house', and 'man'. For the most part, these predicates are not homogeneous but *unitary*: no proper part of an object of type T is itself of type T:

$$T(x) \wedge x' \sqsubset x \rightarrow \neg T(x').$$

Unlike mass-term predications, count-term predications are not susceptible to change. If a table is converted into a chair then I shall say that the table ceases to exist and a chair comes into being; the continuity of matter is handled by means of the $\mathcal{Q}$ ontology. The formal coupling between the two ontologies is provided by a predicate $Comp$ (introduced as C in Burge 1979), which relates an object to a quantity of matter at a time: thus $Comp(x, q, t)$ says that the object x is composed of the quantity of matter q at time t. This enables us to say whether or not a particular material object exists at a given time: x exists at t if and only if $\exists q Comp(x, q, t)$. We may allow the temporal argument to be either an instant or an interval. If a table is converted to a chair, we have a situation which can be described formally in the terms

$$Table(x) \wedge Chair(y) \wedge Comp(x, q, t) \wedge Comp(y, q, t') \wedge t < t' \wedge \\ \neg\exists q', q'', t''(Comp(x, q', t'') \wedge Comp(y, q'', t'')),$$

the final conjunct ensuring that the table and chair never exist simultaneously.

The versatility of this formalism can be illustrated in a series of examples as follows:

- Heraclitus said that one can never step into the same river twice. What he meant was that the water in the river when you step into it for the first time is not the same as the water in the river when you step into it the second time. This can be expressed as follows:

$$(River(x) \wedge Comp(x, q_1, t_1) \wedge Comp(x, q_2, t_2) \wedge (t_1 \neq t_2)) \\ \rightarrow (q_1 \neq q_2).$$

Here q_1 and q_2 refer to the entirety of the water making up the river at times t_1 and t_2; we could refine our analysis so that we only refer to the water in the stretch of the river that is being stepped into, but the analysis above should already suffice to dispel any notion that there is a deep logical puzzle here.[4]

- Archimedes reputedly shouted 'Eureka!' when he realized how he could determine whether the gold in King Hiero's crown had been adulterated with silver. How can we express the fact that the crown has been so adulterated? The quantity of matter now constituting the crown is the fusion of two quantities of matter, one of which was formerly gold and the other silver:

$$Crown(x) \wedge Comp(x, q, t) \wedge (q = q_1 \sqcup q_2) \wedge \\ \exists t' < t(Gold(q_1, t') \wedge Silver(q_2, t')).$$

- 'We are what we eat'. That is, all the matter in us was once in food we have eaten, drink we have drunk, or air we have breathed. By no means all the matter we have eaten, drunk, or breathed is still part of us, though. Thus what we want to say is something like:

$$Person(x) \wedge Comp(x, q, t) \wedge (q' \sqsubset q) \rightarrow \\ \exists y, q'', t'(t' < t \wedge Ingest(x, y, t') \wedge Comp(y, q'', t') \wedge q'' \circ q').$$

This says that every part of a person has a part whose constituent matter was once part of the matter of something ingested by that person, where 'ingest' is here being used comprehensively to include eating, drinking, and breathing.

It should be understood that all I have proposed here is a vehicle for describing specific facts concerning the material constitution of objects. I have not proposed a theory of how objects are to be identified or re-identified; in particular, I have laid down no criteria of identity for different kinds of object, or

[4] This analysis is essentially similar to that proposed by Dummett (1973: 574–5).

even said anything about what form such criteria might take. All this would be necessary for a complete account, but for my purpose here what is important is the availability of a notation with sufficient expressive capabilities; identity criteria may be introduced on the fly as and when they are needed.

A further limitation in my account is that it treats the 'upper' level of material objects as unstructured, providing no analogue to the mereological structure of the lower level. Link (1983) proposed a lattice structure here, generated by the part–whole relation on material objects. I shall not pursue this further.

3.3. Parasitic objects

Not all things that we refer to as objects are composed of matter. Some good examples are *holes* and *surfaces*; the former have been the subject of a detailed study by Casati and Varzi (1994), the latter of a similarly detailed study by Stroll (1988). Stroll notes that we have a number of different conceptions of surface, but among them is what he calls the *interface* conception of a surface, which he also calls a 'Leonardo surface', quoting from Leonardo a passage which ends:

> [A] surface is the common boundary of two bodies which are not continuous, and does not form part of either one or the other, for if the surface formed part of it, it would have divisible bulk, whereas, however, it is not divisible and nothingness divides these bodies the one from the other.
>
> Leonardo da Vinci, quoted in Stroll 1988: 41

Interpreting 'two bodies' liberally to include the case of a body and the outside world, we arrive at the notion of the surface of a body as its interface with the outside world. We can characterize it in terms of where it is situated in relation to the body: if the body occupies exactly some three-dimensional spatial region R, which we may assume to correspond to a regular open subset of $\mathbb{R}^3$, as discussed in the previous chapter, then its surface occupies the two-dimensional region ∂R, the topological boundary of R. This region has no volume, and hence the surface cannot be composed of matter, since all matter occupies volume, any sample of matter having finite mass and density. I shall adopt this conception of surface here, but should note in passing an alternative conception of a surface, also discussed by Stroll, as the *outermost part* of an object: this kind of surface is a material object, since it is part of a material object, although the boundary between the surface, in this sense, and the interior of the object can only be drawn sharply in the case of objects in which there is an outermost layer (a skin, crust, or rind) that is of different composition from the rest.

In my chosen sense of the term, the surface of an object has no existence independently of that object; there is no such thing as a surface which is not the surface *of* something. Following Casati and Varzi (1994), I shall express this by calling a surface 'parasitic' on its *host*, the object it is the surface of. We might say, perhaps in contradiction to Leonardo's notion, that when two bodies are in contact, each retains its own surface intact, but the two-dimensional regions occupied by the two surfaces overlap.

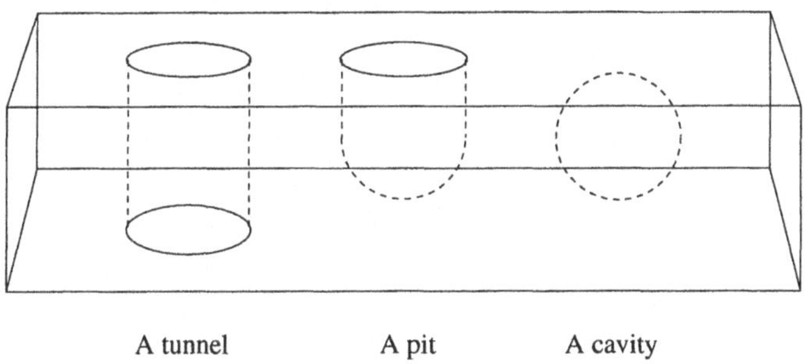

Figure 3.1. Three kinds of holes (adapted from Casati and Varzi 1994)

A hole is similarly parasitic on a host. There are three kinds of holes, namely tunnels, pits, and cavities, as illustrated in Figure 3.1. They are distinguished by the number of entrances they have: a tunnel has two or more entrances, a pit has only one entrance, and a cavity has no entrances at all. Tunnels and cavities, in particular, can assume very complex forms, as illustrated, delightfully, in Casati and Varzi's book. Holes are philosophically puzzling because they seem to be objects but are plainly non-material; nor can a hole be regarded as just a bit of space, since when an object moves, all its holes move with it, whereas bits of space cannot move, since to move is to change position, and bits of space do not *have* positions, they *are* positions.

Casati and Varzi categorize holes as 'superficialities' because they so to speak live on the surfaces of their hosts. This does not mean that a hole is part of the surface: it cannot be, because a hole has volume whereas the surface does not. But the hole shares at least some of its surface with its host—the shared surface forms what Casati and Varzi, adapting a term introduced in a slightly different sense in Lewis and Lewis 1970, call a *hole lining*. We can say that the hole exists by virtue of the configuration of its host's surface.

There are other kinds of superficiality living on the surfaces of objects. Some examples are cracks, wrinkles, waves, grooves, rims, faces, edges, and vertices. All these objects owe their existence to the fact of an object having a particular shape, resulting in its surface being configured in a particular way. But these are not the only kinds of parasitic object.

An interesting case is *shadows*. A shadow lives on a surface, and resembles a superficiality in not being made of matter, but it is not a feature of the surface in the same way that superficialities are. It can move about on the surface and even jump from one surface to another; it can be split between two surfaces. None the less it is a perfectly real object: it can be measured, photographed from different angles, and manipulated (by manipulating the object casting it, the object whose surface it lies on, or the light source), and its presence on a surface can have real effects (when the sun is eclipsed, the birds stop singing). Like holes, shadows have hosts, the objects on whose surfaces they lie, but any shadow also depends for its existence on two other objects: the object which *casts* the shadow, whose shadow it is, and the source of illumination. Thus the true logic of shadows involves a four-place relation: x is a shadow cast by y on surface z in the light coming from w.

Although I have treated shadows as essentially two dimensional, this being the visible form of a shadow, a closely related three-dimensional notion is found in sentences such as 'Come into the shade of this tree' or 'During a lunar eclipse, the moon moves into the earth's shadow'. This conception treats a shadow as a three-dimensional object defined in terms of an object y whose shadow it is and a light-source w. It occupies that portion of space which receives no light, or reduced light, from w on account of the presence of y. An ordinary two-dimensional shadow occurs when part of the surface of an object is situated within the three-dimensional shadow.

Inverse to shadows are patches of light; these include images projected onto a screen. Suppose I have a transparency of, say, a flower, and I project it onto a screen. It is instructive to compare the status of the two flower-images I now have. The flower-image in the transparency is a material object: it is made of the pigments on the surface of the slide. The flower-image on the screen is not a material object; it is parasitic on the screen, being formed as a result of the screen's being situated in the three-dimensional cone of variously coloured light emanating from the projector.

3.4. Perspectival objects

Some things which we speak of as objects are even further removed from bare matter than the parasitic objects discussed in the previous section. A perspectival object is something which depends for its existence on a particular point of view. Some examples are silhouettes, reflections, rainbows, and the horizon. These are all in a certain sense illusory, although it would be misleading to call them illusions; perhaps a better term is *virtual*, though since this term has recently been coming to acquire a host of new associations from the world of computers (virtual machine, virtual reality) it is perhaps better avoided in this context.

A *silhouette* is the outline of the visual appearance of something from a particular viewpoint. It is thus like a surface or boundary, but is further removed from the object itself by virtue of being defined in terms of a viewpoint. A silhouette has a quite definite shape which can be photographed or drawn---indeed it is the shape of a photographic image of the object. It has a position in the visual field but it does not seem to have a position in space, nor does it have a definite size. If one changes one's point of view, then we can either say that the silhouette changes, or that a continuous series of different silhouettes appear. The logic of silhouettes is expressed by a three-place relation: x is the silhouette of y as seen from point z.

It is possible to treat silhouettes as being parasitic rather than perspectival objects. From a particular point of view, the silhouette of an object is defined by the set of straight lines from the observer's eye that just graze the surface of the object. Each such ray contains at least one point on the object's surface; if more than one, we choose the nearest such point to the observer. The collection of all these points forms a possibly disconnected one-dimensional subset of the points on the object's surface, which could be regarded as a superficiality parasitic on the object. It has many of the same properties as the silhouette, and for some purposes we could perhaps identify the silhouette with it. But we do not normally think of silhouettes in this way, so I am inclined to prefer the interpretation of the silhouette as a perspectival object.

The *horizon* can be regarded as a kind of silhouette: the silhouette of the earth as seen from a particular viewpoint. Essentially similar remarks can be applied to the horizon as we applied to silhouettes in general. The fact remains, though, that it enjoys a rather special status in our everyday conception of the world, and most people would quite reasonably want to pause for thought on being presented with the claim that the horizon is a silhouette.

Reflections can also be photographed, and they are conventionally regarded as having size, position, and shape. Your reflection in a flat mirror is treated, in optics, as being located at a position behind the mirror that exactly corresponds, in the geometrical sense of 'reflection', to your own position in front of the mirror, and its size and shape are identical to yours apart from its opposite 'handedness' (you and your image are 'enantiomorphs'). However, all this is merely a convenient fiction: there is not really an object behind the mirror with these characteristics. Optical terminology recognizes this by calling it a 'virtual image' (as opposed to the 'real image' thrown by a projector, which can be focused onto a screen). The reflection does not really have a size or position, though arguably it does have a shape. It is thus quite different from a shadow (though small children take a while before learning the difference). Looking at the still surface of a pond, I may see both the reflection of a tree and its shadow cast over the water; moving round to the other side of the pond I still see the shadow, but not the reflection. If we were to take seriously the convenient fiction of optics, we should say that the reflection is still there, but now hidden from me.[5]

Three factors determine the existence and nature of a reflection: a source (what the reflection is *of*), a reflector (the reflective surface in which the reflection is seen), and a viewpoint. Thus the logic of reflections involves a four-place relation: x is the reflection of y in z as seen from point w. When you look at yourself in a mirror you are the source of the reflection, the mirror the reflector, and the position of your eye the viewpoint. When you look at the reflection of a cloud in a puddle, the cloud is the source, the puddle the reflector, and the position of your eye the viewpoint. Where I have referred to 'your eye', I should really have said 'your eyes'; and since your two eyes occupy different positions, they see different reflections, but because the two reflections are related to each other in exactly the same way as the separate views that your eyes have of any object, the psychophysical mechanisms of stereopsis are brought into play, thus reinforcing the illusion that there is a real object in the position to which the virtual image is assigned by optical theory.

Highlights are a type of reflection, but they need not be *of* anything in particular, merely of the ambient light, coming from the sky, a window, or

[5] Seeing the reflection of something can often be regarded simply as a way of seeing the thing itself: it is, indeed, the only readily available way you have of seeing yourself.

some other light source. Although these are perspectival objects, the reflected light, if strong enough, can appear as a bright patch on a wall or ceiling, and this patch is parasitic, not perspectival.

A *rainbow* is another optical effect, belonging to a whole class of wonderful phenomena caused mainly by atmospheric water and ice. Other examples are halos, glories, and the *Heiligenschein* (a halo-like glow seen round the shadow of the observer's head under certain conditions, believed by Benvenuto Cellini to be a sign of his own genius). We refer to all these things using count nouns; we can point to them and photograph them. Thus far, they can be regarded as objects. But they are not made of matter and they are not parasitic on matter in the sense discussed in the previous section, though they are certainly *caused* by matter in interaction with light. They are very different from silhouettes, and second cousins to reflections. I shall have little more to say about them.

3.5. Natural and conventional objects

Dummett draws attention to two views of reality: on the one hand, the 'picture of reality as an amorphous lump, not yet articulated into discrete objects', and, on the other hand, the 'naïve conception' which 'presupposes that the world presents itself to us as already dissected into discrete objects, which we know how to recognize when we encounter them again, in advance of our acquiring any grasp of language at all' (Dummett 1973: 577). A concomitant of the 'amorphous lump' view is that, since it is we who, through our cognition, and in particular through our language, articulate the world into discrete objects, there must be an element of arbitrariness about how we do this. In particular, it is to be expected on this view that people from very different cultural backgrounds will split the world up in very different ways.

There are several issues tangled up here which need to be sorted out. To begin with, we must distinguish the carving up of the world into discrete objects from the allocation of those objects to different categories. Cultural relativism with respect to the latter need not imply cultural relativism with respect to the former. For example, Mark (1993) shows that the English words 'lake', 'lagoon', and 'pond' divide up the semantic field of *standing water bodies* quite differently from the French words 'lac', 'lagune', and 'étang', but it is not suggested that, in this area at least, the two languages recognize different individual objects. Most of the now quite extensive literature on Prototype Theory and Cognitive Linguistics (Lakoff 1987; Taylor 1989;

Langacker 1991) focuses more on the issue of categorization, how objects are assigned to categories, than on the issue of what objects there are in the first place.

We must also distinguish the effects of language, culture, and cognition. The difference between English and French just alluded to is a linguistic difference; it may or may not have cultural repercussions—only an extreme Whorfian would suppose that every linguistic difference betokens a cultural difference. And there seems to be no reason at all why a linguistic or cultural difference in categorization should necessarily have any effect on the way English and French speakers perceive and interact with standing bodies of water, as would be entailed by the existence of a cognitive difference.[6]

A further important point is that we are not, in fact, faced with a sharp dichotomy between 'natural' and 'arbitrary' divisions. Adherents of the 'amorphous lump' picture sometimes speak as if the carving up of the world into discrete objects is somehow arbitrary, as if, had our culture or our brains been very different, we might, for example, have regarded the saddle as belonging with the rider rather than with the bicycle. A more sober attitude might be to say that there are degrees of naturalness, and that to the extent that a particular division is not natural, it may be more or less conventional, and that to the extent that it is conventional, it may be more or less arbitrary. I suspect that most conventions are well motivated in a way that makes them quite unarbitrary. But the motivation may come from a complex set of cultural or cognitive considerations rather than directly from nature itself.

One might wonder whether our recognition of objects is ever totally arbitrary. Consider the case of *rivers*. A river-name typically attaches to a connected, linear (non-branching) part of a complete river system. Travelling upstream from the mouth of the Mississippi, for example, one encounters various branch points at which two streams come together; a choice has to be made as to which of the two is to share with the downstream portion the name 'Mississippi'. In most cases one of them is much larger than the other, so we can regard the smaller branch as a tributary, and the larger branch as the upstream continuation of the main river. This is clearly the case where the Arkansas joins the Mississippi, for example. But at some of the branches, for example where the Ohio joins, the issue is not so clear. The Missouri is the most problematic case; indeed, the Missouri exceeds in length that part of the Mississippi upstream from their confluence, which is why in lists of the

[6] See Montello (1995) for a broadly sceptical statement against the idea that there are significant cultural differences in spatial cognition.

world's longest rivers we find the 'Mississippi-Missouri' rather than just the Mississippi—something which always used to puzzle me as a child. If we regard a river as an object, as our conferring a proper name upon it would seem to suggest, then there is clearly an element of convention as to which connected linear sub-stretches of a river-system we single out as rivers, although this convention is usually far from arbitrary. In a sense the most natural object to single out would be the entire river-system, including all the rivers which can be traced downstream to a single mouth or delta. We do, to be sure, *have* the concept of a river-system (and the closely related notions of catchment and watershed), but it seems to come much less naturally to us than that of a river.

The geographical domain provides a good number of instances where there seems to be an element of convention in the identification of individual entities. Some of this comes from the matter of *boundaries*. Smith (1995) distinguishes *fiat* boundaries from *bona fide* boundaries. Bona fide boundaries exist independently of human cognitive acts; they are 'boundaries *in the things themselves*'. Smith cites as instances river-banks and coastlines. Fiat boundaries, by contrast are 'delineations which correspond to no genuine heterogeneity on the side of the bounded entities themselves'; they do not exist independently of human cognitive acts. Smith cites as instances 'the borders of practically every single political and administrative unit of the North American continent', and 'medical divisions, such as that between the upper, middle and lower femur'. Smith admits that there are problematic cases of boundaries that are not clearly classifiable as of either type, and mixed cases of boundaries which are partly bona fide and partly fiat; but these, he claims, are not sufficient to detract from the essential validity of the distinction. Smith goes on to say that an *object* is bona fide or fiat according as its boundaries are one or the other.

The distinction between the bona fide and the fiat is clear enough in some cases, but inevitably there are problematic cases as well. One problem is how to characterize a 'genuine heterogeneity'. Consider the division of the shore into upper shore, middle shore, and lower shore. These divisions can be given seemingly clear definitions based on tidal ranges. The shore is the area bounded above and below by the *extreme* high and low water marks; the middle shore is the area bounded above and below by the *mean* high and low water marks. The upper and lower shores are the areas of shore respectively above and below the middle shore (Yonge 1949). There are some difficulties in identifying the boundaries between these regions; I doubt if they can be fixed to within a metre. But that is not the main issue, which is whether the regions are natural or conventional. In support of the former view, we might note that, first, the tides themselves are entirely natural, and the definitions depend only on the tides, and, second, the divisions can to some extent be identified by the different composition of littoral flora and fauna, many seaweeds, for example, being quite selective about which area of the shore they will grow on. In support of the conventional nature of the division, we might note that even though once we have laid down our definitions we must turn to nature to determine where the boundaries are, it is entirely a conventional matter that we decided to use those definitions in the first place, the criteria in terms of tides being in no way forced on us by nature itself.

This example shows us that even though an object may be defined by convention, the consequences of that conventional definition are not themselves conventional. That is, we lay down a definition, and must then accept whatever consequences follow from that definition by the laws of nature or logic. Having defined the shore and its various parts in the way described above, we are then not free to determine the properties of what we have defined. Moreover, we may find that amongst the objects we have defined, seemingly by pure convention, there are some whose properties are highly significant from a naturalistic point of view.

Consider for example the lines of latitude on the surface of the earth. Surely these are conventional? As regards, say, the line at 50°N, one must agree that there is a strong element of conventionality about recognizing this as an object of any kind, although it is worth noting that having once established the conventions by which lines of latitude are to be defined we then have no choice but to recognize the existence of this particular line—we cannot draw lines at random on the globe and call them lines of latitude. But there are at any rate five lines of latitude which have some claim to being more than merely conventional, namely the Equator, the Tropics of Cancer and Capricorn, and the Arctic and Antarctic Circles. These can be identified independently of any decision about lines of latitude in general, as the loci of points having certain properties in relation to the yearly apparent motion of the sun. The fact that the lines are somewhat abstract in character, corresponding to no visible features on the ground, reflects not so much on their conventionality or otherwise but on their position along the material/parasitic/perspectival axis.

Let us see if we can apply the distinction between bona fide and conventional objects to some of our earlier examples. A pebble is thoroughly bona fide: its boundary (that is, its surface) is a natural discontinuity between stone and non-stone. What about two pebbles? Do they constitute a single object? If it is an object, it is a *scattered object* (Cartwright 1975). The boundary of the two pebbles is disconnected, consisting as it does of two individual pebble-boundaries, but it is still quite bona fide in the sense of being a natural discontinuity between stone and non-stone. Just as we can count the pebbles in a bucketful of shingle, so we can count the pebble-pairs (if there are n pebbles there will be $\frac{1}{2}n(n-1)$ possible pebble-pairs, though in any actual pairing off there will be at most $\frac{1}{2}n$). Yet it does not seem natural to regard a pair of pebbles as a single object. Does the reason for this lie in our language, our culture, our cognition, or in the nature of things themselves? This is quite a different case from a pair of shoes, which also has a disconnected bona fide boundary. Shoes *come* in pairs; they do so because feet come in pairs, which is a fact of nature. One might still claim that attributing the status of an object to a pair of shoes is a human cognitive act, sanctioned though not necessitated by nature. Given a set of identical left shoes and the same number of identical right shoes, how they are paired does indeed seem to be arbitrary, subject only to the condition that left is not paired with left, or right with right.

Is my bicycle a bona fide object? Its being man-made does not in any way prevent it from being bona fide. Indeed, so many of its properties—its mass, shape, size, chemical composition, and efficiency as a mechanism—are intrinsic to it, capable of being investigated by anyone competent to do so, without reference to any human conventions or transactions or stipulations, that there would seem to be a good case for regarding it as bona fide. Not *all* the properties of the bicycle are bona fide, of course: the fact of its being *my* bicycle and not anyone else's, is a matter of convention, a fiat property of the bicycle. I paid for the bicycle, I have not given or sold it to anyone else, I have not abandoned it: by our conventions of ownership, therefore, it is my bicycle, yet there is nothing about the bicycle itself, as a material object, in virtue of which this is so.

It might be claimed to be a matter of convention that we single out the particular collection of components that constitute my bicycle as a unitary object. People with a very different view of the world, so the argument goes, might simply not recognize this particular collection of components as an object at all. But is there really any scope for doubt that my bicycle constitutes a single object? One reason for this is that the components are attached together so that they move around together, and this fact is not merely a matter of convention. Some leeway can, it is true, be granted to convention as regards the status of various attachments such as lights, panniers, and perhaps even mudguards, which, though attached, can be detached without interfering with the functioning of the machine: thus we can readily grant that it is a matter of convention whether or not these attachments should count as parts of the bicycle, without thereby granting that this conventionality extends to the wheels, frame, and handlebars. Note, however, that here we have appealed to the functioning of the bicycle: but objects in general cannot be intrinsically associated with a unique functionality. There are many uses to which a bicycle or other artefact *could* be put, with varying degrees of effectiveness. In calling something a bicycle we are considering it from the point of view of its functioning in a particular way; we may consider it natural to consider it from this point of view for a number of reasons, for example (*a*) because that is what it was made *for*, or (*b*) because that is how it is, in fact, being used, or (*c*) because it is more effective when used in that way than in any other way (though it is hard to see how anyone could be sure of this). Of these reasons, only (*a*) can be regarded as conventional, and only (*c*) is concerned with intrinsic, rather than extrinsic, properties of the object itself. There seems to be enough leeway here to accommodate both those who would prefer to regard the bicycle as a bona fide object and those who would prefer to regard it as conventional.

What about the status of organisms? With higher animals, the matter is generally rather clear-cut. An animal moves about in a co-ordinated fashion as a single entity, separated from other individual animals; it is not a matter of convention where one individual ends and the next begins, and indeed biology and common sense are largely in agreement in these cases. Problematic cases are few ('Siamese' twins are a case in point). With some lower animals, biology and common sense part company. A well-known case is the Portuguese Man-of-War (*Physalia*), which to the layman seems obviously to be a single complex creature not unlike a jellyfish but from a biological point of view is better regarded as a colony of many comparatively simple individuals. Originally, all multicellular organisms evolved from colonies of single cells; and even those single 'eukaryotic' cells, with nucleus and organelles, are believed to have evolved as symbioses amongst what were originally little colonies of simpler free-standing 'prokaryotic' cells (Margulis 1970). Presumably the full set of evolutionary pathways would provide us with many maddeningly intermediate examples where it could only be a matter of convention whether to regard a particular aggregation of living matter as a single organism or a colony.

A somewhat different case is provided by fungi. What the layman sees as a fungus, growing on the ground or a tree-trunk, is from a biological point of view only the fruiting body of a larger organism consisting of a complex tangle of mycelial threads hidden from view under the ground or inside the wood. This does not mean that the layman's fungus is not a bona fide object: it is as bona fide as an apple, while the biologist's fungus, mycelia and all, is somewhat comparable to an apple-tree, though it has to be said that the apple-tree has greater unity of organization than the fungus. Plantlife often seem to present itself in the form of relatively unindividuated living matter—living stuff rather than living things, appropriately referred to using mass terms rather than count terms. We speak of 'grass' and 'moss', for example; we can, if need be, refer to individual grass plants, possibly consisting of many shoots connected by underground runners, but this does not come easily to us, and given a uniform expanse of grass in a lawn it would be a major undertaking to sort it out into a set of biologically distinct individuals. From the layman's point of view the natural way to individuate the small-scale components of a lawn is as individual blades of grass, which biologically are proper parts of organisms rather than organisms themselves.

The clearest examples of conventional objects remain the divisions of territory into administrative units: countries, counties, boroughs, parishes, and the like. Although it is common to run the boundaries of these units along natural linear features such as mountain ranges and rivers, the majority of these boundaries are (as already noted above) without any overt physical correlate of this kind. An interesting consequence of the conventional nature of such objects is that they can come into existence, undergo changes, and cease to exist in a discontinuous fashion. When Czechoslovakia was divided into the Czech and Slovak republics, there was no gradual change involved. Abruptly, at the moment the division came into force, the old country disappeared and two new ones took its place (cf. §7.6.6). This was a conventional change, in that nothing happened 'on the ground' in virtue of which the split took place; rather such changes as subsequently became manifest were consequent on the stipulative division of the country rather than the other way about. Of course, there was a natural event—at least the signing of a document—by which the stipulated division was brought into effect.

This kind of behaviour is in stark contrast to that of natural objects, which tend to undergo continuous rather than abrupt changes. A case in point is Great Britain, the island. This is a natural, not a conventional, entity, which must have come into existence as the culmination of a continuous, cumulative process, forcefully envisaged by Jacquetta Hawkes:

> It was during the Middle Stone Age that an event took place which still appeals to the imagination: the isolation of our country from the continent. It is tempting, though shockingly unwarranted, to think that there may have been one dramatic hour when it ceased to be possible to walk dry-foot from France to Britain. I like to imagine a stormy winter night when a wave overtopping the precursors which had prepared the way, smashed the last tenuous ridge of chalk somewhere between Kent and the Pas de Calais and swept over it to make Britain an island.
>
> (Hawkes 1946: 12)

The distinction between natural and conventional does not only apply to material objects. It seems odd in any case to regard administrative units as material objects. We could perhaps identify France, say, with a certain area of land together with at least some of the material objects distributed over its surface; but if so, how do we decide which material objects are *part* of France rather than merely being *in* France? It seems better to regard France as parasitic rather than material; or if that sounds somehow degrading, perhaps the philosophical term *supervenient* is more dignified. Some geographical objects are undoubtedly parasitic and at the same time at least partly conventional; river-mouths are examples. What exactly is a river-mouth? It is not a stretch of river (unlike an estuary, say); it is more like a line separating the river itself from the sea. But there is an element of convention about exactly where that line should be drawn.[7]

The ontological status of territories becomes even more problematic in the case of constellations, which we might regard as 'territories in the sky'. What, exactly, is the constellation Leo? Originally, no doubt, this term referred to a particular collection of bright stars (pre-eminently the first magnitude star Regulus), conceived as pinpricks of light etched on the dark background of the night sky, and arranged in a configuration which to the fanciful human imagination was somewhat suggestive of a recumbent lion. That particular collection of stars has no significance beyond the context of human observation from the vicinity of the earth; from a viewpoint a hundred light-years away, there would be no more reason to single out those particular stars than any other set. To the modern astronomer, 'Leo' refers to a particular col-

[7] Note that when applied to people, the word 'mouth' is somewhat ambiguous: sometimes we mean the visible external aperture framed by the lips, sometimes the internal cavity stretching back from the lips towards the throat—the latter sense is what we mean when we speak of the front or the back of the mouth.

lection of celestial co-ordinates enclosed by boundaries drawn on the purely imaginary celestial sphere. When we say that the star Regulus is 'in Leo', we mean merely that its celestial co-ordinates fall within those boundaries. By the same token, we might say that at a particular time the planet Mars is in Leo, or that a certain galaxy is in Leo, although the former is orders of magnitude less distant from us than Regulus and the other bright stars making up the lion configuration, while the latter is orders of magnitude *more* distant. If all these objects are 'in Leo', then perhaps Leo, properly considered, is a three-dimensional chunk of space extending outwards from the earth indefinitely far. The Leo of the astrologers is something different again, a mere symbol, laden with entirely fanciful significance, whose connection with the astronomical constellation becomes progressively more unreal with the continuing precession of the equinoxes.

3.6. Objects with indeterminate boundaries

While Smith may be right in supposing that an object should be regarded as bona fide or fiat according as its boundaries are one or the other, we must recognize the existence of a large class of objects—geographical features prominent among them—which are undoubtedly bona fide but for which there is inevitably an element of convention in the exact delineation of the boundaries.

Consider a marsh, for example, an area of ill-drained ground characterized by the presence of certain types of vegetation, and the absence of others. A particular marsh, we may suppose, is surrounded by some dry heathland. The heathland has gorse and heather and scattered birch trees, whereas the marsh is dominated by sedges, rushes, and sphagnum moss. As you walk from the heathland into the marshy area, you will notice a succession of vegetational changes, determined in part by the various drainage tolerances of the component species, but also no doubt by other factors such as the nature of the underlying rock and the micro-climatic variations resulting from uneven topography. There will be corresponding changes in the fauna as well. Where does the marsh itself begin? It is impossible to answer that exactly without refining our criteria. For any of the characteristic marsh plants, we might place the boundary at the outer limit of the zone inhabitable by that species; or, for any of the plants characteristically absent from the marsh, we might place the boundary at the limit of the zone uninhabitable by that species; or we might turn to hydrology and define the limits of the marsh in terms of the level of soil saturation; and so on. In a sense there are as many different marshes there as there are cogent reasons for choosing this or that boundary; yet in another sense they are all the same marsh in slightly different guises. The term 'marsh' is not so much vague as somewhat indeterminate, but this in no way means that the marsh itself is merely a conventional object; its bona fide character becomes evident as soon as the water gets into your shoes.

A plausible approach to the problem of boundaries was suggested by Wittgenstein (1975), who noted that '[w]hat corresponds in Euclidean geometry to the visual circle isn't a circle but a class of figures, including the circle, but also, e.g., the hundred-sided regular polygon', and went on to remark that 'it seems as though an exact *demarcation* of the inexactitude is impossible'. Addressing this 'cardinal difficulty', Wittgenstein suggests, in effect, that any curve wholly contained within an annulus of radii $r \pm \delta$ appears as the visual circle of radius r, and any curve which strays outside an annulus of radii $r \pm \Delta$ (where $\Delta > \delta > 0$) does not so appear. This leaves open an indeterminate zone, consisting of those curves which are wholly contained within the latter annulus but not the former. Wittgenstein remarks that

> The boundaries . . . are still only like the walls of the *forecourts*. They are drawn arbitrarily at a point where we can still draw something firm. – Just as if we were to border off a swamp with a wall, where the wall is not *the* boundary of the swamp, it only stands around it on firm ground. It is a sign which shows there is a swamp inside it, but not, that the swamp is exactly the same size as that of the surface bounded by it.
>
> (Wittgenstein 1975: §211)

If we are to follow Wittgenstein's suggestion, we should not be trying to fix boundaries between the area that is clearly marsh and the area that is not clearly marsh, or between the area that is clearly not marsh and the area that is not clearly not marsh; but rather one boundary within which all the marsh lies, and another boundary within which *only* marsh lies. The former is not intended to be minimal, nor the latter maximal.

A related approach has been suggested by Cohn within the framework of his RCC calculus of qualitative spatial relations (see §2.5.2). The *egg-yolk theory* (Cohn and Gotts 1996*a*; Cohn and Gotts 1996*b*) represents an indeterminate region by means of two sharp boundaries, an inner boundary separating the area that is clearly within the region from the area that is not clearly within it, and an outer boundary separating the area that is clearly outside the region from the area that is not clearly outside it. Thus between the two boundaries lies a 'buffer zone' for which it is not clear whether it should be included within the region or not. This indeterminacy may arise for various reasons. At one extreme there is the possibility that the buffer zone is genuinely intermediate in character between the inner and outer zones; this might be the case with our marsh, where the vegetation in the buffer zone contains some heathland species and some marshland species, or perhaps some species that are precisely characteristic of such intermediate zones. At the other extreme, the indeterminacy may be purely epistemic, arising from our own uncertainty as to where the boundary should be drawn. This may itself arise from two different sources: an uncertainty concerning what criteria should be adopted, and an uncertainty concerning the application of known criteria.

One might naturally think of refining the egg-yolk theory still further by interpolation of boundaries intermediate between the inner and outer boundaries. In the limit we arrive at a 'fuzzy' theory in which the feature we are modelling becomes represented as a function from points in space to real numbers in the interval [0,1]. A point is assigned the number α as a measure of the goodness of the attribution 'marsh' to that point. The inner boundary of the egg-yolk theory is the outer limit of points assigned the value 1 in the fuzzy theory, while the outer boundary is the inner limit of points assigned the value 0.

There is a deep-seated paradox at the centre of fuzzy theory. The aim of the theory is to eliminate the spurious precision that it finds in the sharp boundaries of classical theories; but the only way in which it is able to do this is by importing far greater precision. Whereas the classical theory need only determine for each point whether or not it lies in the marsh (or, exceptionally, on the boundary), in the fuzzy theory we must determine the exact value to be assigned to that point as a measure of its 'marshiness'. Thus the paradox of fuzzy theory is that it tries to model less precision with more. The obvious ploy to circumvent this difficulty is to admit higher-order fuzziness. Instead of saying that point P is marshy to the degree of exactly α, we might introduce a function which assigns to each degree of marshiness the degree to which point P is marshy to that degree. Clearly this is a desperate expedient: once again, to model the imprecision, yet greater quantities of precise information are brought into play. In the limit we must have an infinite amount of precise data concerning each point.

This is related to the problem of higher-order vagueness discussed at length by Williamson (1994). Williamson argues cogently that all attempts to model vagueness or indeterminacy by means of a proliferation of values intermediate between 'true' and 'false' must ultimately prove unviable, even incoherent. His view is that vagueness is an epistemic phenomenon. We find it difficult to fix the boundary of the marsh, not because the marsh is itself some kind of indeterminate entity, but because of our own ignorance. We do not know what criteria to adopt; as soon as we adopt a clear criterion for what constitutes a marsh, we can in principle fix the boundary determinately. Our difficulty arises from the existence of so many competing criteria all seemingly compatible with our understanding, enshrined in our normal use of language, of what a marsh is.

3.7. Alternative bases for classifying objects

The classification described above is by no means the last word in the classification of objects. In this section I briefly review two alternative bases for classifying objects, one with venerable philosophical roots, the other perhaps more modern in spirit.

3.7.1. Matter, form, function, and origin

If we consider an object such as a table we can classify it with respect to four different kinds of characteristics. The *matter* of the table is what it is made of: wood, metal, stone, or plastic, or some combination of these, as the case may be. This may be thought of generically, as a particular *kind* of material, or specifically as the particular quantity of matter which constitutes this particular table. The *form* of the table concerns the way the matter is put together to form the table: it has a flat rectangular top about 2 centimetres thick, supported by straight legs 70 centimetres long, of cross-section 3 centimetres square. The *function* of the table concerns the way it behaves, how it can be used, and what it is for. Its function is to act as a support to hold objects at a suitable height above the ground for ease of use by humans. Note that the three aspects of function—how something behaves, how it is used, and what it is for—can be separated; the mark of appropriate design and use is that the three aspects are in harmony. Finally, the *origin* of a table tells us about how it came into being: it was manufactured in a workshop somewhere out of wood that came from trees growing in such-and-such a plantation, and so on.

While any individual object can be classified under all four headings, *categories* of objects tend to be defined in terms of just one or two of them.

Artefacts are normally defined primarily in terms of function: the definition of a table need not refer to its matter, or origin, and although form is important it is function that is paramount. Of course matter is important, but only in so far as it conduces to the realization of the function: a table cannot be made of a liquid, for example, not because solidity is part of its definition, but because it is a necessary condition for something to function as a free-standing support that it be solid. Similarly, the form is subservient to function, which is what makes it possible for radically new designs of familiar artefacts to appear from time to time, as designers discover new forms that can implement the same old function.

For natural objects the other headings can play a more significant role in classification. Rocks, for example, are classified as sedimentary, igneous, or metamorphic, which are defined in terms of origins. In modern biology, organisms are also classified in terms of origin, that is, in terms of evolutionary history. This is in contrast to 'folk' biology, where form plays a much more important part—classifying whales and dolphins alongside sharks and goldfish, and bats alongside birds, for example. In astronomy, the classification of bodies as stars, planets, comets, galaxies, and various kinds of nebulae is largely based on matter and form, origin here being much less important. But function, in the sense of behaviour rather than use or purpose, can play a role here, too, for example in the classification of variable stars on the basis of the form of their light curves. (On the other hand, one might argue that behaviour is an aspect of form—*temporal* form, as it were—rather than function.)

These four bases for classification formed a central element in the philosophy of Aristotle, who introduced them under the name of 'causes' (*aitiai*). For Aristotle they applied more broadly than just to objects; they could apply to events as well. In Aristotle's terminology, they are the 'material' cause (*hē hylē*), the 'formal' cause (*to eidos*), the 'final' cause (*to kinēsan*), and the 'efficient' cause (*to hou heneka*). This terminology seems outdated to us nowadays, because in modern English what we generally mean by the term 'cause' is the efficient cause—and specifically the efficient cause of an event rather than an object. Thus we would not regard the shape of a table as in any sense causing the table, although for Aristotle it is indeed part of the formal cause of the table. We can to some extent reconcile ourselves to Aristotle's usage by thinking of all four elements as contributing to a complete *explanation* of an object: the table is what it is *because* it has the particular matter, form, function, and origin that it does. If it differed in any of these respects it would be in important ways a different table.

A weakness of Aristotle's classification is that it provides but a single category—the final cause—to cover a range of different, albeit related notions such as function, behaviour, use, and purpose. To some extent the English word 'function' shares this ambiguity. Modern jargon partly mitigates this by means of the notion of *functionality*, which refers to behaviour in so far as it is relevant to use or purpose, as distinct from those aspects of an object's behaviour which are merely incidental from the point of view of the use to which it is, can be, or is meant to be put.

When we speak of the functionality of a device we are referring to the way in which it is designed to interact with the outside world. The functionality tells us that if a certain input signal impinges on the device from outside, then the device will respond by sending a certain output signal back. The functionality of a calculator, for example, can be expressed by describing what output will appear on the display panel in response to each possible sequence of button-pushes. This functionality can be described quite independently of what actually goes on inside the calculator, although what goes on inside is what allows it to exhibit the functionality that it does.

From an abstract point of view, the functionality of a device is a mapping from the set of possible inputs to a set of outputs. It is thus a function in the strict mathematical sense. For the purpose of describing its functionality, not all possible impingements of the external world onto the device need count as inputs; for example, if you drop a calculator from the tenth floor of a tower-block, it will respond by breaking into irregularly shaped fragments. But we do not count this as part of the functionality of the calculator: it has to do with its behaviour as a physical object, but not with its behaviour as a device.[8]

If you want someone to construct a calculator for you, you will specify what kind of calculator you want. Some of what you specify will have to do with 'cosmetic' features such as the layout of the keypad, the colour of the casing, and the overall shape and size; but the essential part of your specification will be to do with the functionality of the calculator as a device. You will say what functions you want it to calculate, what sort of error messages it will give, whether you want it to have a memory or to be programmable, and so on. All these aspects together constitute a *functional specification* of the calculator. It is then up to the designer to work out what electronic circuitry is needed to implement this specification. You say *what* behaviour you want,

[8] As Dennett (1989) would put it, in considering the mathematical functions computable by the calculator, we are adopting towards it the *design stance*, whereas in considering the effects of subjecting it to various different forces, we are adopting the *physical stance*.

the designer works out *how* to achieve this.

The use of the word 'function' in everyday life is different from, though related to, the use of the word 'functionality' as just described. In everday English, the 'function' of a device can mean either 'what it does' or 'what it is for'. Both these meanings are present when we say that the function of the brake is to slow the vehicle down or the function of the accelerator is to speed it up. Clearly this notion of function is related to functionality, although the added dimension of *purpose* cannot be expressed in the abstract language of mathematical functions, since it has to do with people's intentions, which lie right outside the sphere of mathematics. Of all disciplines it is perhaps computer science in which both aspects, the abstract mathematical and human, are most thoroughly intermingled. Within computer science the need to systematically characterize objects in terms of their purpose and behaviour has given rise to the paradigm of *object-orientation*, the underlying philosophy of which is deeply pertinent to the issues discussed in this chapter.

3.7.2. Mobiles, fixtures, features, and fields

In this section I present an alternative basis for classifying objects, which is to some extent independent of the categories discussed so far. I begin by presenting the four new categories, which I shall then discuss in more detail:

1. A *field* is an extended object forming an environment within which objects of the remaining three categories have their existence.
2. A *feature* is an object occupying a fixed location within a field and inseparable from it.
3. A *fixture* is an object occupying a fixed location within a field but separable from the field.
4. A *mobile* is an object that is free to move within, or even outside, a field.

The prototypical objects of everyday life are mobiles. They include apples, bicycles, chairs, dishes, elephants, fax machines, guns, hats, and ice-creams. The objects we regard as mobiles are typically on a human scale, not more than two orders of magnitude larger or smaller than an individual human body. This does not mean that nothing larger than this can be a mobile, but rather that we as humans tend not to interact with such very large objects in their capacity as mobiles. The earth itself is a mobile, but in non-astronomical contexts we seldom adopt a perspective within which it is natural to treat it as such; on the other hand the sun, moon, and planets other than the earth do normally present themselves to us as mobiles, the relevant scale being their apparent rather than actual sizes. For us, mobility is especially salient in the case of animals, artefacts designed to be handled, and vehicles designed to be travelled in. As well as the animals, nature also provides us with mobiles in the form of chunks of matter (such as a pebble).

Looking about my room I see plenty of mobiles: books, pens and pencils, mugs, a desk lamp, a stapler, a telephone, my coat. These things are mostly placed on the surfaces of various larger objects: a table, a desk, a filing cabinet, a bookcase, the door. These larger objects can be regarded as mobiles in the sense that they could be moved, if I wished to move them, albeit with greater effort than is required for moving the smaller mobiles. But in fact they do not move. They stay in the same position for years on end, forming a fixed framework within which the mobiles proper move. These objects are fixtures. Some, like the bookcase, are in fact physically fixed, being attached to the wall by means of screws, although they could be moved if necessary. Others, such as the table and desk, owe their fixity to custom rather than physical attachment. Not all fixtures are large: the lights hanging from the ceiling, the light-switches, and the door-knob are all on the same scale as the typical mobiles.

The room itself constitutes the field within which all these mobiles and fixtures are situated. It includes various subfields such as the floor, walls, and ceiling, relative to which the fixtures in the room can be located: the larger items of furniture mostly standing on the floor adjacent to one or two walls, the light-switch on one wall, the lights on the ceiling. The door is attached to the door-frame, and this could be regarded as another fixture. The door-frame itself surrounds the doorway, which cannot be separated from the wall in which it is situated and therefore is neither a fixture nor a mobile: it is a feature, parasitic on the wall and hence on the room itself.

The categories considered here are all to some extent fluid and relative to a point of view.[9] I have already said that from one point of view the earth is a mobile, but from the normal everyday human point of view it is a field: more precisely, the surface of the earth is the field within which most human activity is located, the interior of the earth having at most a rather shadowy existence from this point of view. The features of this field are such things as

[9] The same point was made by Lewin (1936: quoted in Smith 1994) in relation to the categories of *object* and *region*: 'A hut in the mountain has the character of a thing as long as one is trying to reach it from a distance. As soon as one goes in, it serves as a region in which one can move about.'

oceans, islands and continents, lakes and rivers, mountains and plains. Each of these can be regarded as a field in its own right, with its own features (although oceans and plains are typically characterized as featureless). On this scale, forests and towns are perhaps best regarded as fixtures since in principle they are separable from the ground they occupy, although in practice any such separation will result in their destruction. On a smaller scale, forests and towns can themselves be regarded as fields. In a town, considered as a field, a street may be regarded as a feature, with the individual buildings along it as fixtures. Or the street itself may be regarded as a field, its features being such things as bends, humps, and dips.

In general the surface of any object can be regarded as a field, with its parasitic superficialities (bumps, holes, wrinkles) as features, and objects placed on or moving about on it as fixtures and mobiles. If a human face is regarded as a field then its features are precisely the things we normally call facial features—eyes, nose, mouth, ears—whereas accoutrements such as spectacles and earrings are fixtures. In this field there are usually no mobiles since on the whole things do not move about on a face (though the features themselves are capable of limited movement, giving rise to a wide range of facial expressions). But an insect crawling about on someone's cheek would be an example of a mobile temporally located within the field of that person's face.

The distinctions introduced here are related to various distinctions that have been discussed in the literature. The distinction between fields and features on the one hand and fixtures and mobiles on the other corresponds quite well with a fundamental distinction between two ways of representing geographical data in Geographic Information Systems (GIS); the distinction in question has been variously labelled 'raster-based' versus 'vector-based' (Peuquet 1984), 'field-based' versus 'object-based' (Worboys 1994), or 'location-centred' versus 'entity-centred' (Yuan 1994). Couclelis (1992) has an interesting general discussion of some of the issues involved in the debate between proponents of these two representational styles.

To illustrate the distinction, on the right-hand side of Figure 3.2 are shown two different representations of the two-dimensional figure shown on the left. In the *vector-based* method, the figure is represented as a polygon which can then be described in terms of the sequence of vertices, each represented by a vector giving its co-ordinates:

$$(2.5, 5),\quad (3, 7),\quad (4, 9),\quad (6, 10),\quad (7, 9),\quad (8, 8),\quad (9, 8),$$
$$(9.5, 6),\quad (9, 4),\quad (7, 4),\quad (6, 3.5),\quad (5, 3),\quad (4, 3),\quad (3, 4).$$

Here the resolution is half a unit; by increasing the resolution and increasing the number of points on the polygon, one can obtain arbitrarily close approximations to the true figure, at the cost of a corresponding increase in the amount of storage space required. Note that this is essentially a representation of the *boundary* of the figure.

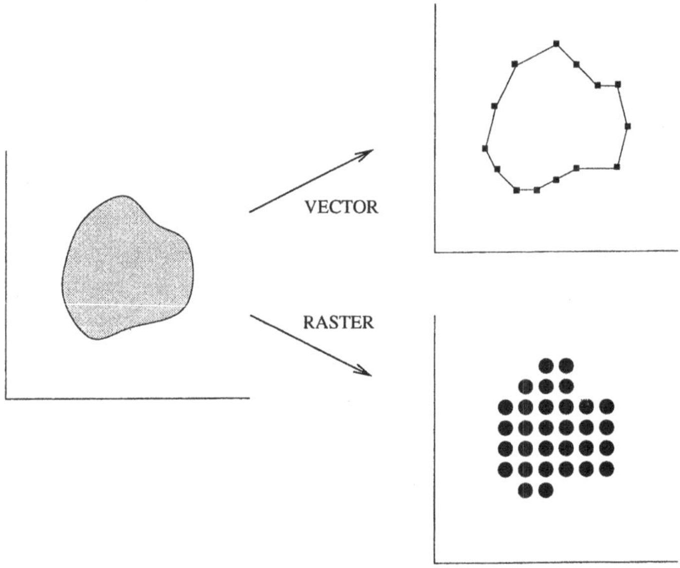

Figure 3.2. Vector-based and raster-based representations of the same two-dimensional figure

In the *raster-based* method, on the other hand, it is the area covered by the figure that is represented. Space is imagined to be divided up into a grid of squares (other divisions are possible, of course) and the figure is represented by a set of these grid squares, namely those squares more than half of which fall within the figure to be represented; rather than list all 31 squares one can adopt a more compact form of listing, such as a 'run-length encoding':

Column	Rows
4	5–8
5	4–9
6	4–10
7	5–10
8	5–8
9	5–8

Here too, by increasing the resolution (choosing a smaller-sized grid square), one can approximate the original figure as closely as one likes.

One may say as a general rule that the raster-based method is most suitable for representing the kinds of objects that I have called fields and features, whereas the vector-based method is most suitable for representing fixtures and mobiles (Couclelis 1992). Geography tends to be dominated by fields and features rather than fixtures and mobiles, so it is easy to concur with Couclelis's advice to choose raster-based representations for GIS if in doubt, but, better still, to 'demand a system that supports both raster and vector equally and that allows the purposive creation of temporary objects [i.e. fixtures or mobiles] out of features'.

To an extent, the raster/vector distinction in GIS can be aligned with the distinctions between mass terms and count terms and between continuous and discrete distributions. Raster-based representations are well adapted to the depiction of *kinds of terrain* such as woodland, grassland, marshland, and desert, for which, as in the present sentence, the natural designation is by means of mass terms. To each raster cell are assigned the particular attributes defining the nature of the terrain at the spatial location represented by that cell. Vector-based representations are rather adapted to the depiction of individual pieces of territory, each clearly delineated by means of a boundary, which may be more or less artificial: individual forests, fields, marshes, deserts, as well as man-made territories such as countries, counties, or parishes. For these, as illustrated in the last sentence, the natural designation is by means of count nouns.

The fourfold distinction between fields, features, fixtures, and mobiles is also related to a body of work concerned with discriminating a range of spatial scales from the psychological point of view. The underlying idea here is that as human beings our conceptualization of space is highly scale-dependent: we interact with, and hence apprehend, spatial entities much larger than ourselves in ways quite different from how we interact with spatial entities of comparable size to or smaller than ourselves. This leads to the idea of a series of cognitively dissimilar spaces defined in terms of the way in which we humans apprehend and interact with them.

Typical of such classifications is that of Zubin, cited in Couclelis 1992, from which the quoted passages in the following enumeration are taken:

- *A-space* consists of manipulable objects contained in a static visual field.
- *B-space* consists of larger, non-manipulable objects whose visual appear-

ance is 'pieced together from many non-simultaneous views'.

- *C-space* consists of 'the vast landscapes seen from some vantage point—apprehended through a sweeping glance, but otherwise not directly accessible to sensorimotor experience'.
- *D-space* is 'beyond the range of direct experience' and must be 'pieced together by means of a very diverse range of spatial and other knowledge, information, and belief'.

What Zubin calls 'A-space' is sometimes called *table-top space*, while 'D-space' is what is often called *geographic space*. Researchers with a strong cognitivist bias often claim that the formal space of mathematics and science has no objective validity, being merely a projection to all scales of the properties of A-space; the mathematicians and scientists themselves might counter with the observation that all of Zubin's spaces are the result of viewing objective physical space through the filter of human cognition, which reveals different facets of the same underlying reality under different circumstances.

A similar classification, drawing on earlier work including that of Zubin, was proposed by Montello (1993):

- *Figural* space, which is or appears smaller than the body and whose properties 'may be directly perceived from one place without appreciable locomotion' (divided into two-dimensional *pictorial* space and three-dimensional *object* space).
- *Vista* space, which is or appears larger than the body but 'can still be apprehended from a single place without appreciable locomotion'.
- *Environmental* space which 'is larger than the body and surrounds it' and cannot be apprehended directly without 'considerable locomotion'.
- *Geographical* space which 'is much larger than the body and cannot be apprehended directly through locomotion'.

Montello's geographical space corresponds well with Zubin's D-space; and likewise figural space (or at least object space) corresponds with A-space. But Montello's vista space is Zubin's C-space, which leaves B-space and environmental space unpaired.

How do these spaces relate to our classification of objects? There is no direct correspondence, but it is tempting to suggest that the spaces earlier in the two lists are dominated by fixtures and mobiles, whereas the later spaces are dominated by fields and features. Having considerable sympathy with the mathematical and scientific view that there is but one space in nature, I believe that the whole issue here is most satisfactorily handled in terms of objects rather than spaces. Spaces are defined by the objects they contain, and these objects can play different roles in accordance with the different ways in which we as humans identify and interact with them. In different situations we may conceptualize the same objects differently.

In support of this view we may adduce the fact that the most fundamental components of our spatial language, the prepositions, operate in much the same way across all the scales. The differences that do appear relate to the different kinds of object. For example, the preposition 'at' almost invariably applies to features and fixtures, which serve as landmarks relative to which the position of a mobile can be specified. It is thus an asymmetric relation, despite the fact that its 'ideal' meaning is 'for a point to coincide with another' (Herskovits 1986); we can say that the train is at the station, but it would take an exceptional set of background assumptions to render intelligible the statement that the station is at the train. 'At' can only be used with a mobile in a context in which the mobile can be regarded as a fixture—in particular, not moving with respect to its immediate environment. We can only say that someone is 'at the car' when the car is parked, thereby defining, *qua* fixture, a relatively fixed location in space.

3.8. Summary

In this chapter I have discussed a number of different principles by which the variety of kinds of object can be classified.

First, I noted that objects possessing spatial attributes must be dependent in some way on matter, and I used the nature of this dependence as one principle of classification, furnishing us with the categories *material*, *parasitic*, and *perspectival.* Within the category of material objects, I noted that object identity may be more or less strongly coupled with the identity of quantities of matter, giving us a two-level ontology that can provide the basis for a workable formalization in the predicate calculus. The degrees of coupling between object and matter provided us with another principle of classification, using the categories *piece of matter*, *heap*, *assembly*, *organism*, and *flow*. I do not claim to have a complete classification here.

A separate principle of classification concerns the extent to which the individuality of an object is determined by nature itself rather than imposed on nature by human cognitive acts. This leads us to a broad distinction between *natural* (or *bona fide*) and *conventional* (or *fiat*) objects. This distinction cuts across the categories previously listed.

A fourth principle of classification is provided by a consideration of the various kinds of explanation we can give for an object's being what it is, in terms of *matter*, *form*, *function*, and *origin*, this fourfold division reflecting Aristotle's system of the four 'causes' (material, formal, final, and efficient). I noted in particular that the category of 'function' seems to conflate a number of distinct elements such as behaviour, use, and purpose.

Finally, I introduced a classification based on the role played by objects in relation to each other and to human experience. I distinguished four categories of *field*, *feature*, *fixture*, and *mobile*, and related these to some previous discussions of different kinds of psychological space.

After all this discussion we have not arrived at a final definitive classification. The main role of this chapter is to draw attention to the extraordinary complexity of the notion of an 'object'. The lesson to be learnt from this is that we must tread carefully in our subsequent discussion, where inevitably objects of various kinds will play an important part. The crucial thing is not to forget that there are many different kinds of objects, differing in respect of their logic and ontological status, so that almost any claim about objects in general is sure to have abundant counter-examples.

4

Spatial Attributes and Relations

It is perhaps not possible to identify a clearly delineated subset of the attributes of an object which are to count as the spatial ones. Attributes may be more or less spatial. At one extreme are those properties of an object, such as shape, size, and position, that can be expressed in purely geometrical terms; they can be regarded as properties of the space occupied by the object just as well as properties of the object itself. Next come physical properties which include a spatial element as an essential component of their definition, but which are not purely geometrical; an example is density, which involves both volume (geometrical) and mass (non-geometrical). Third, there are attributes which are spatial only in the sense that they are dependent on more immediately spatial attributes; an example is colour, which can only be attributed to extended, and hence spatial, objects. In this chapter I draw the line between the second and third categories: I shall regard density as a spatial attribute, but not colour. One reason for this is that *change* in density always involves an overt spatial change, whereas change in colour need not.

Amongst purely geometrical attributes we distinguish those which require measurement for their specification from those which do not. The former are metrical properties, the latter topological. Topological attributes are those that are preserved under arbitrary continuous deformation: examples are dimension and connectivity, which are discussed in the first two sections below. Metric attributes are those that involve some form of measurement, either of length or of angle, in their specification. We distinguish between intrinsic metric attributes such as size and shape, which involve only measurements within the object itself, and extrinsic ones such as position, which involve measurements relating the object to its environment.

It would be possible to interpolate a number of other categories between the metric and the topological. 'Projective' properties are those which are preserved under arbitrary projections: they include collinearity, incidence, and cross-ratio. 'Affine' properties are preserved under *parallel* projections: in addition to all projective properties, they include parallelism, and division of a line in a given ratio. For further details, see Brannan *et al.* 1999.

4.1. Dimension

Under this heading we distinguish between zero-dimensional or *pointlike* objects, one-dimensional or *linear* objects, two-dimensional or *areal* objects, and three-dimensional or *solid* objects. We must further distinguish between 'strict' and 'apparent' dimension. In three-dimensional space, a strictly n-dimensional object has extension in n dimensions and is therefore extensionless in the remaining $3 - n$ dimensions. An apparently n-dimensional object is extended in three dimensions, but its extension in n of the dimensions is very much greater than its extension in the remaining $3 - n$, with the result that in these latter dimensions it can be regarded as effectively extensionless.

Since all material bodies are extended in three dimensions, strictly n-dimensional objects, where $n < 3$, can only be parasitic or virtual, and can moreover only be recognized as part of an idealized representation of reality. A perfect cube has faces which are strictly two-dimensional, edges which are strictly one-dimensional, and vertices which are strictly zero-dimensional. At the macroscopic scale of everyday experience, many objects, especially manufactured ones, come close to exemplifying perfect polyhedral forms with well-defined faces, edges, and vertices. The table at which I am working, for example, has a pretty nearly smooth rectangular top whose edges are very well defined. We know, of course, that at a microscopic scale, roughnesses appear which make it harder to characterize the strictly lower-dimensional parasites on the three-dimensional form, and once we are down to the level of individual atoms, the band of uncertainty in the definition of these parasites becomes relatively enormous.

A somewhat different form of scale-dependency is illustrated by those objects whose lower dimensionality is apparent rather than strict. Consider a road, for example (Hobbs 1985). To the maker of a road map, a road is essentially a linear object, since the true width of the road does not show up on the scale of such maps—the fact that the road is none the less portrayed on the map as having some width is a concession to usability, a convenient fiction that is at variance with a strict interpretation of the map's scale. On a larger scale, to the traveller on the ground, the road becomes areal, since what one travels on is a surface, having significant width—only at this scale does the injunction to drive on the left (or right) side of the road make sense. From the point of view of the builder of the road, the third dimension becomes salient: although all the traveller needs is a surface to travel along, the builder can only provide this by providing a solid object whose surface it is. Thus the road has depth, and the road-builder must devote much of his efforts to the road's constitution as a material body.

Landau and Jackendoff (1993) provide a special place in their categorization of objects for 'objects best described as surfaces', which are 'principally extended in two dimensions, with a relatively negligible thickness'. They draw attention to the existence of a set of general terms for what they call 'thickened surfaces', such as *slab, sheet, layer, slice, lamina*, and *stratum*. These are good examples of objects which in our terminology are apparently two-dimensional. They point out that for such objects the notion of *enlargement* only applies to the two non-negligible dimensions; thus an enlargement of a photograph will have a larger area than the original, but is not expected to be thicker.

The difference between strict and apparent dimensionality can be readily portrayed in the context of a raster-style representation. Figure 4.1 shows a road running from east to west across the border between two neighbouring countries. In (*a*), where the scale is small, both the road and the border are portrayed as linear features one raster cell wide. In (*b*), the scale is larger, and now the road is shown as having a width of four raster cells; the border, however, remains of minimal thickness. In (*c*), an alternative form of representation is shown in which the two countries are indicated by a difference in shading, and now the border between them becomes represented as the strictly linear feature that (in idealization at least) it is.

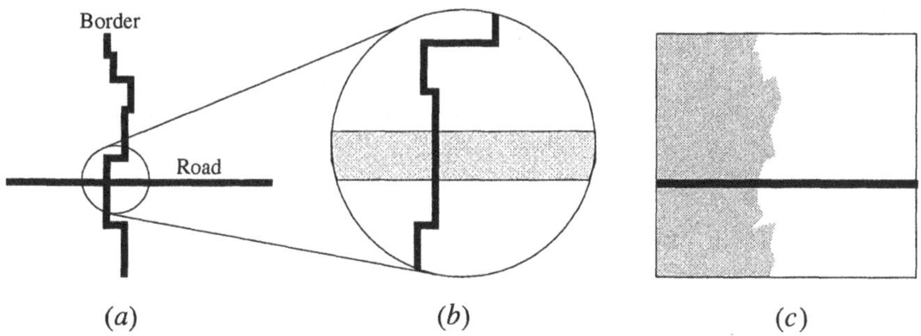

Figure 4.1. Three representations of a road crossing a border

In Chapter 2, I mentioned the idea of spatial regions having mixed dimension such as a sphere with linear 'spikes' (p. 35). Objects of mixed strict dimension tend to be rather contrived. In Figure 4.2 is shown a series of three idealized buildings, each consisting of a circular tower attached to a section with a gabled roof. Below each building is illustrated, in plan view, the object named by the expression 'the topmost part of the building'. It will be seen that the middle building has a top which consists of a linear edge (the top of the gable) and a circular face (the top of the tower). Can we conceive of any situation in which we might wish to speak of such an object? It is possible: at any rate, a theory which explicitly rules out mixed dimension leaves any such object unrepresentable.

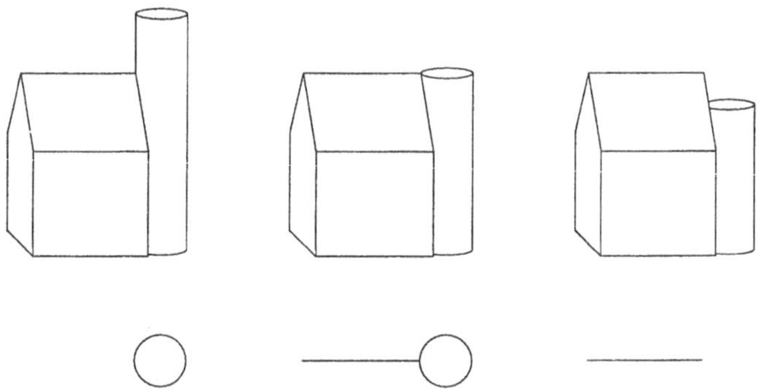

Figure 4.2. Three buildings and their tops

If natural objects of mixed strict dimension are hard to come by, objects of mixed apparent dimension are, perhaps, more frequent. A river might broaden into a lake, which then narrows again, thereby forming an object with two linear sections and an areal section. Similarly, an idealized model of a sailing-boat might consist of a three-dimensional part representing the body of the boat, a two-dimensional part (the sail), and various one-dimensional parts (the mast and cables). A cactus can be thought of as a three-dimensional succulent body with many one-dimensional spines attached.

4.2. Connectivity

The topological definition of connectedness given in §2.3.1 is that a point-set is connected if, and only if, whenever it is contained in the union of two disjoint open sets, it is wholly contained in just one of those sets. A more intuitive rendering of this definition, in the case of sets in $\mathbb{R}^n$, is that the set S is connected if, and only if, between any two points in S there is a continuous path lying wholly within S. A continuous path is the image of the unit interval [0,1] under a continuous one-to-one transformation. To see why these definitions are equivalent, suppose first that $S \subseteq S_1 \cup S_2$, where S_1 and S_2 are disjoint, open, and non-empty, and consider a path from $p_1 \in S \cap S_1$ to $p_2 \in S \cap S_2$. Let this be the image of [0,1] under the continuous map f, where $f(0) = p_1$ and $f(1) = p_2$. As we move along the path from $f(0)$ towards $f(1)$, we start in S_1 and eventually leave S_1 (since we have to end up in S_2, which is disjoint from it). By continuity, there must be either a greatest value of x such that $f(x) \in S_1$ or a least value of x such that $f(x) \notin S_1$. The former case is not possible since S_1 is open. Let x_0 be the least value for which $f(x_0) \notin S_1$. Then $f(x_0) \in \partial S_1$, and hence cannot be in S_2, since S_2 is open but disjoint from S_1. Hence $f(x_0) \notin S_1 \cup S_2$, so $f(x_0) \notin S$. Any continuous path from $S \cap S_1$ to $S \cap S_2$ has to pass outside S.

When we come to consider objects rather than point-sets or regions, the issue is complicated by the difficulty of finding a single physical analogue of connection. The problem is that objects and matter are not just occupiers of space; they have their own dynamics. Consider the difference between two sheets of A4 paper lying side by side flat on the table, and one sheet of A3 paper lying flat on the table in exactly the same position. As far as spatial occupancy is concerned, these situations can be virtually indistinguishable: in both cases the region occupied by paper is a very thin cuboid measuring approximately 296mm×420mm×0.1mm. But physically what we have is different. If the paper at the left-hand edge is pulled away from the centre, in the first case only half the paper moves (the left-hand A4 sheet), whereas in the second case it all moves (the whole A3 sheet).

At a microscopic scale the difference between these two situations is easy to account for, at least in principle. Paper essentially consists of a tangled mass of densely matted fibres. If a fibre traced along its length meets the edge of the paper, it abruptly comes to an end or bends back into the interior of the sheet. When two sheets are placed side by side, there is a sharp discontinuity between them: no fibres cross from one sheet to the other, whereas between any two points in the same sheet many paths can be traced along chains of overlapping fibres. In this way we can also distinguish between an intact sheet of paper and one which has a cut in it without being disconnected.

One way of illustrating this is to introduce a crude 'atomistic' level of representation to act as an intermediary between the complex physico-chemical picture and the idealized abstract topological picture. At this level, matter is represented by discrete 'atoms' linked together by inter-atomic forces. These forces are all or nothing in the sense that any two atoms either are or are not linked by a force. A chunk of matter is represented as a collection of atoms between any two of which can be traced a path from atom to atom following the inter-atomic links. For convenience we assume that the atoms in a chunk of matter are arranged in the form of a cubic lattice (represented in diagrams by a square lattice).[1]

The topological picture is derived from the atomistic picture as follows. Round each atom is inscribed an open cube whose sides are equal in length to the distance between two adjacent atoms. When two atoms are linked, the union of the corresponding cubes is replaced by the interior of its closure. Thus the points on the face separating the two cubes are added. The piece of matter is identified with the region thus produced.

In Figure 4.3 are illustrated a number of phenomena as they appear in the atomistic and the topological representations. The atomistic picture simply shows atoms and their linkages; the topological picture is superimposed on this in the form of dashed lines representing the surfaces of bodies. In (*a*) and (*b*) are portrayed the two situations discussed above: the two A4 sheets compared with the single A3. In (*c*) is portrayed an A3 sheet which has been cut more than half way across; in (*d*) is an A3 sheet with an internal cut. Note that in the topological representation these last two cases are not regular: parts of their surfaces are adjacent only to their interiors. By contrast the *regions* occupied by paper in all four cases is the same: the interior of the large A3-sized rectangle (ignoring the third dimension for the moment). This nicely points up the dissociation between the matter and space. The illustrations here are highly schematic in that the 'atoms' are shown as relatively very large; this is for convenience only. Note that there is a rough correspondence between the atomic/topological representations on the one hand and the raster-based/vector-based distinction on the other (see p. 137).

This model allows us to distinguish between mere adjacency on the one hand and complete physical coherence on the other. But these two conditions are merely the extreme points of a range of physical conditions which can be described as forms of connection, any one of which could be used as the

[1] This atomistic picture could be modelled in an Incidence Space (see §2.6.2). The lower-dimensional elements—edges and vertices—can then be seen as the 'glue' which binds together the atoms (cells). The relationship between the atomistic and continuous models then becomes clear: the former is a quotient topology of the latter.

(a) 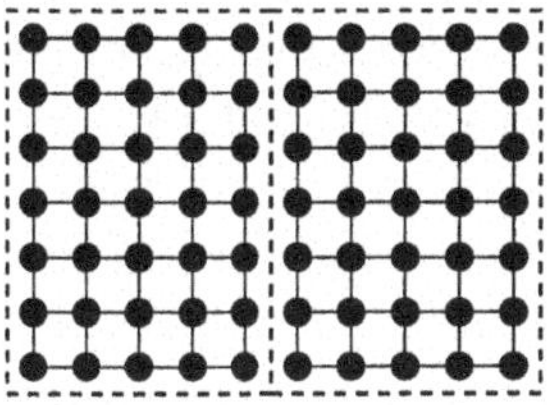 (b) 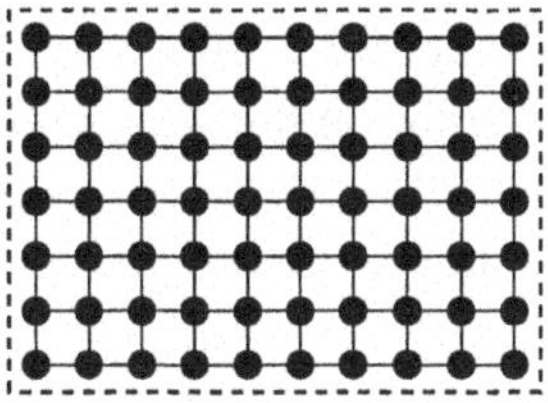

(c) 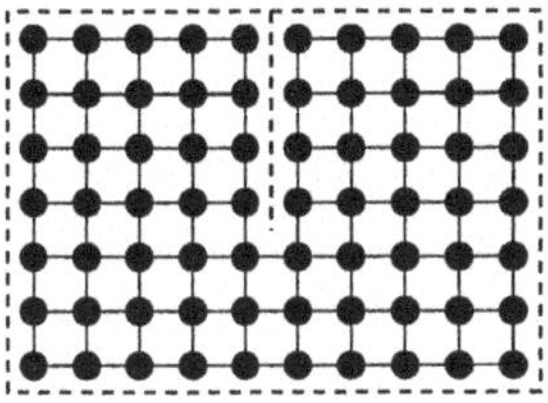 (d) 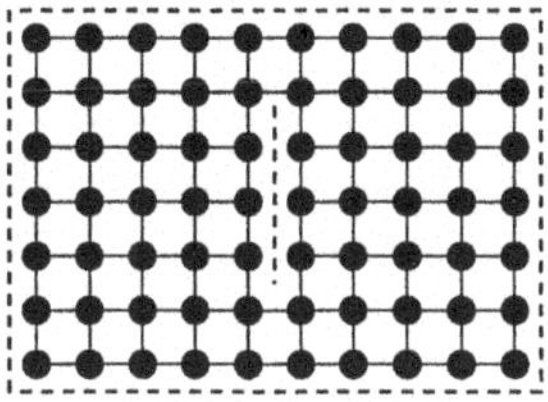

Figure 4.3. Atomistic and topological representations of material bodies: (*a*) two sheets of A4 paper lying side by side; (*b*) a single sheet of A3 paper; (*c*) a sheet of A3 paper partly cut in two; (*d*) a sheet of A3 paper with an internal cut

basis for an account of physical connectedness. In the following enumeration, only connection between disjoint objects is considered; non-disjoint objects are trivially connected by virtue of their possessing a common part. Running in order from the weakest form of connection to the strongest, we may distinguish:

- *Adjacency.* This is a purely geometrical notion, there being no other form of physical connection involved. It is the RCC notion of 'external connection' (EC), as a purely geometrical relation. Our two A4 sheets lying side by side are connected in this way.
- *Support.* Support involves adjacency together with some force tending to preserve it. The force results from the configuration of the two objects; it can be supplied by gravity, friction, or air pressure, amongst other possibilities. A book resting on a table is connected to the table by virtue of being adjacent to it, but in addition the book exerts a force downward on the table, which the table reciprocates with an equal upward force. The table is responsible for the book's remaining where it is; if the table is moved without being tilted, the book moves with it (contrast this with the case of the A4 sheets: one sheet can be moved without disturbing the other), but if the table is tilted too far then the book will slide off. In fact the book and the table can be separated with no more effort than

is required to move the book. Other examples of support are a picture hanging on the wall, clothes on a person's body, and a fly clinging on to the ceiling by means of the suction pads on its feet (the force in this case being provided by the pressure of the surrounding air). In all these cases the support relation is asymmetric: one object is supported by another. The preposition 'on' is characteristically used, in English, to express this physical relation. But there can be symmetric support as well, as when two planks of wood are leaned against one another to form a triangular arch.

- *Interlocking.* Two objects are said to interlock when their shapes and relative position are such that under a wide range of conditions they cannot become separated. This includes topological linkage such as we see between consecutive links in a chain: here separation can only be effected by breaking open one of the links. This kind of linkage is simulated by the relationship between a key and a key-ring; but in this case the severance of the link is accomplished by a clever trick dependent on the tight helical form of the key-ring. Interlocking of a different kind is seen in the way in which a screw is embedded in a piece of wood; here another clever trick—anticlockwise rotation about the axis of the screw —is required to separate them. Human ingenuity has produced a wide range of different devices whose purpose is to secure attachment by means of interlocking: bolts, locks, hinges, anything with a screw thread, dovetail and mortice-and-tenon joints, buttons, hooks, and so on.

There is no sharp cut-off between interlocking and support. When a wedding-ring is placed on a slim young finger, from which it can be readily removed with only a very little effort, we have a case of support; as the finger grows older and stouter and the joints swell, ever more effort is needed to remove the ring. Eventually, perhaps, it can only be removed by cutting; support has been converted to interlocking. The example of a person wearing clothes, given above as a case of support, could also be construed as a kind of interlocking. There are other problematic cases: is holding hands support or interlocking? It depends on how firm the grip is! The clearest cases of support are when the objects can be separated without otherwise affecting them in any way (a book on the table); the clearest cases of interlocking are when one or both objects have to undergo a topological change of form before they can be separated (the links of a chain). Cases involving an intermediate degree of disruption are appropriately regarded as intermediate between support and interlocking. In the case of a jigsaw puzzle the interlocking is weak, but there is no support involved in the relationship between adjacent pieces. They are only interlocked with respect to horizontal movements; all that is needed to separate them is a small relative vertical displacement.

- *Adhesion.* Adhesion involves the existence of some physical attachment between the two objects, which must be broken in order to separate the objects. Adhesion is not dependent on the existence of outside forces: objects that adhere to one another will continue to do so in a vacuum or in free fall. The paradigmatic case is two objects that have been glued together. Adhesion can be closely simulated by connections which on close analysis prove to be examples of support or interlocking. For example, if a cabinet is attached to the wall by means of screws, there is no adhesion between cabinet and screws, screws and wall, or cabinet and wall; what holds the cabinet in place is the forces transmitted from the wall to the cabinet by way of the screws; they arise from the mutual configuration of cabinet, screws, and wall. This involves both support and interlocking. Viewed on a larger scale, the cabinet effectively adheres to the wall (the importance of granularity again). Similarly, when the fruits of plants like burdock (*Arctium*) and goosegrass (*Galium aparine*) attach themselves to the fur of passing mammals (or the clothes of people), what looks (and feels) like adhesion is, when viewed on a sufficiently small scale, seen to be interlocking between the hooked barbs of the fruit and the individual hairs or threads of the coat to which it attaches.
- *Cohesion.* The strongest form of connection occurs when there is no physical discontinuity between the objects that are connected; we shall say that the objects cohere to form a single, coherent whole. The human head is connected to the neck which, together with the limbs, is connected to the torso. All these connections are examples of coherence.[2] Because there is no physical discontinuity, the boundary between the cohering objects can only be specified geometrically; as a result, there is an element of conventionality about just where the boundary should be drawn (recall Barry Smith's remark, cited on p. 123, about the division of the femur). Similarly, the two A4-sized halves of our sheet of A3 paper cohere. Because adhesion typically involves a sharp physical discontinuity whereas cohesion involves no such discontinuity at all, there is the possibility of a

[2] On the other hand, when a body is reduced to a mere skeleton, the component parts—individual bones—do not cohere, and the structure is now readily disarticulated.

gradation between the two forms of connection, according as the physical inhomogeneity is more or less complete. Body parts such as hair and nails are made of a different substance from the skin to which they attach, so there is a fairly sharp discontinuity there; but they grow out of the skin, so on close inspection there is an element of continuity too: we seem to be dealing with something intermediate between adhesion and cohesion. Similarly, the leaf of a tree starts off as a continuous outgrowth from the twig, and is thus coherent with it; but by the end of the season it has become merely adherent, separated from the twig by a thin layer of small fragile cells allowing the leaf-stalk to snap off readily at this point. When we glue together the pieces of something we have broken we are using adhesion to simulate the original cohesion.

With these various forms of connection in mind, we can now proceed to discuss the topological attribute of connectivity as applied to objects in space. To the question whether a given object is connected, one must first respond by asking which sense of connection is intended. Is a teapot a connected object? Yes and no: it hinges on the nature of the connection between the container and the lid. Yes, it is connected if by 'connected' you mean adjacent or supporting; but no, it is not connected if you mean interlocking, adherent, or coherent. Often several different forms of connectedness are employed together in the same structure. In a brick wall, there is support, interlocking, and adhesion in varying degrees.

The most drastic kind of unconnectedness is when an object has parts which are not even adjacent: in this case we call the object 'scattered'. An example is the teapot when the lid has been separated from the container. It would be perverse to deny that the teapot still exists in this case, so for the duration of the separation we must regard it as a scattered object. More generally we can regard an object as disconnected when it is 'scatterable', that is, it has parts whose connection is at most that of support. The parts of such an object will be sometimes scattered, sometimes adjacent, without compromising the integrity of the object itself. In these cases convention often plays a larger role than nature in determining the identity of the object. Some examples are: a newspaper, whose pages, not being stapled together, can be readily scattered and then brought together again; a crowd, consisting of individual people who are at most connected by support; an altarpiece consisting of several panels which have been distributed amongst different art galleries; a jigsaw puzzle (as mentioned above, there is an element of interlocking here when the pieces have been assembled correctly).

In the case of objects not capable of movement, the distinctions amongst the various modes of connections tend to collapse. This is the case with many geographical objects, for which connectedness can reduce to mere adjacency. The old Welsh county of Flintshire was disconnected because it consisted of two disjoint, non-adjacent parts.[3] After the break-up of the Soviet Union and the formation of the Baltic states of Lithuania, Latvia, and Estonia, Russia retained a small enclave beside the Baltic sea, centred on the city of Kaliningrad, thereby becoming disconnected.[4]

There is more to connectivity than just the distinction between connected and scattered objects. Connected objects themselves exhibit a variety of configurations in which their parts are connected to each other. Important phenomena here are tunnels and cavities.

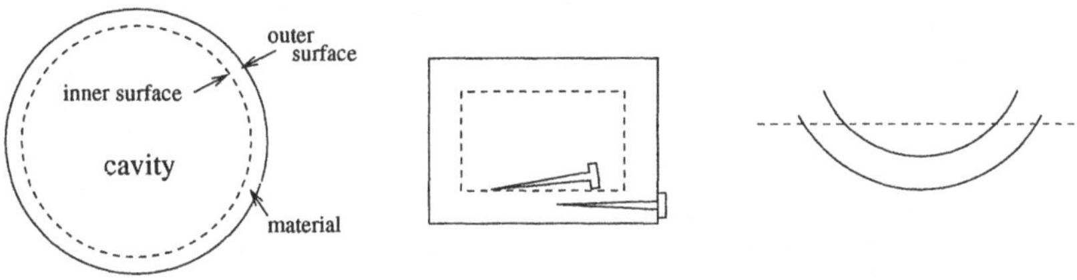

(*a*) A hollow ball (*b*) Two nails in a box (Herskovits) (*c*) A bent tube as a container (Cohn)

Figure 4.4. Connectivity and containment

Cavities appear when the surface (or boundary) of an object is disconnected. Think of a hollow ball, with an outer surface and an inner surface (Figure 4.4*a*). It is impossible to move continuously from a point on the outer surface to a point on the inner surface without passing through the material of which the ball is made. In the case of objects like this, the terms 'interior' and 'exterior' are ambiguous. 'The interior of the ball' can refer to the cavity which is bounded by the inner surface of the ball, or to the region

[3] As a child I had a wooden jigsaw puzzle with pieces in the shape of the counties of England and Wales; every county appeared as a single connected piece of wood apart from Flintshire, which was represented by two pieces, and Hampshire, which is divided into a mainland part and the Isle of Wight.

[4] Smith (1999) cites an extreme example of 'Baarle Nassau and Baarle Hertog, a pair of mutually embrangled (Dutch and Belgian) inland archipelagos in the region south of Turnhout'. See http://wings.buffalo.edu/philosophy/faculty/smith/baarle.htm

occupied by the matter of which the ball is made. This ambiguity extends to the use of the preposition 'in'; Herskovits (1986) and other authors such as Lyons (1977: 967) have drawn attention to the ambiguity of sentences like 'The nail is in the box', which can mean either that the nail is embedded in the wood of which the box is made, or that it is in the interior cavity (Figure 4.4*b*).

Although this discussion naturally arises at this point it applies more broadly to any object that can be used as a container, such as a bowl. In the earth's gravitational field, objects do not normally move upwards, so a container for liquids or solids can be open at the top. Cohn *et al.* (1995) characterize the 'containable inside' of an object as a region of space completely bounded by a portion of the object's surface and a plane. If the object is so oriented that the plane is horizontal, then the containable inside can be filled with liquid up to the level of the plane. An interior cavity (as in our hollow ball) is the limiting case of a containable inside in which the plane is no longer needed. Cohn's characterization neatly explains why, for example, a length of straight tubing does not constitute a container (since no one plane, together with any portion of the tube's surface, completely bounds any part of space), whereas a bent tube can do so (Figure 4.4*c*). Containable insides are, of course, parasitic objects, like holes, to which they are closely related. Several authors (Miller and Johnson-Laird 1976; Hayes 1985; Landau and Jackendoff 1993) have drawn attention to the importance of the general class of *containers* in our ontology of everyday objects.

Whether an object is regarded as having a cavity may depend on the nature of the connections between its parts. A container with its lid in place has a temporary cavity which disappears when the lid is removed. But if the lid is permanently fixed to the container (by adhesion) then the cavity likewise becomes permanent. 'Containerhood' is also sensitive to granularity—a fact exploited by sieves, colanders, filters, and similar devices, which are designed to function as containers at a coarse granularity but not at a fine one.

An object can have any number of cavities, and the cavities themselves can have complex connectivity. Complex connectivity can also arise in the case of objects whose boundaries are connected. The simplest example in three dimensions is a torus, which has a connected surface but still manages to have a hole. The difference between a torus and a spherical shell is that the hole of the torus is connected to the outside, whereas the hole of the spherical shell is a cavity, disconnected from the outside. Objects with connected surfaces can be classified according to how many tunnels they have. This is their topological genus. Any region that is continuously deformable into a sphere with n non-intersecting holes has topological genus n. When the holes intersect, things can be more complicated, but any reasonably well-defined object in three dimensions must be topologically equivalent to a sphere with some number of non-intersecting holes. Topological equivalence does not necessarily imply equivalence in any more useful sense; for example, a block with a Y-shaped tunnel is topologically equivalent to a block with two separate simple tunnels (Casati and Varzi 1994: fig. 5.17), but for most purposes one would regard these as quite different figures. It is well known that a teacup is topologically equivalent to a doughnut, but only the former can function as a container.

In two dimensions, connectivity exhibits less variety. The region enclosed between two concentric circles, an *annulus*, has a cavity, namely the interior of the inner circle, whose perimeter forms the inner boundary of the annulus itself. Some geographical regions take this form. For example, South Africa completely surrounds Lesotho, its boundary thus having two components which are disconnected from each other. The two-dimensional analogue of a tunnel is a passage separating two disconnected components. The only way of bringing in further variety in the connectivities of objects in the plane (or on the surface of a sphere) is by either increasing the number of components or increasing the number of holes. The 'connectivity type' for a planar object can therefore be exhaustively specified by a finite non-increasing sequence $(n_1, n_2, \ldots, n_m)$ of natural numbers, indicating an object consisting of m components of which the ith has n_i holes. Planar connectivity space consists of all possible sequences of this type. Figure 4.5 shows representative examples of the eleven connectivity types for which the number of components plus the number of holes does not exceed four. The structure of this connectivity space is discussed below in §6.2.

Whereas for planar objects the number of boundary components is exactly correlated with the number of 'holes', the latter being one less than the former, in three dimensions we have to be careful to specify what kind of hole we are talking about. The number of boundary components of a connected three-dimensional object is always one more than the number of internal cavities; but the number of tunnels is independent of this. Moreover, the internal cavities can themselves be tunnelled. Thus it is obvious that the connectivity space for three-dimensional objects is much more complex than for two. I shall not attempt to map it out here in more detail.

All the forms of connectivity discussed here are sensitive to variation in

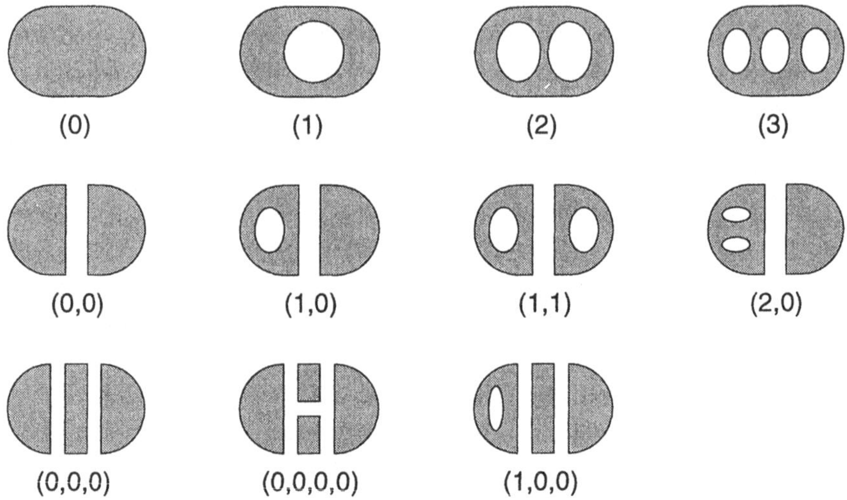

Figure 4.5. The simplest connectivity types in two dimensions

the interpretation of connection. But they are also sensitive to granularity. At a coarse level of granularity a human being can be treated as simply-connected, like a sphere, but on a finer scale the digestive tract must be regarded as a tunnel, increasing the connectivity to that of a torus; at a finer granularity still many more bodily orifices become apparent, leading to vastly increased connectivity. Similarly, a wood, considered as a piece of territory vegetated in a particular way, might be connected, but if it is considered as an aggregate of individual trees then it appears as scattered.

4.3. Position

The position of an object provides an answer to the question 'Where is it?'. Aristotle defined the position of an object to be that portion of space that is occupied by the object; a possible position for an object is a region of space geometrically congruent to some configuration of the object itself, into which the object can 'just fit'. In the formal representations, I shall use *pos* to represent a function which maps each object to its position. Since the position can vary over time, $pos(x)$ is a function of time (a *fluent*—see §5.4) whose range of values R is whatever set of regions we decide to accept. The statement $pos(x) = r$ can be written as $at(x, r)$; this holds just when the object x exactly occupies the region r.

In one sense this analysis seems obviously right, but in another sense it is rather unhelpful, in that it simply converts the problem of describing where an object is to the no less difficult one of describing where a certain region of space is. Moreover, it presupposes an absolute notion of space as a fixed framework within which objects move; but in practice we have no way of defining regions in space except in relation to the position of objects. Accordingly, what we generally do in practice is to describe the position of an object in relation to the positions of other objects. How we do this depends on what system of spatial relations we want to use. I shall discuss various possibilities below. Note that even if we set up a framework of spatial co-ordinates, it is itself ultimately grounded in the positions of objects, so it is still true that we are describing the positions of objects relative to other objects.

The attribute of position can be factored into two components: location and orientation. A clean way of effecting the separation is to identify a set of reference axes on the object, intersecting at a 'centre' which might, for example, be taken as the object's centre of mass, or as its geometrical centroid. The location of the object is then defined by the position of its centre, and the orientation is defined by the directions of the axes. This is a particularly useful procedure in physics, where the motion of a rigid object can be separated into a translational component, given by the motion of its centre of mass, and a rotational component, given by its angular motion about the centre of mass. Taken separately, these two components obey relatively simple dynamics under ordinary conditions; combined together, they look more complex. This is straightforward for rigid bodies; specifying the position of a body which can change shape is more complicated. Discussion of this is deferred to §4.7. I turn now to separate discussions of the location and orientation components of position.

4.4. Location

When the location of one object is specified with respect to another, the first object is called the 'target' object and the second the 'reference' object. In general, the location of one object relative to another has two components, a distance and a direction. The reference object may be regarded as generating a continuous field of locations in the space around and/or within it, which may be partitioned into zones of qualitatively similar locations, where similarity might be judged with reference to distance, location, or both. I shall discuss three different systems which I shall call 'earth-based', 'object-centred', and 'observer-centred'. Within each system differences arise according to the apparent dimensionality of the reference object. Thus there are many different cases to discuss. I shall attempt a broad but by no means comprehensive survey of the possibilities.

4.4.1. The earth-based system

The earth-based system is used for locating objects relative to one another in geographical space, that is, when the objects are situated on the earth's surface at distances greater than the immediate reach of a human being. The earth-based system is distinguished by the fact that the earth itself provides a natural system of directions determined by the cardinal points north, south, east, and west. Directions on the earth's surface are expressed as bearings such as 40° west of north, otherwise designated 320°. The earth also provides an intrinsic measure of distance, based on its circumference; the metric system is supposed to reflect this, in that the metre was originally intended to represent one ten millionth part of the distance from the North Pole to the equator, but there is in fact a slight discrepancy between this and the standard metre constructed in Paris in 1799, which for many years was taken to be definitive.

To locate one point on the earth's surface relative to another, one can provide a distance and a bearing; for example, Birmingham is approximately 160 kilometres from London on a bearing of 320°. It is not always appreciated that this can be interpreted in two different ways. If instructed to go to a point 160 kilometres distant on a bearing of 320°, there are two ways you could set about following the instruction. On the one hand you could turn to face in the direction 320° and then travel in a straight path (more exactly, a great circle) for 160 kilometres. On the other hand you could face in the direction 320° and then travel for 160 kilometres maintaining a bearing of 320° throughout the journey. These two methods will take you to different points. The further you have to go, the further apart the points will be. The reason for this is that as you follow round a great circle your bearing changes constantly (unless it is the equator or a meridian). If you set off north-eastwards on a great-circle path, by the time you are halfway round the world your bearing will be south-east. If you follow a constant bearing, your path will only be a great circle if your bearing is due north or due south; if your bearing is due east or due west, it will be a line of latitude, but in all other cases it will be a loxodrome or rhumb-line, along which you will eventually find yourself winding in a tight spiral around one of the poles. It follows that if we want to give the distance and bearing of point B from point A, we will get different results according to whether we choose the great-circle path from A to B or the loxodromic path.

We can thus describe two separate quantitative location spaces using distances and bearings on the earth's surface. The simpler of the two is based on the great-circle interpretation. If R is the radius of the earth (assuming for simplicity that the form of the earth is perfectly spherical), then the co-ordinates $(d, \alpha), (d+2\pi R, \alpha), (d+4\pi R, \alpha), \ldots$, as well as the co-ordinates $(2\pi R-d, 180+\alpha), (4\pi R-d, 180+\alpha), \ldots$, all represent the same point. Thus we can restrict the co-ordinates to the ranges $0 \leq d \leq \pi R$, $0 \leq \alpha < 360$, and the whole of the earth's surface will be covered. Note that this assumes that the reference object is not at either of the poles, at which points bearings become uninterpretable since from the North Pole all directions are south, and from the South Pole, north. Note also that for any α, the co-ordinates $(0, \alpha)$ pick out the same point, the position of the reference object; and the co-ordinates $(\pi R, \alpha)$ also pick out the same point, exactly antipodal to the reference object.

In the loxodromic interpretation, the location space varies with the latitude of the reference object. If you are on the equator, then maintaining a bearing of due north, south, east, or west will take you along a great circle; anywhere else, only north and south will do so, but due east and west will take you along a closed path, a circle of latitude. In any case, there is always a maximum distance you can travel due north before hitting the North Pole, and likewise for south. These distances depend on the latitude of your starting point. Again, if you travel due east, you will eventually arrive back at your starting-point, and the distance you must travel before this happens is a function of latitude. On any bearing other than due north, south, east, or west, you can travel an indefinite distance along the loxodromic path at that bearing. In particular, on any given bearing there is a maximum distance you can travel before reaching a point that could have been reached in a shorter distance by setting out on a different bearing from your starting-point. An obvious choice is to select the shortest loxodromic path from reference to target as defining the location of the target relative to the reference. We shall end up with a space that looks very different from the one based on great circles.

Although the two interpretations look very different when we consider objects distributed widely over the earth's surface, on a more local scale the differences become much less significant, both converging in the limit to a 'flat earth' model in which the portion of the earth's surface in the vicinity of the reference point is represented by a portion of the Euclidean plane. Considered as *qualitative* designations, the cardinal points obey a logic that is approximately Euclidean. For example, moving more or less eastwards from western Europe, we encounter the regions designated the Near East, the Middle East, and the Far East, and after that we arrive at the vast extent of the Pacific Ocean which from a European perspective counts as neither east nor west. Moving westwards on the other hand, we reach the New World in the west and then beyond it the Pacific Ocean again, which is neither east nor west. Thus 'east is east, and west is west, and never the twain shall meet'. In effect, it is as if we were simply to omit a large chunk of the earth's surface antipodal to us, so that what remains can be represented as a circle centred on our own position. Positions in this circle can then be assigned qualitative bearings according to one of the schemes illustrated in Figure 4.6*a–c*. The exact location of the boundaries between the radial zones could be varied if necessary. Qualitative distances, on a scale of *near–mid–far*, generate concentric zones as shown in Figure 4.6*d*.[5] A radial zonation for bearings can be combined with a concentric zonation for distances to give the qualitative location space shown in Figure 4.6*e*.

The systems of qualitative bearings illustrated here do not exhaust the possibilities. A different scheme is discussed by Frank (1996), here illustrated as Figure 4.6*f*. In this scheme there is a 'neutral zone' of points to which no cardinal direction is assigned. One reason why this might be appropriate is if the reference object is not, after all, pointlike, but an extended roughly rectangular region. Then the neutral zone simply comprises the locations designated 'here' as opposed to 'north of here', 'south of here', and so on. Frank is particularly concerned with what happens when qualitative cardinal directions are composed; for example, if A is north of B and B is west of C, where is A in relation to C? With only directional information given, and qualitative directional information at that, it is not in general possible to give a single unambiguous answer; but one can at any rate narrow the possibilities. In the example cited, the answer depends on which scheme we adopt: according to Figure 4.6*a*, A could be either north or west of C; if we use Figure 4.6*b*, then A can be north, north-west, or west of C; while with

[5] Note however that Vandeloise (1986) regards *accessibility* to be more important than *distance* in the meanings of 'near' and 'far'— this has the result that the nearness relation might end up asymmetric, whereas a nearness relation based solely on distance must be symmetric. Cf. also Lyons (1977: 692).

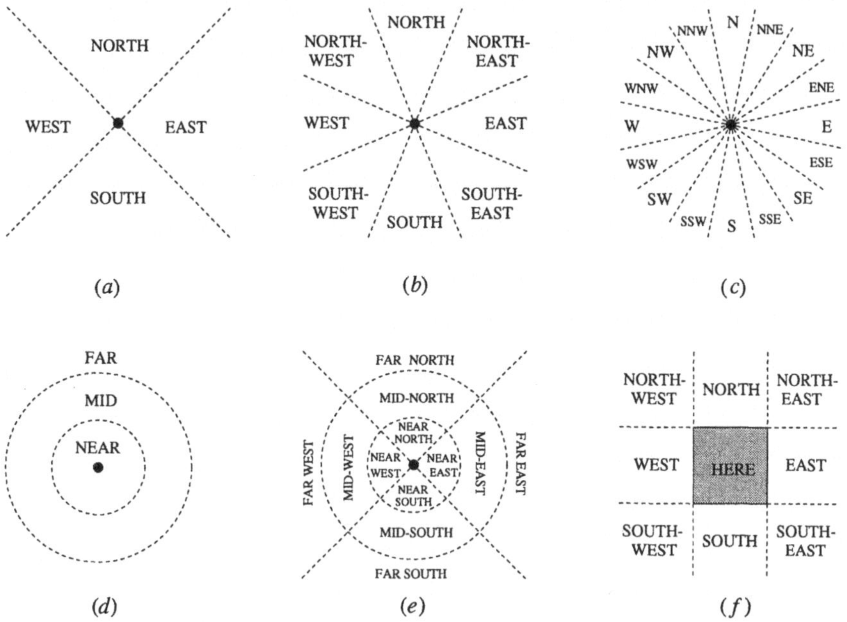

Figure 4.6. Qualitative distance and bearing on the earth's surface

Figure 4.6*f* we see that A must be either north-west or west of C. A complete set of such results could be collected together into a composition table for qualitative bearings analogous to the table for the RCC relations mentioned above (see §2.5.2).

If the reference object is linear then we have a different situation again, varying with the direction along which it runs. Given a river running from west to east, we can locate an object to the north or south of it, but hardly to the east or west of it. 'East of the Thames' would be out in the North Sea, but we do not usually describe positions in this way. Thus linear reference objects typically divide their surroundings into just two qualitative directional zones based on the compass points complementary to those defining the direction of the object itself (Figure 4.7).

If the reference object is an extended field within which the target object is located, then we can still use the cardinal points but in a somewhat different way. The reference field itself is divided into zones. Division by latitude gives *north* and *south* (as in 'London is in the south of England'), division by longitude gives *east* and *west*. These can be combined, as in 'London is

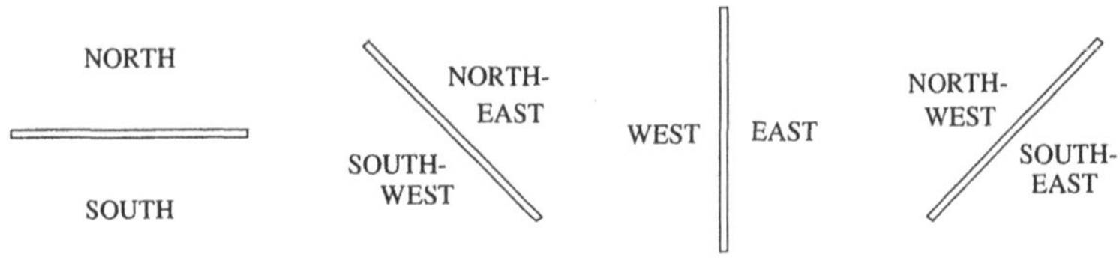

Figure 4.7. Locations relative to a linear earth-based reference object

in the south-east of England'. They can be qualified by 'far' or 'extreme', as in 'Berwick-upon-Tweed is in the far north of England'. We also use *middle* or *centre*, as in 'Hawaii is in the middle of the Pacific Ocean' or 'Piccadilly Circus is in the centre of London' (Figure 4.8*a*). In practice the way we use these divisions does not always follow lines of latitude or longitude exactly: a wealth of complex cognitive factors intervene. In England, for example, the boundary between places felt to be 'north' and places felt to be 'south' slants markedly in a south-west to north-east sense. This has partly to do with the effect of geology on the character of the various regions, but also to do with the strong distorting effect resulting from the position of the capital in the far south-east. A similar effect exists in the case of Italy, but in this instance the explanation has more to do with the fact that the main axis of the country has a marked north-west to south-east slant.

If the reference field is the whole of the earth's surface, then the system becomes problematic because there are no parts of the earth's surface that can be called the west or east of the earth. All places are both east and west of some other places. The standard system of locating objects on the surface of the earth by latitude and longitude is better regarded as a special case of location with respect to two linear reference objects. Thus the co-ordinates of Birmingham, at 52°30′ N, 1°55′ W, pinpoint the city by simultaneously locating it with reference to the equator and the Greenwich meridian. Angular measurements are in effect used here as a measure of *distance*, not direction. Thus the northing tells us that Birmingham is north of the equator and at a distance which subtends an angle of 52°30′ at the centre of the earth; the easting tells us that it is west of the Greenwich meridian, at a distance which subtends an angle of 1°55′ on the earth's axis. The correspondence between angle and distance is complicated by the fact that the distance corresponding to a given longitudinal displacement varies with latitude.

For smaller areas it is impractical to use latitude and longitude, and al-

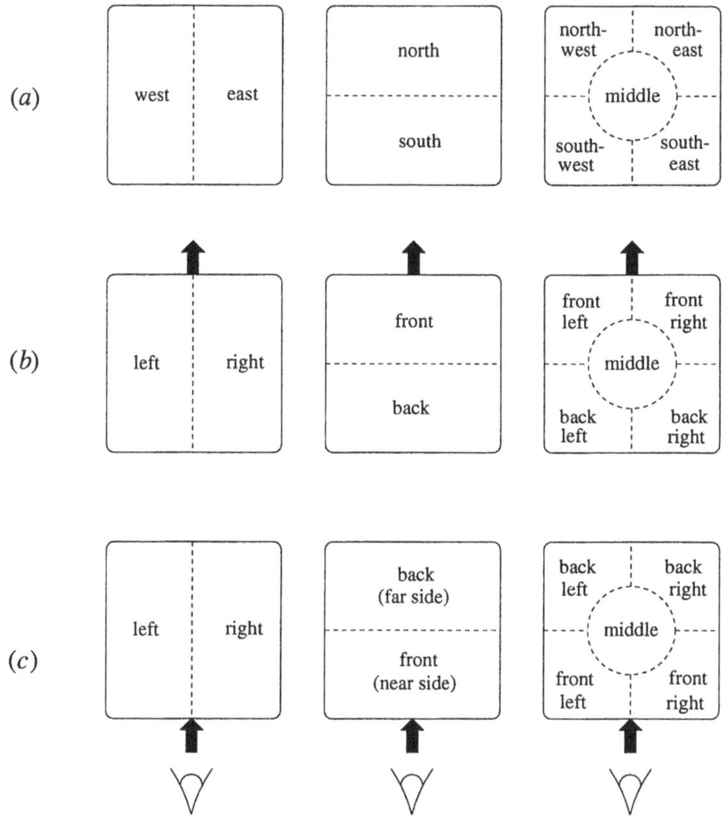

Figure 4.8. Division of a field into zones: (*a*) earth-based system, (*b*) object-centred system, (*c*) observer-centred system

ternative grid-systems can be set up. An example is the British National Grid. On this system the centre of Birmingham is located in the grid square SP075870. On a smaller scale still, the numbering of houses along a street provides a simple numerical index to the location of individual buildings, as does the numbering of flats in an apartment block or the numbering of rooms in a hotel; postcodes provide a somewhat coarser scale of resolution. Such methods for locating objects on the earth's surface are known collectively as 'geo-referencing'. Although such numberings are usually reasonably systematic, in the sense that they respect, at least up to a point, the natural ordering of the numbers, there is inevitably an element of arbitrariness about them.

4.4.2. The object-centred and observer-centred systems

For many purposes outside an explicitly geographical context it is impracticable to use the earth-based system for specifying the locations of things. Instead we use a more local scheme, based on the immediate participants in the situation being described. There are essentially three such participants: the target object, the reference object, and the observer.[6] The target object itself can hardly be used as the basis for the system of locations for then there would only be one possible location ('here'); that leaves the reference object and the observer, and either of these can be used as the basis for defining locations relative to the reference object. I shall consider the two systems in turn.[7]

The object-centred system

The object around which this system is centred is of course the reference object, not the target object. The idea is that the reference object imposes a co-ordinate system on the space around it by reference to which the target object can be located. In order for this to be possible, a preferred direction must be determined by the object to act as a baseline analogous to due north on the earth's surface. The preferred direction will typically be defined by an axis of symmetry in the horizontal plane. It is characteristic above all of animals such as ourselves which have a distinguished forward direction of vision and locomotion; but many other objects similarly have the requisite reflectional symmetry about a vertical plane, but no symmetry about any other plane. Examples of such objects are cars, boats, and other vehicles, many buildings, desks, televisions, computers, control panels, chairs and benches, and statues.

All such reference objects generate a system of intrinsic directions, as illustrated in Figure 4.9*a*, where the reference object is shown as a stylized motor car. These directions constitute what Miller and Johnson-Laird (1976) call the 'intrinsic' system of spatial reference; Herskovits (1986) refers to this as the 'coincidence schema'. Note that in the coincidence schema the sequence left–in front–right–behind follows a clockwise path round the object as seen from above.

Freksa (1992*b*) has investigated the composition of relations in an object-

[6] Note the suggestive analogy here with the three times invoked by Reichenbach (1947) in his analysis of the tenses of the verb: event time, reference time, and speech time.

[7] A more detailed classification of such systems, or 'frames', is given by Jackendoff (1996).

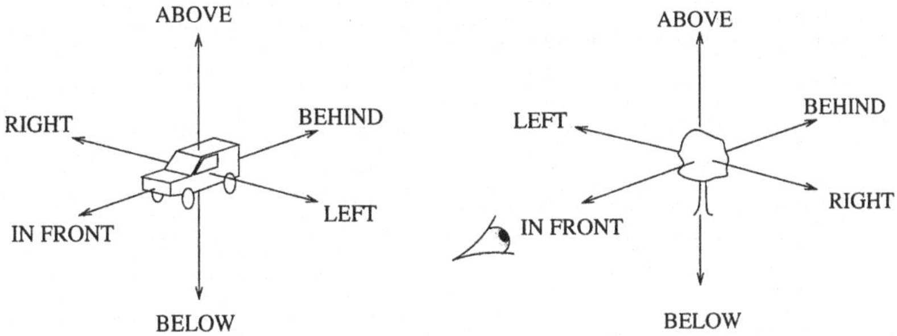

(*a*) Object-centred system (*b*) Observer-centred system

Figure 4.9. Two schemes for relative spatial location

centred system that is somewhat different from ours. This work is analogous to Frank's work on composing cardinal point directions discussed above (see Figure 4.6*f*).

If the reference object is a field within which the target object is to be located, then as with the earth-based system we can generate a set of zones within the field using the intrinsic system relative to a point in the centre of the field. Corresponding to the relations 'left of', 'right of', 'in front of', 'behind', 'above', and 'below', we refer to the zones as the left side, right side, front, back, top, and bottom of the field. Thus one might sit at the top of the bus, at the front right. In the case of an effectively two-dimensional field, we do not need 'top' and 'bottom'; the system we obtain is as shown in Figure 4.8*b*.

The observer-centred system

If the reference object does not have the requisite symmetry to allow the use of the intrinsic system of spatial co-ordinates, then instead we will normally use an observer-centred scheme, as shown in Figure 4.9*b*, where the reference object is a tree. Since there is no intrinsic front, back, left, or right, it is the position of the observer that determines how the space around the reference object is to be divided up and categorized. To do this we make use of the 'encounter schema' (Herskovits 1986), or 'deictic' system of spatial reference (Miller and Johnson-Laird 1976). In this scheme we imagine, up to a point, that the reference object does have an intrinsic front–back direction, and that its front is facing directly towards us, the observer. This determines that the space between us and the reference object is *in front* of the latter, and that the space on the further side of the reference object from us counts as *behind* it. For left and right, however, we retain our own intrinsic scheme, so that the space we say is to the *left* of the reference object would be to the *right* if we used the front–back system we imposed on to it. Thus the sequence left–in front–right–behind now follows an anticlockwise path round the object as seen from above.

For a target object located within an extended reference field, the observer-centred system imposes a division of the field in a similar way to the object-centred system, the resulting divisions differing from one another exactly as the systems themselves do. The zonation generated by the observer-centred system is shown in Figure 4.8*c*.

Ambiguity between the object-centred and observer-centred systems

Ambiguities can arise when it is not clear whether the system of spatial reference being used is object-centred or observer-centred. Imagine you are on a railway platform at which there is a train standing; the phrase 'behind the train' could refer to the space on the far side of the train as viewed from the platform; or it could refer to the space at the rear end of the train, past the hindmost carriage. The former interpretation uses the observer-centred system, the latter uses the train's own object-centred system.

Similar ambiguities arise when the reference object is a field. An interesting example here is provided by the stage and auditorium of a theatre or concert-hall (Figure 4.10). Both audience and performers agree as to which parts of the stage and auditorium are the 'front' and the 'back', but for different reasons. From the point of view of the audience, the auditorium is described using the coincidence schema, and the stage by the encounter schema; from the point of view of the performers on stage, it is the other way round. Although this leads to agreement as regards 'front' and 'back', it leads to differing interpretations of 'left' and 'right'; as a result, expressions like 'the left side of the stage' are ambiguous, and indeed one encounters considerable uncertainty about their meaning.

As Herskovits points out, a reference object which has no intrinsic front–back axis can acquire one as a result of motion. In this case, the leading edge of the moving object can be regarded as the front, allowing the use of the coincidence schema rather than the encounter schema. Herskovits cites a clever example from Kimball: *The red billiard ball was in front of the white*

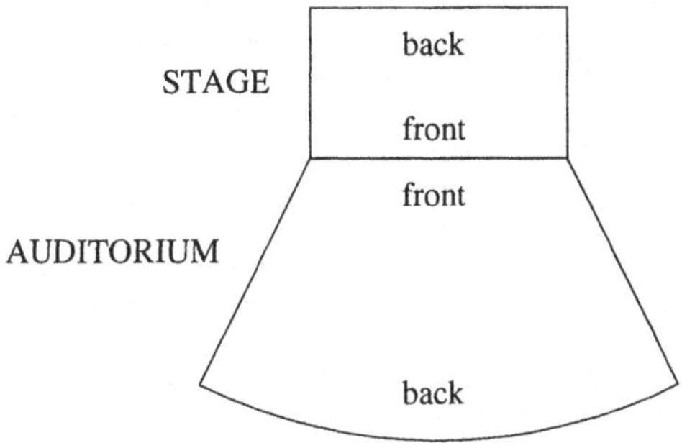

Figure 4.10. 'Front' and 'back' in the theatre

one. As Herskovits says,

> if the white ball is rolling, one may think of it as some 'observer' with line of sight in its direction of movement. This is a metaphorical equivalent of the coincidence situation, with an effective point of observation coincident with the reference object (i.e., the white ball). This point of observation is **virtual**, since it does not correspond to an actual observer.
>
> (Herskovits 1986: 160)

Two interpretations of this example are shown in Figure 4.11.

(*a*) Encounter schema (*b*) Coincidence schema

Figure 4.11. 'The red billiard-ball is in front of the white one' (after Herskovits)

4.4.3. Topological relations

So far I have considered systems for location based on distance and direction; the different systems we have encountered have been characterized in terms of how they describe direction. I have also only considered the cases when the target object is either wholly outside or wholly inside the reference object. More generally we may wish to consider location based on *topological* relations, specifying the position of an object in terms of whether, and if so in what way, it is connected to the position of the reference region. These relations can be represented using the RCC calculus introduced in §2.5.2.

Since the RCC relations are defined as relations between regions, the strictly correct way to describe one object a being wholly outside another object b is $DR(pos(a), pos(b))$, where the relation DR—which, it may be recalled, is the disjunction of DC ('disconnected from') and EC ('externally connected to')—is said to hold between the positions of the two objects rather than between the objects themselves. For simplicity, I shall from now on abbreviate expressions of the form '$R(pos(a), pos(b))$', where R is an RCC relation other than EQ, as '$R(a, b)$'. That is, I shall describe an *object* as bearing a certain spatial relation to another, when what I should say, within the strict terms of the formalism, is that the *position* of the first object bears that relation to the position of the second. Having decided to abuse my notation in this way, I might as well also allow mixed expressions of the form $R(a, r)$, in which a is an object and r is a region, when what is really meant is $R(pos(a), r)$. As already noted, I shall use $at(a, r)$ for $pos(a) = r$ (or equivalently $EQ(pos(a), r)$).

The RCC system is only concerned with topological relationships and hence provides only a crude description of the relative positions of objects, but this may suffice for some purposes. Objects which are separated from one another by a positive distance stand in the relation DC; objects which are touching one another are EC. For material bodies which cannot interpenetrate these are the only possibilities. But we can also locate a body relative to a room, for example, which we may conceive of as the space bounded by the walls, floor, and ceiling. A person entering a room passes through the sequence of relations DC, EC, PO, TPP with respect to the room; if she then gets into bed, say, then her relation to the room becomes $NTPP$.

In general, not all the RCC relations will be realizable for a given pair of objects, even when interpenetration is possible. This is particularly obvious in what is perhaps the paradigmatic use of such relations, the relation of a rigid object to a region. The possible relationships depend on the shape and size of both the object and the region. There are in fact six possibilities to consider (Galton 1993):

1. They are *congruent*: that is, r is a possible value of $pos(a)$.
2. a can just fit inside r: that is, $TPP(a, r)$ is possible, but $NTPP(a, r)$ is not. An example would be a hemispherical body and a spherical region

of the same radius.

3. a can just cover r: that is, $TPP(r,a)$ is possible but $NTPP(r,a)$ is not. Example: a spherical body and a hemispherical region of the same radius.
4. a can fit right inside r: that is, $NTPP(a,r)$ is possible. Example: a spherical body and a spherical region of greater radius.
5. a can more than cover r: that is, $NTPP(r,a)$ is possible. Example: a spherical body and a spherical region of smaller radius.
6. None of the above holds: we say a and r are *incommensurate*. Example: a pole 3 metres long and a cubical region measuring 1 metre along each side.

Each of these possibilities defines a subset of the eight RCC relations that can hold between a and r, as follows:

1. DC, EC, PO, EQ.
2. DC, EC, PO, TPP.
3. $DC, EC, PO, TPPI$.
4. $DC, EC, PO, TPP, NTPP$.
5. $DC, EC, PO, TPPI, NTPPI$.
6. DC, EC, PO.

These are illustrated in Figure 4.12. Each possibility in turn gives rise to a relevant subset of the conceptual neighbourhood diagram, as portrayed in Figure 4.13.

4.4.4. Natural language

There has been much interest in the expression of relative spatial location in natural language. This was the focus of the work of Miller and Johnson-Laird (1976) and Herskovits (1986) already cited. The words most closely implicated in the expression of spatial relationships are the prepositions, both simple (*in, on, at, over, under, above, below, behind, inside, outside, beside, along, across, ...*) and compound (*next to, in front of, on top of, left/right/north/south/east/west of, at the edge of, in the middle of, ...*). Detailed discussion of these is to be found in Herskovits 1986. Perhaps the most important observation to be made in relation to these spatial prepositions is

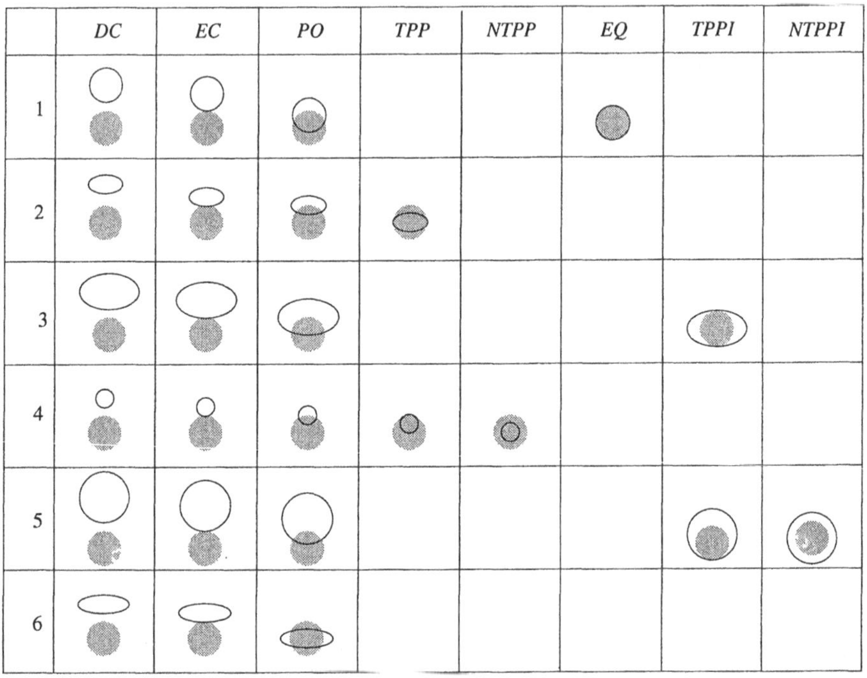

Figure 4.12. The six possible sets of RCC relations between a rigid object (outline) and a region (shaded circle)

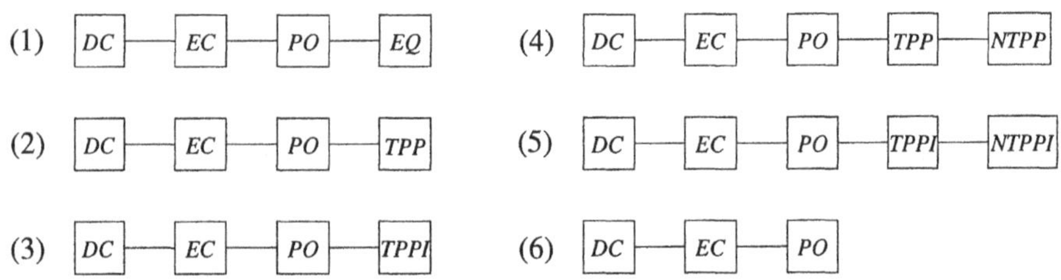

Figure 4.13. Conceptual neighbourhood diagrams for the six possible sets of RCC relations between a rigid object and a region

that their meanings cannot be characterized exclusively in geometrical terms. For example, the preposition *on* signifies not just adjacency but also support: 'X is on Y' typically means that X is adjacent to, and supported by Y. The prototypical case of support is when X is resting on top of Y; this leads to the mistaken idea that *on* means 'on top of'. (Indeed, etymologically, the French *sur*, which behaves very much like the English *on*, is derived from the Latin *super*, which means 'above'). Examples where *on* cannot be replaced by *on top of* are a ring *on* someone's finger, a picture *on* the wall, a fly *on* the ceiling, and washing *on* the line. This does not mean that the geometrical component of the meaning of spatial prepositions is unimportant. Indeed, Herskovits attempts to derive all the aspects of the meaning of a spatial preposition from an underlying 'ideal meaning' which is essentially geometrical in character. For the three basic topological prepositions she gives the ideal meanings:

> *at*: for a point to coincide with another
> *on*: for a geometrical construct X to be contiguous with a line or surface Y
> *in*: inclusion of a geometric construct in a one-, two-, or three-dimensional construct

To the ideal meaning of *on* Herskovits appends an additional clause: 'if Y is the surface of an object O_Y, and X is the space occupied by another object O_X, for O_Y to support O_X'. Figure 4.14 illustrates the various realizations of these ideal meanings. Many actual uses of the prepositions fit these realizations quite well, but by no means all; this is particularly the case with 'at'.

Not all authors agree with Herskovits in treating the ideal meanings of spatial prepositions as geometrical in character. Vandeloise (1986), for example, regards the *functional* characteristics of prepositional meanings as primary, with the geometrical aspects derivative from these. Thus for Vandeloise, 'in' primarily connotes containment, 'at' connotes localization, and 'on', support. Similarly, 'near' and 'far' are defined in terms of relative accessibility rather than distance, and 'behind' in terms of inaccessibility to perception. A similar point of view is expressed by Coventry (1998), who states that

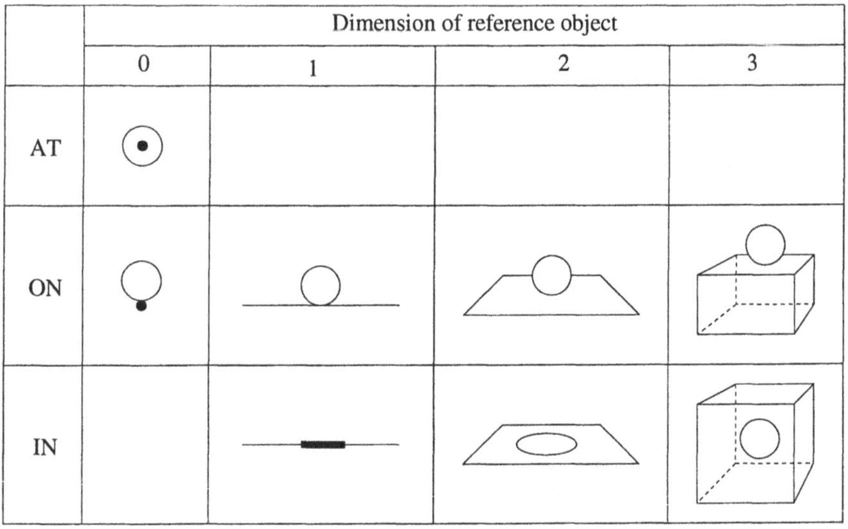

Figure 4.14. The three basic topological prepositions in English

> ... spatial language use is under-determined by geometric relations in the input, and better predicted by functional relations reliant on object-specific knowledge.
>
> (Coventry 1998: 247)

Of particular interest in the current context is how systematic formal representations such as we have been discussing might be related to the expressions we find in natural language. In particular, we may ask to what extent the meanings of the spatial prepositions can be defined in terms of logically characterized relations such as those of the RCC set.

To begin with, it can only be possible to capture the *topological* components of preposition meanings using RCC; this means that those prepositions whose meanings have an irreducibly non-topological component cannot possibly be given a complete definition in RCC terms. Thus, for example, the prepositions *over*, *under*, *above*, *below*, and *on top of* all include in their meanings a reference to the vertical up/down axis defined by the local gravitational field. This axis cannot be distinguished from any other axis by purely topological means, and so none of these prepositions can be completely defined using RCC. They can, however, be partially characterized; for example, we can say that for X to be on top of Y it must be in contact with Y, hence the

RCC relation must be EC; this contrasts with X being above Y, which rules out such contact, so the corresponding RCC relation must be DC.

The best chance for purely topological definitions lies with the so-called topological prepositions shown in Figure 4.14. But, as already noted, *on* denotes more than just adjacency (EC); in addition the notion of support is involved. What about *in*? A plausible candidate is the proper-part relation PP, which is the disjunction of TPP and $NTPP$. Certainly whenever this relation holds, it would seem to be appropriate to use *in* to describe the situation. But there are other situations where *in* is appropriate even when the RCC relation is not PP. In this case we have to do a certain amount of reinterpretation to make the natural language meaning fit the RCC analysis.

This is well illustrated by the case of containment. When there is an apple in the fruit bowl, the apple is not embedded in the material of which the bowl is made, but rather is situated within the space that is regarded as contained by the bowl, Cohn's containable inside (see p. 154). Thus the phrase 'in the fruit bowl' would have to be analysed as 'PP the containable inside of the fruit bowl'. If we now wish to derive this meaning as a composition of the separate meanings of 'in' and 'the fruit bowl', we have two choices. Either we reconstrue 'in' as meaning, in this context, 'PP the containable inside of', leaving 'the fruit bowl' alone; or we reconstrue 'the fruit bowl' as meaning, in this context, 'the containable inside of the [normal] fruit-bowl', with 'in' meaning simply PP. In different contexts the reconstruals may have to go differently; in the long run there may be nothing to choose between reconstruals of the first type and reconstruals of the second. One might easily be led to conclude that therefore it is utterly indeterminate which is the 'correct' construal. Langacker (1991) points out that traditionally linguists have tended to assume that a complex phrase must either have a meaning that can be completely predicted in every detail from the meanings of its components and the way they are put together, or else be a completely opaque idiom whose meaning has to be given as a whole without reference to the meanings of its components. This is an instance of what he calls the 'exclusionary fallacy'. One of the tenets of Langacker's cognitive approach to grammar is that there can be, and indeed are in abundance, intermediate cases, complex phrases whose meaning is well motivated but not wholly explicable in terms of the meanings of the individual components. It may be that phrases involving prepositions are predominantly of this type. If this is the case, then any decision as to how to construe such phrases analytically is bound to have an element of arbitrariness about it, without, however, being wholly arbitrary.

The fruit bowl example raises a further problem, which is that even the analysis involving the containable inside is not adequate. As pointed out by Herskovits (1986: 14), a pear perched on top of a pile of fruit in the bowl will still be said to be *in* the bowl even if it lies wholly outside the containable inside. One way to explain this might be to allow the containable inside to stretch beyond its normal limits under certain conditions such as the situation just described. Alternatively, and perhaps more plausibly, we may say that the pear inherits the property of being in the fruit bowl from the pile of fruit of which it is a proper part, that pile as a whole counting as being in the fruit bowl because part of it is in the containable inside. This fits in with many other instances in which an object is said to be in something when in fact only part of it is, for example a bather can be *in the water* even when not wholly submerged, the key is *in the lock* even when part of it is sticking out, and we speak of a man *in a hat* even though only a very small fraction of the man's body is strictly within the hat.

Another set of cases where reconstrual of some kind might seem to be called for are those in which 'in' has the sense of embedding, for example a nail in a plank. A pedant might insist that the nail is not in the plank itself, but in a nail-shaped hole in the plank. If we take this line then one possibility is to regard the plank as functioning as a container, with the further option of two different reconstruals as in the case of the apple in the bowl. An alternative is to regard the space occupied by the plank as that space which would have been occupied by the plank had the nail not been there.

This brief discussion of 'in' should suffice to convince the reader that the enterprise of relating the meanings of English prepositions to the RCC relations is fraught with difficulty. I could have given similarly problematic examples involving 'on' and 'at'. It is not necessarily a completely fruitless task, but certainly one which must be pursued circumspectly, if at all.

4.5. Orientation

The other component of the position of an object is its orientation: roughly, the direction in which it is pointing. If we choose some directed line through the centre of the object as a reference axis, then the orientation of the object can be specified by specifying the direction in which this line points. Some objects have a natural reference axis, although this is usually determined by function and behaviour and not purely geometrically; for others it would have to be imposed arbitrarily. Having fixed a reference axis the problem of specifying an orientation reduces to that of specifying a direction: we can use angular measure, either quantitative (in degrees or radians) or qualitative (using 'north', 'south', 'east', 'west', or 'right', 'left', 'up', 'down', 'forward', 'back').

The orientation of an object plays a crucial role in determining the appearance of that object from a given viewpoint; as its orientation changes, so does its appearance as different parts of the surface come into view or are hidden. Orientation has played an important role in some psychological studies in spatial cognition. Shepard and Metzler (1971) investigated the phenomenon of *mental rotation*, whereby, it is claimed, human subjects can transform their mental images of objects, in a way that corresponds in a precise fashion to rotation in space, in order to determine whether or not two objects are congruent; the precision of the correspondence is demonstrated by the fact that the time taken to reach a decision is roughly proportional to the angle through which one of the objects has to be rotated in order to determine whether it is congruent to the other.

Shepard has gone on to investigate some of the abstract properties of 'orientation space', which turns out to be a three-dimensional closed manifold since, as Shepard explains,

> With respect to a particular reference orientation, any possible orientation of an object can be specified in terms of the direction of an axis of rotation through the centre of the object (a direction that we can represent by the direction of a vector issuing from the origin in a three-dimensional space), and in terms of the angle of the rotation about that axis (an angle that we can represent by the length of that vector). Because rotations exceeding 180 degrees by an angle Θ are identical to rotations around the oppositely directed vector falling short of 180 degrees by that same angle Θ, the manifold of possible orientations corresponds to the interior and surface of a sphere (of radius 180 degrees), with diametrically opposite points on the surface identified.
>
> (Shepard 1988: 92–3)

If the object is portrayed as a cube having faces labelled F(ront), B(ack), L(eft), R(ight), T(op), and U(nderside), then the space of possible orientations of the cube can be partitioned into twenty-six qualitatively distinct regions characterized by which of the faces is visible to a fixed observer. These twenty-six regions can be arranged on the surface of a cube as portrayed in Figure 4.15, in which the straight lines join together regions between which it is possible to pass by continuous change in orientation of the cube; elements thus linked are conceptual neighbours in qualitative orientation space.

The significance of the arrowheads will be explained below, in Chapter 8. A more detailed view of part of this qualitative space, with illustrations of the variously orientated cube itself, is shown in Figure 4.16.

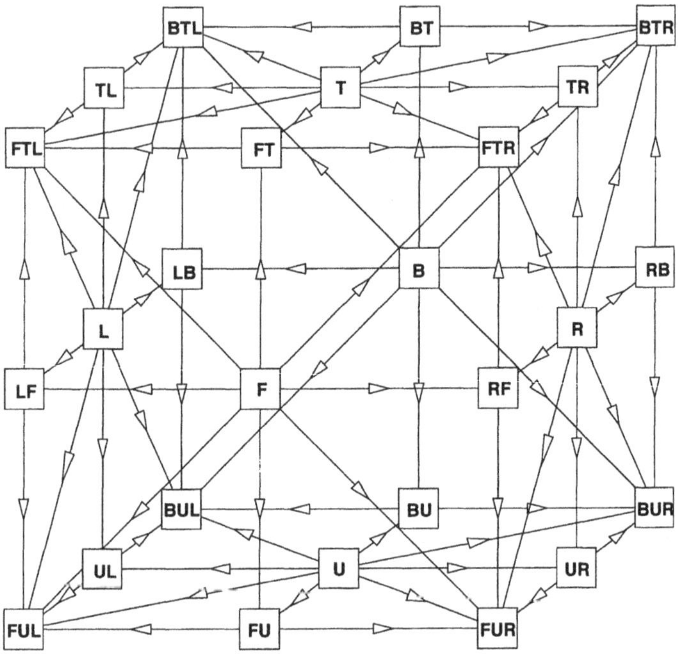

Figure 4.15. The space of qualitative orientations of a cube

Before leaving orientation we should point out that this term has been used in a different sense by Freksa (1992*b*) to refer to the *direction* in which one object is situated relative to another; this is an aspect of location, but is clearly related to orientation as discussed in this section. The orientation, in Freksa's sense, of one object as seen from another, is the orientation, in our sense, that an observer must adopt in order to look straight at the object.

4.6. Size

Size is in some ways the least complex of the spatial attributes we have to deal with, but, even so, it too has its pitfalls. Size in the sense of sheer extent has a separate measure in each different number of dimensions: length

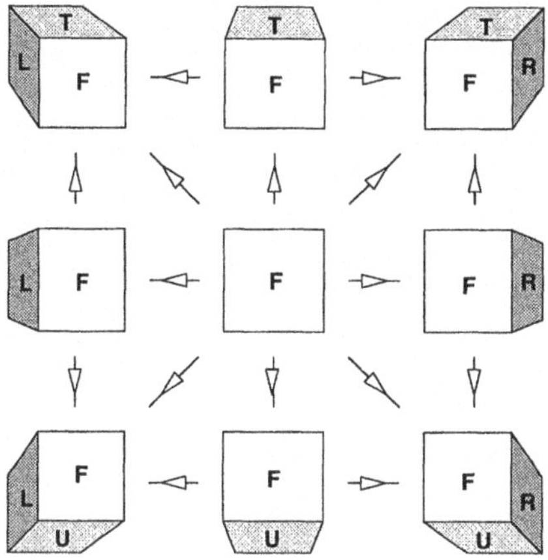

Figure 4.16. A closer look at some of the qualitative orientations of a cube

in one dimension, area in two, and volume in three. Area and volume are useful measures for many purposes, but they can be frustratingly uninformative because each of them is in effect compressing more than one dimension of variation into a single linear measure. For that reason it may be more informative to describe the size of something in terms of a set of lengths. At a minimum, this would be two lengths for a two-dimensional object, three for a three-dimensional one. There are several ways of doing this.

One method is to specify the dimensions of a minimal bounding rectangle (in two dimensions) or cuboid (in three). This is the method used by airlines when they restrict the size of an item of cabin baggage to $20'' \times 15'' \times 10''$, a more stringent restriction than simply requiring the volume not to exceed 3,000 cubic inches. The rectangle or cuboid may or may not be required to align with a predetermined set of co-ordinate axes, yielding different results in many cases. In Figure 4.17, for example, a curved sausage-like figure is shown, together with (A) the minimum bounding rectangle orthogonal to the vertical/horizontal axes, and (B) the minimal bounding rectangle of any orientation. In this example the latter rectangle has an area less than two-thirds that of the former; in other cases the ratio could be made indefinitely smaller. The intersecting dashed lines (C) represent a different way of indicating size, using the lengths of a pair of 'major' and 'minor' axes of the figure.[8]

[8] These can in some cases be defined precisely by means of the notion of *differential symmetry* used by Leyton (1988). See §4.7.2 below.

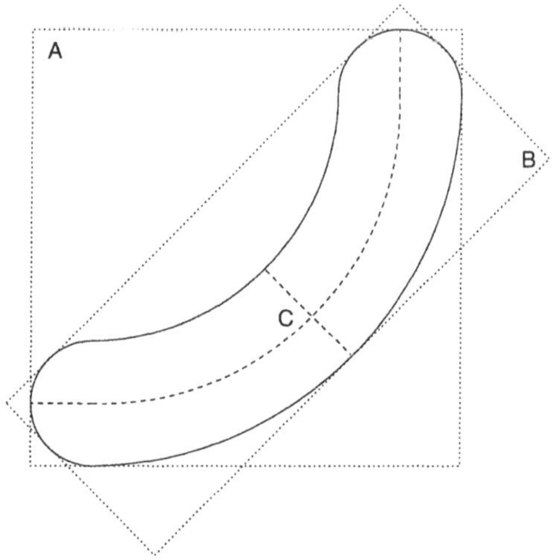

Figure 4.17. The dimensions of a planar figure

From this brief sketch it should be clear that the notion of 'size' is fraught with ambiguity. In particular, phrases like 'twice as big' must be used with great care. If one object is twice as big as another, does this mean that it has twice the volume, twice the area, or twice the linear dimensions? In everyday life, nobody seems to know. Our intuitions of cubic capacity are notoriously fallible: containers generally contain more than we expect, and this is probably because whereas we can in a sense perceive area and length directly, we cannot similarly perceive volume, biasing our intuitive understanding of size in favour of measures which grow as lower powers of the linear dimensions.

Another kind of size is angular size, which must be specified relative to a point of observation. One-dimensional objects have an angular length, measured in degrees or radians, whereas two-dimensional objects have an angular area, measured in steradians. Angular size is not the same as apparent size, which is a cognitive measure, involving an element of compensation for the rapid decrease in angular size that objects undergo as they recede into the distance. The discrepancy between angular and apparent size forms the basis of many well-known visual illusions, such as those of Ponzo and Lyer-Müller, not to mention the famous 'moon illusion' whereby the full moon appears larger when it is low over the horizon than when it is high in the sky, even though the angular diameter is the same in both cases.

4.7. Shape

Here I shall assume that shape is assigned to topologically self-connected units; a non-self-connected entity can only be assigned a single overall shape by considering it at a coarse enough level of granularity for it to be treated as being, after all, self-connected. I shall further assume that the objects we assign shape to are simply connected: another granularity condition. In effect, then, I am concerned with *outlines*, ignoring cavities, tunnels, and gaps. Even with these simplifications, shape is by far the most complex spatial character we have to deal with. It has literally infinitely many degrees of freedom: the outline of a closed curve in the plane can be varied independently at arbitrary many distinct points. Any theory of shape, to be useful, has somehow to reduce this complexity to a manageable set of principles; as Feldman (1993) puts it, 'the power of a shape theory is in constraining the useful dimensionality of shape without giving up too many desirable distinctions among classes of things'.

The most straightforward approach to shape is in terms of pure geometry. This provides us with many of our most basic shape categories, both in two dimensions (for example *circle*, *square*, *triangle*) and in three (*sphere*, *cube*, *cone*, *cylinder*, *pyramid*). For some purposes, though, a purely geometrical account can be seriously incomplete, even misleading. As usual in our discussion of spatial matters, cognitive factors intervene.

In geometry, a class of shapes is defined in terms of necessary and sufficient criteria for any particular shape to belong to the class. For example, a square is a plane polygon having four equal sides, each perpendicular to its neighbours. Two examples are shown in (*b*) and (*e*) of Figure 4.18. From the human point of view the most salient characteristic of a shape is its appearance. From that point of view (*e*) is a much 'better' square than (*b*), even though they are geometrically congruent. Our visual system is so organized that we perceive axes of symmetry readily if they are horizontal or vertical but only with much greater difficulty if they are not. As a result, in (*b*) we perceive mainly the diagonal symmetry, whereas in (*e*) we perceive mainly the orthogonal symmetry. For this reason (*b*) and (*e*) look quite different, and one has to learn to see that they are in fact the same from a geometrical point of view. Indeed, on the basis of appearance, (*b*) looks more like the non-square rhombuses (*a*) and (*c*), with the same symmetry about the diagonals, whereas (*e*) looks more like the non-square rectangles (*d*) and (*f*), with the same orthogonal symmetry. In everyday language (*a*), (*b*), and (*c*) are classed as 'diamonds', whereas (*d*), (*e*), and (*f*) are classed as 'squares'. This illustrates

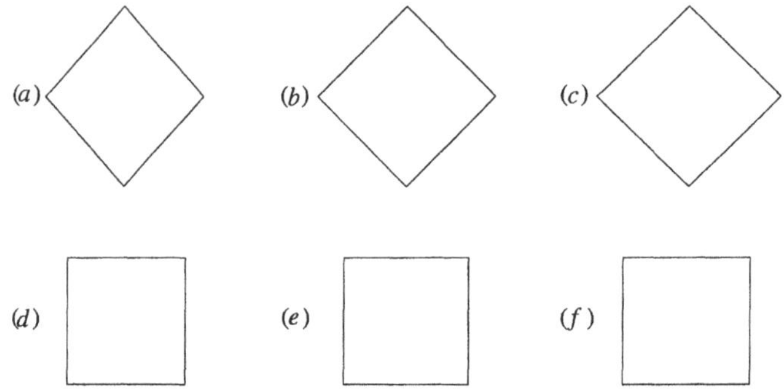

Figure 4.18. Being a square is not the same as looking like a square

the distinction between 'folk' and 'learned' vocabularies which has been the focus of considerable discussion in cognitive psychology and linguistics.

From a cognitive point of view, shape can be approached as a problem either of *recognition* or of *description*. The former has been the province of researchers in the psychology of vision; the latter has tended more to be the preserve of linguistics. In a computational context, similar problems arise in relation to computer vision and computational linguistics. I shall discuss the geometrical, psychological, and linguistic contributions in that order.

4.7.1. Shape in geometry

Geometrical language allows both quantitative definitions of particular shapes and also qualitative descriptions of classes of shapes. For simplicity I shall confine the discussion initially to planar shapes—that is, figures in the two-dimensional Euclidean plane. Many objects have shapes that can be described at least approximately in these terms, for example, the flattish 'thickened slabs' of Landau and Jackendoff (1993), features on the earth's surface, so long as they are small enough for the curvature of the earth to be negligible, and also anything appearing in a picture-plane, such as the diagrams in this book. Geometric classification of such shapes makes use of a number of different dimensions of possible variation, involving such notions as convexity, curvature, and symmetry. I shall discuss these three properties in turn.

Convexity

A shape is convex so long as the straight line segment joining any two of its points lies wholly within the shape. In Figure 4.19*a*, a convex shape is shown together with many pairs of points with their connecting line segments lying wholly inside the shape; in (*b*), a concave shape is shown with a pair of points whose connecting segment contains points outside the shape. A consequence of the definition of convexity is that a *bounded* shape is convex if, and only if, no straight line crosses its boundary more than twice.

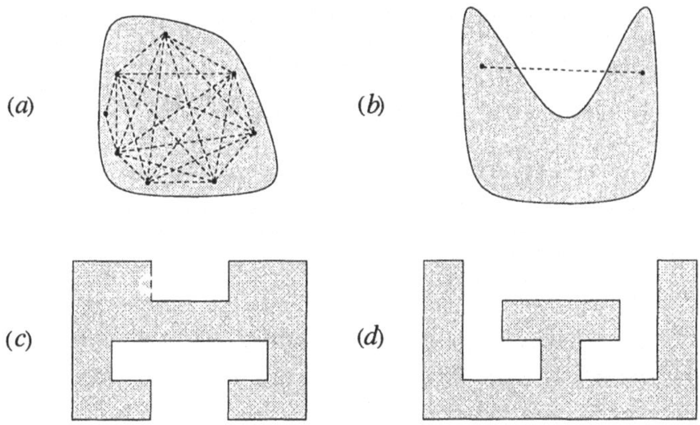

Figure 4.19. (*a*) All points inside a convex figure are joined by line-segments entirely within the figure; (*b*) a concave figure has a pair of points joined by a line-segment partly outside the figure; (*c*) a concave figure with convex and concave concavities; (*d*) a concave figure with a single concavity which itself has a concave concavity

Given any figure we can define its *convex hull* to be the smallest convex figure which contains it. A figure is convex if, and only if, it is its own convex hull. The concavities of a convex figure are the regions which remain when the figure itself is subtracted from its convex hull. Different concave figures can be discriminated by the nature of their concavities. A figure may have a single connected concavity (Figure 4.19*b* and *d*), or it may have two or more scattered concavities (Figure 4.19*c*). Each concavity component can then be characterized in terms of whether *it* is convex or concave, and so on recursively, as suggested by Cohn (1995). Of the two concavities in Figure 4.19*c*, one is convex and the other is concave. Figure 4.19*d* has one concavity which itself is concave, with a single concave concavity whose own two concavities are, at last, convex. Convexity and concavity alone thus already provide a rich language for the description of broad categories of shape, although it is still crude in the sense that within each category that is discriminable in these terms there are many different shapes which look very different to the human eye. For more on this, see §6.6.

From a formal point of view we can think of the convex hull as being given by a convex-hull operator conv mapping spatial regions to spatial regions (here represented as point-sets). This operator has some simple formal properties such as

$$\begin{aligned} \mathrm{conv}(\mathrm{conv}(x)) &= \mathrm{conv}(x) \\ x &\subseteq \mathrm{conv}(x) \\ x \subseteq y &\rightarrow \mathrm{conv}(x) \subseteq \mathrm{conv}(y) \end{aligned}$$

from which may be deduced further properties such as

$$\begin{aligned} \mathrm{conv}(x \cap y) &\subseteq \mathrm{conv}(x) \cap \mathrm{conv}(y) \\ \mathrm{conv}(x) \cup \mathrm{conv}(y) &\subseteq \mathrm{conv}(x \cup y) \end{aligned}$$

These do not, however, come near to providing a complete axiomatization of conv, as can be easily seen from the fact that the properties here given for conv apply equally well to the topological closure operator. Further axioms, still falling short of completeness, are given in the context of the RCC theory in Cohn *et al.* 1997. Convexity itself can be defined in terms of convex hull: x is convex if, and only if, $x = \mathrm{conv}(x)$.

Curvature

From a geometrical point of view it is much easier to handle rectilinear figures, whose outline is composed entirely of straight line segments, than curvilinear figures, whose outline is composed of smooth curved segments. A rectilinear figure is a polygon. Polygons are classified primarily in terms of the number of straight line segments needed to define their outline; this is also the number of vertices. Thus polygons may be triangles, quadrilaterals, pentagons, hexagons, heptagons, octagons, and so on. Polygons with a small number of sides are easy to discriminate from one another perceptually, and their properties have been intensively investigated. Even triangles have many highly unobvious and extremely beautiful properties.

The simplest curvilinear figure is the circle, which has again been the subject of intensive geometrical investigation (Pedoe 1957). Defining curves is in general mathematically much more complex than defining polygons, especially when the curves are the outlines of objects occurring in nature. In computer graphics, much use is made of curves defined in terms of a control polygon (for example, Bézier curves and splines); many of the curvilinear figures occurring in the diagrams in this book have been produced in this way.

Of course there are also mixed figures whose boundary contains both straight and curved segments; a semicircle is perhaps the simplest example. In general the outline of a figure can be characterized in terms of a single numerical variable called the curvature, defined at each point along the boundary apart from at most finitely many singularities where its value undergoes a discontinuous change. This is the reciprocal of the radius of the circle which, in a certain technical sense, best approximates the curve at that point; it can also, equivalently, be defined as the rate of change of the bearing as distance along the curve increases. The curvature can be regarded as zero at any point along a straight segment (a straight line being thought of as a circle of infinite radius), and as infinite or undefined at vertices and cusps, where the tangent to the curve undergoes a discontinuous change in direction. The curvature is negative if the radius of the approximating circle is directed out into the exterior of the figure, positive if into the interior.

There have been a number of attempts to use curvature to provide a basis for the qualitative description of shapes in two dimensions. The idea is to use the variation in curvature along the outline of a shape to divide it into segments, the qualitative description then consisting of a sequence of descriptive labels for the segments. One implementation of this idea is that of Hoffman and Richards (1982), discussed below in §4.7.2, where various similar ideas are also described.

From a qualitative point of view the nature of the outline of the figure at each point can be represented in terms of the following set of seven 'curvature types':

Curving outward	$\supset$	(Positive finite curvature)
Curving inward	$\subset$	(Negative finite curvature)
Straight	$/$	(Zero curvature)
Angle outward	$>$	
Angle inward	$<$	
Cusp outward	$\succ$	
Cusp inward	$\prec$	

A qualitative description of the outline of a figure can then be given by recording the sequence of qualitative curvature types encountered as the boundary is followed round clockwise. This is illustrated in Figure 4.20, where the outline, starting at the outward-pointing cusp at the top of the figure, is of type

$$\succ \subset \supset < \supset \prec \supset / > \subset .$$

This system of recording the sequence of curvature types on an outline is explored further in Galton and Meathrel 1999, where amongst other things, a formal grammar is given for the set of valid curvature-type sequences.

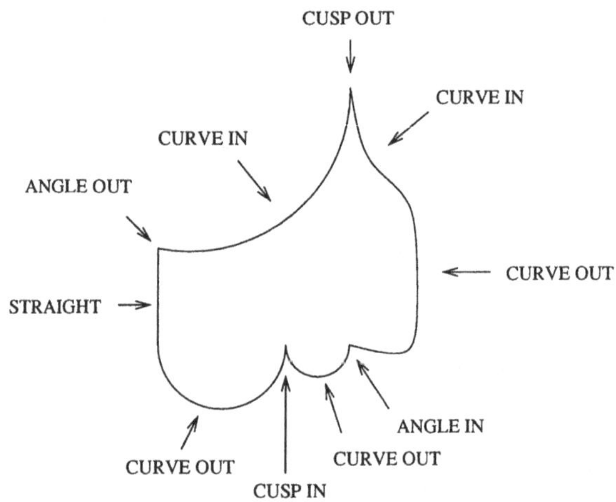

Figure 4.20. An outline characterized by its sequence of qualitative curvatures

Where a 'curve out' segment meets a 'curve in' segment without either a cusp or a vertex, there is a point of zero curvature (a point of inflection), but we do not count this as 'straight' since it is only an isolated point. Although the curvature types are presented as a linear sequence, this must be understood as cycling back on itself, the last member of the sequence being followed by the first. With this understanding, the following constraints apply:

1. Consecutive terms of the sequence must be different from one another.
2. Two consecutive terms of the sequence may contain at most one member of the set $\{<, >, \prec, \succ\}$.

3. Any occurrence of $\prec$ or $\succ$ must be adjacent to an occurrence of $\supset$ or $\subset$ respectively.
4. In a closed outline there must be either at least one $\supset$ or at least three occurrences of either $>$ or $\succ$.

Figure 4.21 depicts representative exemplars of all the possible curvature type sequences of length at most three. Fourteen outlines are shown; eight of them involve sequences of length three, giving rise to a further eight, not shown, obtained by reversing the sequence (which corresponds to reflecting the outline in a mirror).

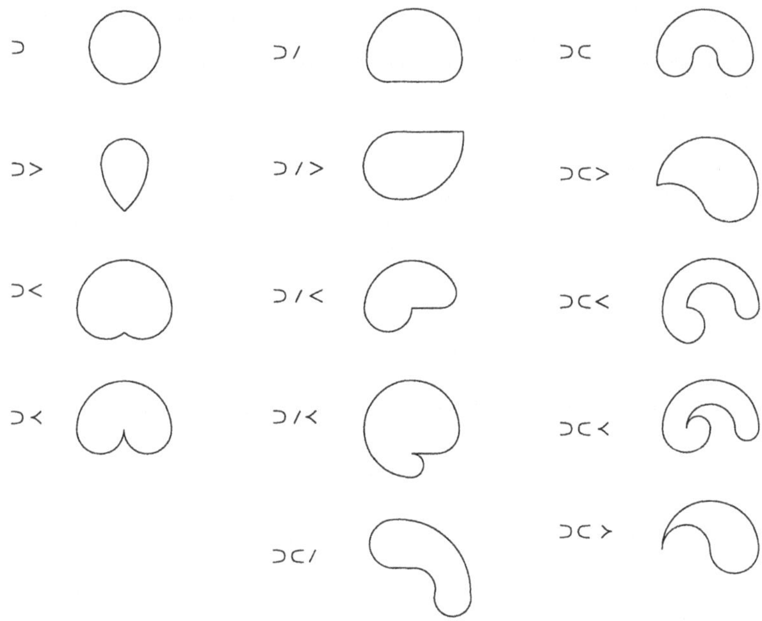

Figure 4.21. The fourteen outlines containing three or fewer curvature types

Various classes of outlines can be characterized in terms of restrictions on the available curvature types. Thus the set $\{\supset, /, >\}$ defines the convex figures, $\{/, <, >\}$ defines the rectilinear figures (polygons), their intersection $\{/, >\}$ defines the convex polygons, while $\{\supset, /, \subset\}$ defines the figures with *smooth* outlines, lacking angles and cusps.

Note that I have been assuming that a figure has at most finitely many changes of curvature type as its outline is followed. A figure whose boundary oscillates infinitely many times in a finite stretch (for example if it contains a segment of the curve $y = \sin(1/x)$ over an interval $0 < x < x_0$) is thus excluded from the analysis. Likewise I exclude fractal-type curves such as Koch's snowflake curve, every point of which might be thought of as either an inward-pointing vertex ($<$) or an outward-pointing one ($>$).

Curvature plays an important role in the work of Leyton, to be discussed in section 4.7.2. The next topic, symmetry, also has a role to play in Leyton's work.

Symmetry

Symmetry plays a very important role both in geometry and in human perception. For bounded figures such as we are mostly considering, there are two kinds of symmetry to consider, rotational and reflectional:

- A figure has reflectional (or mirror) symmetry about a line L if for each point P in the figure not on the line there is another, identical, point P' in the figure such that L is the perpendicular bisector of PP'. The line L is called an axis of symmetry of the figure.
- A figure has n-fold rotational symmetry about a point O (where n is an integer greater than 1) if for each point P in the figure distinct from O there is another identical point P' such that $OP' = OP$ and $P\hat{O}P' = 360°/n$. The point O is called a centre of n-fold symmetry of the figure.

Each type of symmetry is associated with a symmetry operation. For reflectional symmetry, this is most easily described by thinking of the plane as embedded in a three-dimensional space, in which case reflection in an axis can be achieved by rotating the plane through 180° about the axis so that it returns to its original location but with all the figures in it 'reversed'. A figure has reflectional symmetry about the axis exactly if it remains unchanged when the plane is reflected in this way. For n-fold rotational symmetry the corresponding symmetry operation is rotation of the plane through $360°/n$ about the centre of symmetry.

If a figure has both reflectional and n-fold rotational symmetry then it will have reflectional symmetry about n different axes, since corresponding to the points in the figure along any one axis of symmetry there will be identical points along lines making angles of $360i°/n$ (for $i = 1, \ldots, n$) with the first axis, and these will all be axes of symmetry too.

Familiar shapes can be characterized in terms of the types of symmetry they possess; for example, amongst quadrilaterals (Figure 4.22):

- a square has fourfold rotational symmetry as well as reflectional symmetry about both its diagonals and the lines joining the midpoints of opposite sides.
- a rectangle has twofold rotational symmetry, and reflectional symmetry about the lines joining the midpoints of opposite sides.
- a rhombus has twofold rotational symmetry, and reflectional symmetry about the diagonals.
- a parallelogram has twofold rotational symmetry.
- a kite has reflectional symmetry about one diagonal.
- an isosceles trapezium has reflectional symmetry about the line joining the midpoints of one pair of opposite sides.

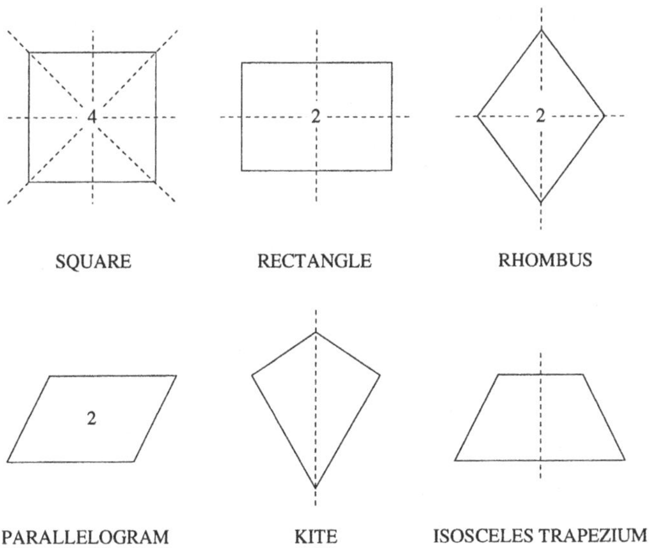

Figure 4.22. Symmetries of quadrilaterals

The individual symmetries of a figure combine to form a closed system of operations known as a *group*. The symmetry group of a square, for example, consists of eight distinct operations: rotations through 90°, 180°, and 270°, reflections in four distinct axes, and also the *identity* operation which leaves everything the same. Any two of these operations combine to produce a third, and each operation has an *inverse* with which it combines to yield the identity operation, which itself has no effect on any operation it combines with. The group of eight operations are all and only those rigid operations on the square which result in the square's occupying exactly the same location as before.

Other combinations of symmetries are possible for polygons with larger numbers of sides, and for curvilinear figures, but all are subject to some rather tight constraints, namely:

1. A finite figure can have at most one centre of symmetry.
2. If n is the largest number for which the figure has n-fold symmetry, then it has m-fold symmetry for each factor m of n, and for no other numbers.
3. If a figure has mirror symmetry about n distinct axes, where $n > 1$, then it also has n-fold rotational symmetry, and all the axes of symmetry pass through the centre of symmetry.
4. If a figure has n-fold rotational symmetry and at least one axis of reflectional symmetry, then it has n axes of reflectional symmetry.
5. A circle has n-fold rotational symmetry for all $n > 1$, and every diameter is an axis of reflectional symmetry.

The upshot of all this is that a complete list of the possible symmetry types for plane figures is

$$1, 2, 3, 4, \ldots, 1', 2', 3', 4', \ldots, \infty'.$$

where integers indicate the degree of rotational symmetry (1 for no rotational symmetry—the only rotation bringing the figure into coincidence with its starting position being 360°), and a dash indicates presence of reflectional symmetry; ∞' represents the symmetry type of the circle. Thus a square is type $4'$, a non-square rectangle or rhombus is $2'$, a general parallelogram is 2, and a kite or isosceles trapezium is $1'$. Any other convex rectangle is of type 1, which is to say, completely asymmetrical.

If we allow unbounded figures then not only must the above constraints be modified (since then a figure can have infinitely many centres and axes of symmetry) but we must in addition recognize two further fundamental symmetry operations, namely translation and glide reflection. Unbounded figures can be assigned to one of seven distinct types of 'frieze pattern' or seventeen distinct types of 'wallpaper pattern' (Cadwell 1966: ch. 11).

Symmetry is important because many objects we see in the world about us are at least approximately symmetrical, particularly in the animal world and amongst human artefacts and symbols. Of the twenty-six letters of the alphabet, in their upper-case forms, only seven (F, G, J, L, P, Q, R) have no symmetry at all.

The notion of symmetry about an axis can be generalized to cases where an object displays a kind of symmetry about an axis which is not straight. This applies both to reflectional symmetry in two dimensions and rotational symmetry in three. The sausage-like outline in Figure 4.17 has an approximate symmetry about the curved principal axis indicated by the longer dotted line, and indeed this kind of symmetry is characteristic of many organic forms, which seldom have absolutely straight principal axes. Various kinds of such 'differential' symmetry have been described; Leyton (1988) mentions three of them. I shall refer to differential symmetry again when I discuss Leyton's work in §4.7.2.

Shapes in three dimensions

All the above considerations apply, with correspondingly greater complexity, to three-dimensional shapes. Convexity, concavity, and convex hull can be defined in the same way, but the range of possible forms is much greater. For example, an hourglass or dumbbell-shaped figure has a single concavity which is topologically a torus.

Curvature in three dimensions is a much more complex business. There is no such simple set of qualitative curvature types; Figure 4.23 illustrates a selection. Apart from the flat and saddle types, all of these have inward and outward pointing forms. In addition, the edge and edge-cusp types have variants according to the form of the line of the edge, which may be curved in a plane which either cuts or merely touches the surface, or may even be non-planar. It is not clear how many of these variants we might want to discriminate. Accordingly the number of three-dimensional qualitative curvature types may be anything from fourteen (those illustrated plus their reversals) to at least thirty-one and quite possibly many more. An additional problem is that whereas the curvature types encountered around the outline of a two-dimensional figure can be arranged in a linear sequence (or more exactly, a circle), for a three-dimensional figure this is no longer the case, the curvature types being distributed instead over a spherical surface.

Symmetry in three dimensions comes in the same varieties as in two, but has to be described differently. Reflectional symmetry is now associated with

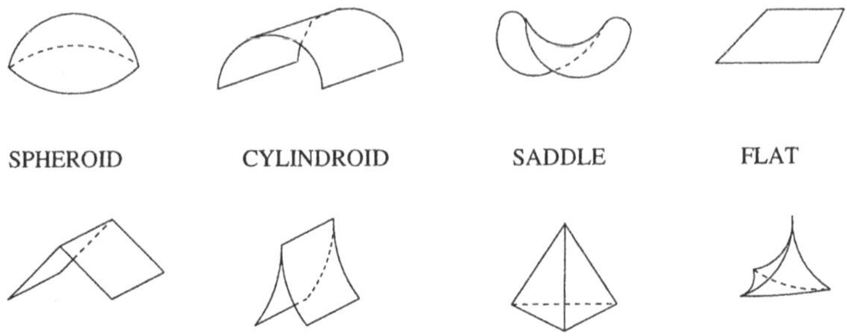

Figure 4.23. A selection of three-dimensional curvature types

a *plane* of symmetry rather than a line (axis), and rotational symmetry with an *axis* of rotation rather than a point. There are many more ways in which these symmetry operations can be combined; the symmetry group of the cube, for example, has forty-eight distinct elements, that of the icosahedron has 120. Note that if one wants to explain reflection in three dimensions in terms of rotation (as reflection in two dimensions was explained as three-dimensional rotation about the axis of reflection), then one has to enter the realms of mathematical abstraction and invoke four-dimensional space, in which it is possible to 'rotate' a figure about a plane; as viewed within the three-dimensional space of the figure, the result of such a rotation is to convert the figure into its mirror image or enantiomorph, as vividly envisaged by H. G. Wells in 'The Plattner Story' (Wells 1927).

4.7.2. Cognitive factors in shape recognition

Although we can to some extent recognize shapes by touch, shape is generally felt to belong to the visual domain. One of the central problems in explaining vision is to give an account of how the visual system is able to convert an essentially two-dimensional retinal image into a three-dimensional cognitive model of the visual scene. The task of the draftsman or painter is the reverse of this: to represent a three-dimensional scene on a flat surface. This shift of dimensionality opens up the possibilities of treating shape as essentially two dimensional or essentially three dimensional. We thus distinguish *contour-based* and *volume-based* approaches to shape representation.

In everyday life we tend to think of shapes as essentially two dimensional.

A popular toy for young children is a *shape-sorter*. This consists of a hollow box with variously shaped holes in its surface, together with a set of solid pieces, each of which can be fitted through exactly one of the holes; the idea is to 'post' the pieces into the box through the correct holes, thereby exercising the child's perceptual, motor, and classificatory abilities. This task is most easily accomplished by visually matching the pieces to the holes; it is much more difficult to perform when blindfolded.[9] Now the point about the shape-sorter is that although the 'shapes' to be sorted are, of course, three-dimensional objects, in every example of the toy that I have come across, they are *prisms*, that is figures of uniform cross-section, generated by sweeping a planar figure along an axis orthogonal to its plane. Thus the 'shapes' in question are distinguished from one another solely by having different cross-sections; in effect, they are two dimensional.

Even so, our visual system is extremely efficient at imposing a three-dimensional 'reading' on a two-dimensional image, a capacity selected for by evolution because of its obvious utility: most retinal images do in fact derive from three-dimensional scenes, and the task of the visual system is somehow to run the derivation backwards. So efficient is it at doing this that it is easily fooled into seeing as three dimensional what is in reality quite flat. This forms the basis of all naturalistic drawing and painting, especially in the *trompe l'œuil* genre, and many delightful illusions and 'impossible' pictures.

'Compositional' approaches to shape seek to describe the full gamut of shapes in the world in terms of some small set of primitive elements, in a way that is somewhat analogous to how the enormous variety of words in a language are built up out of a small selection of phonemes. This analogy is explicit in Biederman 1987. Different systems may differ in the inventory of primitive components and the allowed ways of combining them.

If the emphasis is on the modular structure of complex shapes such as those of animals and machinery, with less concern for the nuances of fine detail, then it may suffice to use a very limited repertoire of primitives. This is exemplified by Marr and Nishihara (1978), who use cylinders of various sizes and proportions as their only primitive. Shapes built out of cylinders can often encode enough information to enable one to identify unambiguously the object that is being represented.

[9] This observation is related to Molyneux's question, whether a blind man, on recovering his sight and being presented with a cube and a sphere, would be able to identify which was which without feeling them. See Locke's *Essay Concerning Human Understanding*, Book II, ch. 9.

Biederman (1987) uses a more complex set of primitives, called *geons*. These are 'generalized cones', volumes 'swept out by a cross section moving along an axis'. Biederman classifies geons along four dimensions of variation:

1. the cross-section may be curvilinear or rectilinear;
2. the cross-section may have both rotational and reflectional symmetry, only reflectional symmetry, or neither;
3. the cross-section may have constant, increasing, or increasing and then decreasing size along the axis;
4. the axis may be straight or curved.

This provides $2 \times 3 \times 3 \times 2 = 36$ different geons. Biederman goes on to classify along a further four dimensions the way in which two geons G_1 and G_2 can be combined:

1. G_1 may be much greater than, about the same size as, or much smaller than the G_2;
2. G_1 may be above, below, or to the side of G_2 (applicable in about 80 per cent of cases);
3. G_1 and G_2 may be joined end-to-end or end-to-side;
4. the join may be on either the long or the short surface of G_1 or G_2.

This leads to 74,649 two-geon objects discriminable on the basis of these principles of classification, and 154 million three-geon objects; as Biederman says, this value 'readily accommodates the liberal estimate of 30,000 objects actually known'. When we reflect further that many objects require more than three geons for their satisfactory representation, we see that Biederman's system is in fact highly redundant. This is, of course, unavoidable in any combinatorial approach to shape, and should not be regarded as a defect of the approach. It merely reflects the obvious fact that there are many more possible shapes than are actually realized by the objects we see in the world.

What is more worrying is that Biederman's principles of classification are—as he himself admits—far too crude: they do not enable one to specify the shapes of even very simple everyday objects without an enormous range of ambiguity. A simple cup with a handle, for example, as illustrated in Figure 4.24*a*, might be specified using Biederman's primitives as follows:

> The body of the cup is a geon with both reflectionally and rotationally symmetrical curvilinear cross-section increasing along a straight axis. The handle is a geon with reflectionally but not rotationally symmetrical curvilinear cross section of constant size along a curved axis. The handle is much smaller than the body, and is attached to the side of the body, end-to-side, with the join on the long surface of the body and the short surface of the handle.

The object in Figure 4.24*b* also fits this description; whatever it is, it certainly isn't a cup!

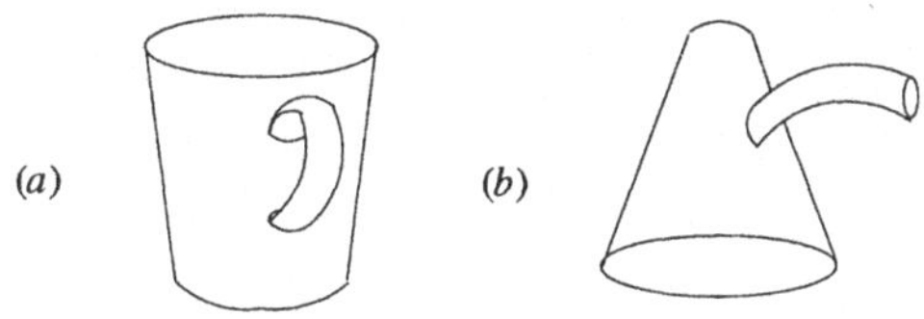

Figure 4.24. (*a*) A cup built out of two geons; (*b*) a non-cup answering to the same description in Biederman's system (see text for details)

In contrast with these volume-based approaches to shape recognition, there has also been extensive work using contour-based approaches. Attneave (1954) noted that 'information is concentrated along contours . . . , and is further concentrated at those points on a contour at which its direction changes most rapidly'. A consequence of this is that the visually most salient parts of an outline are the curvature extrema, an idea which led Attneave to suggest that

> Common objects may be represented with great economy, and fairly striking fidelity, by copying the points at which their contours change direction maximally, and then connecting these points appropriately with a straightedge.
>
> (Attneave 1954)

A practical illustration of this idea is Attneave's now famous rectilinear drawing of a sleeping cat, which is instantly recognisable despite the fact that every contour in the cat itself is curved.

Hoffman and Richards (1982) also used the salience of curvature extrema in order to characterize an outline from the point of view of recognition, but their approach was to break the curve into segments at the curvature *minima*, using maxima and zeros to describe the shape of each segment. This results in four basic types of segments (loosely analogous to our curvature types, but restricted to smooth curves), which they called 'contour codons'; a table is provided to show which types of codons can occur adjacent to which other types. Hoffman and Richards use the asymmetry in their treatment of maxima and minima to explain figure–ground reversal phenomena well known from the Rubin vase which can also be seen as two faces. Maxima and minima are defined relative to one side of the contour being taken as the figure; hence under figure–ground reversal, the maxima and minima are reversed, leading to a completely different description in terms of contour codons.

A rather different approach to the understanding of shapes through the curvature of their outlines is Leyton's *process-based* theory (Leyton 1988). The idea is that shapes are naturally understood in terms of the processes that could have given rise to them. Starting with a stock of very simple basic shapes such as circles and spheres, other shapes are described in terms of the modifications that have to be made to the basic shapes in order to produce them. One looks first for the extremes of curvature in the outline of the shape: points at which the curvature assumes either a local minimum or local maximum value. In a smooth outline, between two consecutive minima there must be a maximum, and vice versa. Leyton now makes use of a generalization of the notion of a symmetry axis: a *differential symmetry axis* is the locus of some midpoint associated with the series of circles touching both of two segments of the outline. Different types of differential symmetry arise from different interpretations of 'midpoint'; possibilities include the centre of the circle, the midpoint of the chord joining the points where the circle is tangent to the curve, and the midpoint of the corresponding arc. We then invoke Leyton's *symmetry-curvature duality theorem*, which says that any segment of a smooth planar curve bounded by two consecutive curvature maxima (minima) has a unique differential symmetry axis, which terminates at a curvature minimum (maximum). Leyton claims that 'each curvature extremum implies a process whose trace is the unique symmetry axis associated with, and terminating at, that extremum' (Leyton 1988: 219). These processes are 'protrusion', associated with a positive curvature maximum, 'indentation' (negative curvature minimum), 'squashing' (positive curvature minimum), and 'internal resistance' (negative curvature maximum). They are illustrated in Figure 4.25, in which the differential symmetry axes of a complex shape are shown along with the processes by which the shape can be produced from an initial circle. Although this example, and most of Leyton's discussion, involves only two dimensions, the underlying principles can be generalized to three dimensions as well, though as usual at the cost of much greater complexity.

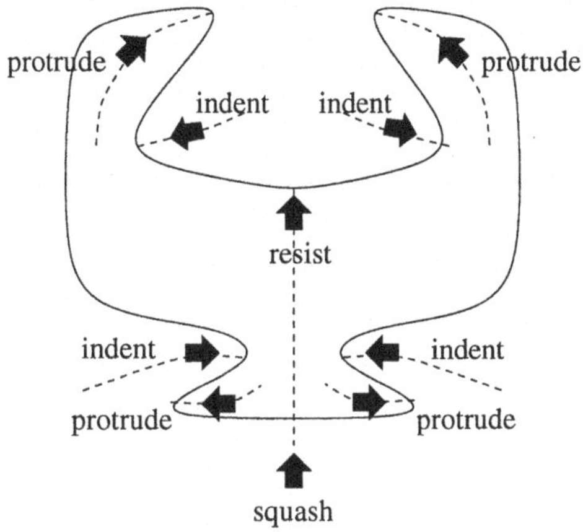

Figure 4.25. A complex outline and the processes that could be used to generate it (after Leyton 1988)

Whereas Leyton constantly speaks of processes that *have acted* upon a shape, thereby licensing the use of his representational system for making inferences about the history of the shape, it would be more realistic to refer to them only as processes that *could have acted* to produce the shape. While it may in some cases be true that a particular outline has been produced by the processes which Leyton's analyses would assign to it, this cannot be assumed in general. Moreover, even in cases where this assumption is justified, it is not obvious that the best way of describing a shape is in terms of the processes that have brought it into being.

A general curvature-based theory of qualitative outlines in two dimensions is presented by Meathrel and Galton (2000). This theory subsumes not only the systems of Hoffman and Richards (1982) and Leyton (1988) described in this section, but also the qualitative curvature-type theory described on pp. 183–6. The systematic relations between these contour-based approaches and the volume-based methods (exemplified by the work of Biederman) remain to be elucidated.

4.7.3. Describing shape in natural language

Although we are very sensitive to nuances of shape when we perceive them, we are on the whole much less adept when it comes to *describing* shape. Here we are referring to the explicit description of shape *as* shape; not the implicit encoding of shape in the common nouns we apply to classes of objects. To emphasize this distinction, consider the noun 'giraffe'. For the most part we have little difficulty with the application of this term; we readily recognize a giraffe when we see one, there being no other animal at all similar. The distinctive characteristics of a giraffe that enable us to recognize it so easily all have to do with shape: the long neck and legs prominent amongst them. Colour and surface texture are not very important, since we recognize giraffes just as readily on the basis of their silhouettes.[10] Implicitly, the word 'giraffe' encodes all this: that is, it serves as a point of access to the web of associations that constitutes any individual's knowledge of giraffes, and in particular to a representation of its shape, which in itself provides a separate such point of access.[11] On the other hand, there is nothing about the word 'giraffe' which explicitly encodes any information about shape: if one did not know that a giraffe had long neck and legs, one could not infer this from the name.

Because the names of objects implicitly refer to the shapes of the objects which they name, they can be used as proxy descriptions for shapes. This provides us with one of the commonest means employed in natural language for the description of shape: a shape is described by reference to some category of objects which have that shape. Thus to revert to our earlier example, if someone were to describe some newly discovered or imagined animal as *giraffe shaped*, we would already have a pretty good idea of the appearance of the creature.

Some objects have such distinctive shapes that their names are often used for the 'X' in locutions of the form 'X shaped' or 'shaped like a(n) X'. The object itself is used as a prototypical exemplar of a class of shapes. In English, a (doubtless very incomplete) list of nouns that are frequently employed in this way is:

> *arrow, barrel, bell, boomerang, bottle, cigar, club, crescent, cross, cup, diamond, dome, doughnut, dumbbell, egg, fan, funnel, heart, horseshoe, hourglass, kidney, kite, lens, lyre, mushroom, paddle, pear, ring, saddle, sausage, sickle, spoon, star, sword, wedge.*

Typical examples of the shape prototypes represented by some of these terms are shown in Figure 4.26.

[10] Indeed, a classic image of the African savannah, repeated endlessly in travel brochures, wildlife documentaries, etc., is a group of giraffes silhouetted blackly against the sunset, with a few acacia trees thrown in for good measure.

[11] This sense of 'implicit encoding' is what underlies the assertion in Landau and Jackendoff 1993 that 'for a large proportion of object categories, shape is amongst the most important criteria for identification, and in particular for judgments of what a thing should be called'.

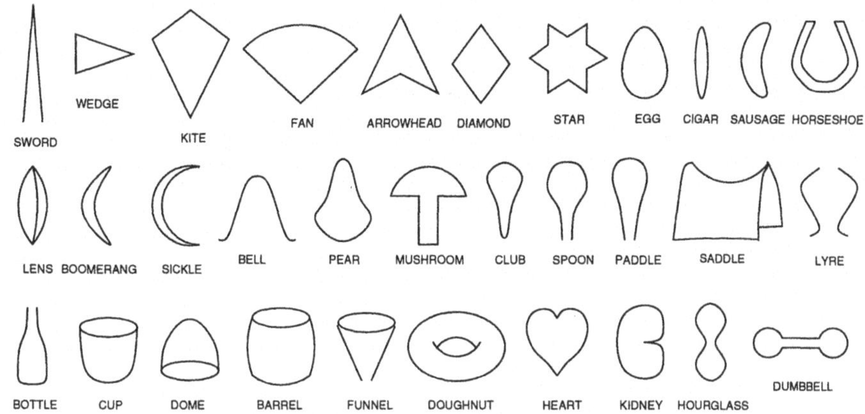

Figure 4.26. Shape prototypes: a selection of familiar (or not so familiar!) objects used as prototypical exemplars of classes of shapes

For some of these examples, such as *heart* and *star*, the object itself does not really have the shape for which it is taken to be the exemplar: in these cases, the shape is that of a conventional representation of the object in question. In other cases, such as *doughnut* and *kite*, objects of the prototype class may themselves exhibit a wide variety of shapes, of which a particular instance may be singled out as the standard. Some of the shapes are essentially three dimensional, others can refer to outlines or two-dimensional shapes.

Also belonging in this category are descriptions based on the shapes of letters of the alphabet. It is always the upper-case letter that is used: *L-shaped, S-shaped, T-shaped, U-shaped, V-shaped.* Note also expressions such as *C-spring, G-clamp, H-girder, T-bar, T-bone, T-junction, T-shirt, U-tube, U-turn, Y-fronts*. Numerals are not employed in this way to anything like the same extent, though we frequently use the expression *figure-of-eight* (its technical analogue, *lemniscate*, being little known by general language users).

The use of prototypical shape exemplars provides a prolific source of terms in technical shape vocabularies. These include, for example, the rich (some would say over-rich) repertoire of terms used by botanists for describing the shapes of leaves and other parts of plants. Some terms refer to the overall shape of the leaf (*linear, lanceolate, elliptic, oblong, ovate, triangular, peltate, palmate, pinnate, pinnatifid, pinnatisect, renate, cordate, ter-*

nate), some to the leaf-margin (*entire, lobate, crenate, dentate, denticulate, serrate, undulate*), some to the base of the leaf (*cordate, cuneate, sagittate, perfoliate*), yet others to the tip of the leaf (*mucronate, apiculate, acute, obtuse, obcordate, truncate*). Taken together they can be used to provide a remarkably exact description of a leaf. Many of these terms derive from prototypes, for example *cordate* (heart-shaped), *cuneate* (wedge-shaped), *ovate* (egg-shaped), *renate* (kidney-shaped), and *sagittate* (arrow-shaped). Heidorn (1998) has studied in detail the technical vocabulary used for describing just one relatively restricted class of leaves, those of the North American oaks, and developed a computer program for generating images from descriptions cast in this vocabulary.

The technical vocabulary of geometry also draws extensively upon prototypes for defining shapes, particularly as regards some of the exotic curves which were investigated with such enthusiasm in the wake of the introduction of Cartesian geometry and the subsequent analytical power conferred by the introduction of the calculus (Lockwood 1963). These include *astroids* (stars), *cardioids* (hearts), *cissoids* (ivy-leaves), *conchoids* (sea-shells), *helices* and *limaçons* (snails), *nephroids* (kidneys), and of course *ovals* (eggs). With the exception of the last, these terms are all pretty recondite; our more basic geometric vocabulary has more diverse origins.

Names for general classes of polygons and polyhedra refer to the number of vertices, sides, or faces they have, giving us the series *triangle* (three angles), *quadrilateral* (four sides), *pentagon* (five angles—literally 'knees'), *hexagon* (six angles), *heptagon* (seven angles), *octagon* (eight angles), and so on,[12] and *tetrahedron* (four faces), *hexahedron* (six faces), *octahedron* (eight faces), *dodecahedron* (twelve faces), and *icosahedron* (twenty faces). Only a few of these terms, of which *triangle* is the commonest, are in general use in non-technical contexts.

A group of other terms which have precise geometric meanings are employed in everyday speech, more or less loosely, for describing shapes. The commonest ones are:

> Two dimensional: *circle, semicircle, square, oblong, rectangle, triangle, oval, spiral*
> Three dimensional: *sphere, cube, cone, cylinder, pyramid*

[12] The first two would be *trigon* and *tetragon* if the Greek basis of the others were extended back to the beginning of the series. The former does indeed show up as a component of *trigonometry*.

Terms such as *parallelogram, rhombus, prism, ellipse, parabola, hyperbola* are less widely used, presumably because less well understood. The only widely used terms for other polygons are *pentagon, hexagon, octagon*, and these are generally used for examples which are more or less regular. *Oval* is often used to mean 'elliptical', although strictly speaking it has a different meaning (the same remark applies to the term *egg-shaped*). Similarly, *rectangular* is used to mean 'oblong', though strictly speaking it means 'oblong or square'. *Spiral* is generally also used for a three-dimensional helix, the term *helix* itself being seldom used in everyday language—its most familiar use being perhaps the description of the DNA molecule as a 'double helix'. Terms such as *cuboid, tetrahedron, octahedron, dodecahedron, icosahedron*, and *parallelepiped* are little used outside specialist domains. Instead of *rhombus*, everyday speech favours *diamond*. The term *kite* is used in geometry, less often in everyday speech, to refer to the shape which stands in relation to a rhombus as a true oval does to an ellipse. The adjective *round* is used as a more familiar version of *circular*, the latter perhaps connoting greater geometrical precision. *Hemisphere* is in general use normally associated with one of the adjectives *northern*, *southern*, *eastern*, and *western*.

Up to this point, the spatial vocabulary I have considered has for the most part consisted of nouns. However, we also make extensive use of adjectives for describing shape. I am not here referring to adjectives such as *triangular* or *spherical*, which are derived from the nouns we have already discussed, but to 'basic' adjectives which are not analysable into separate morphemes.

English contains a small basic set of 'dimensional' adjectives describing linear spatial extent, namely:[13]

> *long, tall, high, deep, broad, wide, thick, fat, short, low, shallow, narrow, thin.*

Around this core list one can collect a number of less basic terms such as *stout, slim, slender*, but we shall not consider these here. The basic terms can be organized as a set of partially overlapping oppositions, namely

long/short	*high/low*	*broad/narrow*	*thick/thin*
tall/short	*deep/shallow*	*wide/narrow*	*fat/thin*

One can convey an idea of the shape of something by pairing adjectives; typically one pairs the first element of one opposition with the second element

[13] Detailed work on the dimensional adjectives of German is reported in Lang 1989 and Lang *et al.* 1991.

of a different opposition. Thus a *long, thin* object is one whose length along one axis considerably exceeds its length along any other axis, while a *short, fat* object is much less extreme in this respect. A building might be described as *long and low*, whereas *long and high* conveys a less definite image. There are exceptions, of course, such as a *long, thick* rope—but arguably this is merely describing the size of something whose shape is known rather than describing the shape itself.

More generally, as pointed out by Lyons (1977: 701–3), *which* oppositions we use to describe something depends very much on its shape. The available oppositions convey information about shape but also about orientation, function, and perhaps other characteristics as well. Some examples are:

- Plank, ribbon, piece of paper: *long/short—wide/narrow— thick/thin*
- Wall: *long/short—high/low—thick/thin*
- Tunnel, mountain-range: *long/short—high/low—wide/narrow*
- Door: *high/low—wide/narrow—thick/thin*
- River, trench, groove, swimming-pool: *long/short—deep/shallow—wide/narrow*
- Alcove, recess: *high/low—deep/shallow—wide/narrow*
- Post, tree-trunk: *tall/short—thick/thin*
- Person: *tall/short—fat/thin*
- Leg, finger: *long/short—fat/thin*
- Log, branch, pencil, needle, wire, rope: *long/short—thick/thin*
- Nose, chin, hand: *long/short—wide/narrow*

Note also that a book *qua* text is *long/short* whereas the correlated dimension for the book *qua* physical object is *thick/thin*. Moreover, the opposition *high/low* is also used to provide positional rather than dimensional information: a high (or low) branch is a branch situated high up (or low down) in the tree; this is not included in the list.

For some objects the situation is more complicated. A container such as a cup or a bowl can be described from different points of view. The vertical extent of its interior is measured by *deep/shallow*, the horizontal extent of its opening (spanning the rim) is given by *wide/narrow*, while the material it is made of may be *thick/thin*. On the other hand a vase would be described as *tall/short* rather than *deep/shallow*.

Finally, we should mention the existence of languages in which shape plays an important part in the classification of nouns. In such languages, nouns may be categorized syntactically in terms of their co-occurrence with a fixed set of lexical elements called *classifiers*, in a way that is roughly analogous to the classification of nouns by gender in many European languages. In many such classifier systems, it turns out that the underlying semantic motivation for the classification includes shape as a significant component: thus there might be one class for long thin objects, another for flattish objects, another for roughly spherical objects, and so on. Friedrich (1970) gives a detailed examination of the classifier system of the native American language Tarascan, as well as a more general discussion of classifier systems in the world's languages.

4.8. Posture

This discussion of the geometrical attributes has for the most part proceeded on the implicit assumption that the geometrical attributes of objects are inherited wholesale from the regions of space which they occupy (their positions, in Aristotle's sense). Thus an object has a certain shape, size, position, dimension, and connectivity because the space which it exactly occupies has precisely that shape, size, position, dimension, and connectivity.

For many objects, however, this gives us a misleading idea of what we mean by shape. When we speak of, for example, the shape of the human body, we are referring to something which is invariant across all the various postures that the body is capable of assuming. This is a more abstract notion than we have been considering hitherto. The shape of a human body is a composite of the shapes of its various parts; it consists of a torso to which are attached limbs and head. These appendages are attached at specific invariant positions on the torso; the limbs themselves can be decomposed into proximal and distal sections, with the hand or foot attached to the ends of the distal sections, the points of attachment again being invariant. Thus the overall articulation of the body remains constant despite the possibility of much repositioning of the individual parts. It is this overall articulation, which is the same whether the body is standing, sitting, crouching, or lying down, which constitutes our abstract conception of the shape of the body. We thus have the idea of a constant shape over a range of different postures. This does not only apply to visibly articulated objects; a snake too has a single abstract shape—long and thin—but is capable of assuming a wide range of different postures, for example coiled, undulating, or straight. In all these cases the regions occupied by the object in its various postures have different shapes, so the shape of the object (invariant across all its different postures) cannot be identified with the shape of the region it occupies (which varies with the posture).

Not all objects are capable of assuming different postures. This is obviously the case for rigid objects, even when they function as representations of non-rigid ones. A set of clay models of a human being in various postures have different shapes according to the postures represented. If the same postures are represented by means of a doll with articulated limbs, head, and neck, on the other hand, then we are more likely to speak of the different postures of the doll, the shape remaining invariant.

We can say in general that both shape and posture arise from the configuration of parts. Given two configurations of the same set of parts, when do we say that we are dealing with different shapes, and when with different postures of the same shape? There are some clear cases of both: the bodies of humans and other animals provide clear instances of the latter. A lump of clay, which can be moulded into almost any shape we like, provides a clear instance of the former. The *word* 'posture' is generally reserved for animals, paradigmatically for humans, becoming less felicitous as we move to species that are less closely related to us (does one ever speak of the posture of a fish?). However, the idea of an abstract underlying shape that is preserved under reconfiguration of parts extends more widely, being applicable, for example, to books (which at least in one sense do not change shape when they are opened), pieces of string (which, again in one sense, remain long and thin even when rolled up into a tight ball), and pieces of paper (which remain rectangular even when crumpled up). This extension stops short of lumps of clay and the like.

4.9. Spatial attributes that are not purely geometrical

In this section I deal with those attributes of an object which are not shared with the space occupied by the object. In fact we have already encountered a number of such attributes, and it is worthwhile here to collect together these scattered encounters in order to form a clearer idea of what sorts of things these non-geometrical attributes are. Of the spatial attributes I have discussed, then, the following depend on more than just geometry:

- Amongst the various forms of connection enumerated in §4.2, only adjacency and interlocking are purely geometrical. Thus support, adhesion, and cohesion all involve non-geometrical effects. One form of support depends on gravity, and hence on the mass of the supported object, as well as the strength of the supporting object. Cohesion and adhesion involve intermolecular forces.
- Support appeared again in our discussion of the meaning of the word 'on' in §4.4. The meaning of 'in' similarly involves the notion of containment, which is not purely geometrical since, as we have seen, we can speak of something's being *in* a container even when only a small part of it is located within the containing space of the container. Again, although I did not mention this above, the preposition 'at' often has a non-geometrical (indeed non-spatial) component to its meaning, since when we speak of someone being 'at' a computer terminal, for example, we imply that they are actually using it.
- The distinction between the coincidence and encounter schemas makes use of the non-geometrical notions of intrinsic 'front' and 'back', the intrinsic orientation that is possessed by objects such as animals, vehicles, buildings, and many machines and articles of furniture.
- The distinction between shape and posture makes use of an abstract notion of shape that is not a property of any region of space, but depends on the reconfigurability of the parts of an object.

To summarize, then, the non-geometrical features which can contribute to the spatial attributes of an object include mass, strength, force, use, intrinsic orientation, and reconfigurability.

In addition to these attributes we may briefly mention *density*, *constitution*, and *state*.

Density is a measure of how much of the material making up an object or distribution fits into a given amount of space. In physics it is primarily associated with substances, and is defined as the ratio of mass to volume. In geographical contexts a more frequently encountered case is the density of a distribution, for example population density (how many people per square kilometre).

By the constitution of an entity I mean the actual material of which it is composed. An entity can retain its identity through changes of constitution;

for example, a river remains the same river despite wholesale replacement of the water it contains over a period of time (cf. §3.2).

By state here is meant the physical state of the component matter: solid, liquid, gaseous. This list should be supplemented by other states such as granular (sand) or dissolved (the salt in the sea). It may be stretching things somewhat to call these *spatial* properties, but they are clearly of great importance in constraining the kinds of spatial change that an entity can undergo. Granular entities partake of both solid and liquid features; they can form heaps like solids but can be poured like liquids (cf. Hayes 1985).

4.10. Summary

In this chapter I have discussed a range of spatial attributes that may be ascribed to objects: dimensionality, connectivity, location, orientation, size, shape, and posture. I have discussed these from a variety of points of view, ranging from the purely geometrical, through the physical, to the cognitive and linguistic.

Over and over again I have returned to the basic idea, introduced in Chapter 1, of a continuous quantitative phase space partitioned into a discrete qualitative mode space. This general picture has been articulated in some detail here for a number of particular cases: connectivity space (§4.2), the various location spaces (illustrated in Figures 4.6, 4.7, 4.8, 2.9, and 4.12), orientation space (Figure 4.15), and one specialized shape space, namely triangle space (Figure 1.2); later I shall illustrate other shape spaces such as convexity space (Figure 6.19), and symmetry space (Figure 6.20). This way of organizing qualitative spatial attributes will prove to be important later, in the discussions of spatial change (Chapter 6) and, especially, of continuity (Chapters 7 and 8).

5

Time and Change

5.1. Absolute and relational theories of time

Time and change are two sides of the same coin: you cannot have one without the other.[1] The passage of time, as we experience it, seems for the most part to consist of a sequence of things happening, of changes in the world, and it is precisely when nothing happens to excite our interest that time seems to flow most slowly, as if change itself were the dynamo that powers the flow of time. Our most intimate knowledge of time comes from the flow of events in our subjective consciousness, with all its vagaries and fluctuations; but we begin to transcend mere subjectivity when we start correlating our experiences with those of others. In this way we can identify intersubjectively invariant features of time and can recognize in what respects our personal experience agrees with or contradicts that of others. From here we can move on to the idea of an objective flow of time, independent of any one individual's experience, and come to see our own experience of time as an imperfect representation—a distortion—of some objective reality. When we are absorbed in some activity, time seems to rush by without our noticing, but when there is nothing to absorb us, it slows down to a snail's pace; and when we are asleep it seems to take no time at all. All these variations we readily learn to put down to our personal misperception of an objective, uniform flow such as Newton described.

It is disputed how far different cultures embrace this notion of an objective flow of time, independent of human experience.[2] What is certain,

[1] This, at least, seems to be the most natural position to take, though some philosophers have argued that time without change is in principle possible (Shoemaker 1969; Newton-Smith 1980); cf. Newton's famous pronouncement that 'Absolute, true, and mathematical time, of itself, and from its own nature, flows equably without relation to anything external . . . '.

[2] For the history of the doctrine of absolute time before Newton, see Čapek 1987, where the source of this doctrine, unknown in the ancient world, is claimed to be a decree issued by the Bishop of Paris in 1277.

though, is that this notion is particularly fostered in a culture, such as ours, which takes seriously the notion of *measuring* time. With our calendars and clocks we do have an independent arbiter against which to compare our own experience of the flow of time. The nested cycles of years, months, days, and hours provide a temporal framework within which all change and events occur. A brand-new diary, acquired at the start of a calendar year, provides slots in which to inscribe the planned or actual happenings of the year; the slots seem to have an independent existence. Whatever I do on 23 July, indeed whether I do anything at all, 23 July itself will surely come and go exactly as the calendar foresees.

We see here a divergence between an 'absolute' or 'technological' view of time, which takes seriously the notion of a fixed absolute temporal framework within which all change and events have their being, and a 'relational' or 'philosophical' view of time, which takes account of the fact that all the temporal cycles of the technologically defined framework are dependent for their existence on phenomena in the world, indeed ultimately on phenomena of our experience.[3] The philosopher Leibniz famously championed the relational view of time (and indeed of space) in argument with Newton and his followers.

Note, incidentally, that despite its title, the Theory of Relativity does not necessarily favour the relational rather than absolute conception of space and time. In relativity, spatial and temporal measurements become relativized to the state of motion of the observer, but these measurements can be combined to give an integrated space–time measurement which is independent of the observer's state of motion and thus is a candidate for absolute rather than relational status. Thus while Newton and Leibniz argued about the statuses of space and time separately (which could allow, in principle, the bizarre option of simultaneously holding, say, a relational view of space but an absolute view of time), for Einstein and his successors the issue is rather one of absolute versus relational space–time (see, for example, Ray 1991: ch. 7).

Apart from the metaphysical issue, there is a methodological choice to be made as to whether our *representations* of space and time should be absolute or relational. An absolute representation of space and time involves first setting up a spatio-temporal framework and then populating it with objects and events. For time, this might go somewhat as follows. First we posit a set T of entities called 'times'—these might be instants or intervals (see §5.3.1

[3] The latter remark suggests that we might further distinguish an objective relational time from a subjective one, an issue I do not pursue further here.

below)—and a temporal ordering relation on T, for example strict precedence ('t_1 is earlier than t_2'). We then introduce some way of referring to states and events, along the lines of 'Event e happens at time t' or 'State s holds at time t'. Relations between events are then paraphrased using statements of these kinds; for example, to say that e_1 precedes e_2 is to say that there exist times t_1 and t_2 such that e_1 occurs at t_1, e_2 at t_2, and t_1 precedes t_2.

In a relational representation, on the other hand, we begin with a set E of events, together with temporal relations of simultaneity and succession defined on E. Thus the statement that e_1 precedes e_2 receives direct expression as a primitive in the representation. Times are then defined as equivalence classes of events under the relation of simultaneity: thus a time might be identified as the time at which a certain event e_1, and hence any event simultaneous with e_1, occurs. Relations on times are then defined in terms of relations on events: to say that t_1 precedes t_2 is to say that an event which t_1 is the time of precedes an event which t_2 is the time of.

So long as the concern is over representation rather than metaphysics, it is appropriate to appeal to purely practical issues in deciding which view to adopt. However philosophically appealing the relational views of space and time may seem, it is undoubtedly true that it is far easier to base a formal account on the absolute view. The plan of this chapter reflects this. I begin by discussing in a relatively informal way some issues concerning the classification of different kinds of change, but when I turn to a somewhat more formal account, I start with the description of time as an absolute framework existing in its own right, and only after that do I embark on a formal account of states holding and events occurring in time.

5.2. States, processes, and events

The things that go on in time can be broadly classified as states, processes, and events. Approximate definitions of these are as follows (further refinements come later):

- a state is characterized by the absence of change;
- a process is characterized by change as it is actually occurring;
- an event is a chunk of change picked out as an individual from the ongoing flux.

These approximate characterizations need explaining and qualifying.

When we say that a state involves no change, we cannot mean no change *at all*. We mean that a state is a situation viewed from a certain perspective in terms of which any change that takes place is regarded as incidental. For example, being asleep is a state that a person can be in. It is not true that when a person is asleep, no changes are occurring. The heart is beating, there is electrical activity in the brain, there is steady breathing, and the sleeper tosses and turns. In describing someone as asleep, however, one is rather focusing on the *constancy* of the situation: all the little changes that are occurring do not add up to one big change, so that as we observe the sleeper at different times we do not discern important differences. The situation is homogeneous with respect to time.[4]

Being awake is also a state, although it is less obviously homogeneous, in that a person who is awake is likely to indulge in many different activities which involve observable changes. However, in contrast to being asleep, being awake involves a higher level of activation of most bodily systems, including a higher level of mental alertness, and it is on grounds of this consistently higher level of arousal that one might reasonably make a claim for homogeneity here.

We may describe a situation as a state if, at some level of description, its holding results in no systematic net change. A process, by contrast, should result in net change. An example is growth: when something grows, it gets bigger. This is not homogeneous, since the thing which is growing is discernibly different at different times during the process, and these differences are ordered in a systematic way. The ordering is important: compare sleep and growth. If a sequence of snapshots of a sleeping person are jumbled up and presented in a different order, the resulting sequence will not look markedly different from the original; this is not true of a growing person.

The distinction between states and processes is not as clear-cut as one might wish. It is dependent on the point of view from which one is trying to classify situations; linked to this, it is dependent on the granularity or level of detail at which the situations are described.

The 'jumbled snapshot' test is particularly sensitive to temporal granularity. Consider the movement of the earth around the sun. Suppose we have a sequence of ten snapshots of this situation taken over a period of six months. These will show the earth at a sequence of positions along its orbit, with a clear correlation between the ordering of the positions and the ordering of

[4] Temporal homogeneity is discussed further below, see p. 212; for the formal expression of homogeneity, see §5.4.

the snapshots. If they are jumbled up, the resulting sequence will look quite different, with the earth appearing to jump back and forth along its orbit in an erratic manner. Now consider instead a sequence of ten snapshots taken over a period of ten years. If they are taken at regular intervals, a year apart, these will all show the earth in the same position; if jumbled up, the resulting sequence will be equally credible. But even if taken irregularly, the sequence of positions they show will not correlate in a systematic way with the ordering of the snapshots, so that again if the snapshots are disordered, the resulting sequence will be equally credible.

We might say that when viewed at a sufficiently fine granularity, the motion of the earth around the sun appears as a process; but when viewed at a much coarser granularity, it is more like a state. This is typical of repetitive, cyclical processes in general. We often call them 'steady-state' processes for exactly this reason. A steady-state process is one which, when viewed at sufficiently coarse granularity, appears homogeneous, and hence as a state. It is a process which in the medium or long term leads to no net change.

We may contrast these with 'cumulative' processes, which involve an accumulation of net change. The example of growth, which we considered earlier, is a good example. Here the 'jumbled snapshot' test behaves in the opposite way to steady-state change. If a growing child is photographed ten times over a period of a week, and the resulting snapshots disordered, the chances are that one would not notice the incorrect sequence, since except when very young a child does not grow discernibly over a week. But if the period is extended to a year, the changes become obvious.

Compare the flow of water in a river with the progress of a boat down the river. In the former case, although there is change in the sense that there is a systematic displacement of water downstream, the process looks the same from one moment to the next; it is a clear case of a steady-state process. The motion of the boat, on the other hand, is an equally clear case of a cumulative process: a sequence of snapshots will show the boat at a sequence of different positions along the river, which will become disrupted if the snapshots are jumbled up. Notice, however, the importance of point of view: consider a passenger on the boat, who is unfamiliar with the topography of the banks; a sequence of snapshots taken by this passenger might indeed be disordered without its becoming apparent that they have been so.

The example of the river shows that steady-state processes need not be periodic, involving the cyclical repetition of some sequence of changes. Nor for that matter need they be regular. An example of an irregular steady-state process is the succession of weather-fronts blowing in eastwards off the Atlantic and afflicting the British Isles with that peculiar climate which so many people love to hate and which King Charles II described as the best climate in all the world (Manley 1952). This process continues much the same (apart from a certain seasonal periodicity) year in, year out, yet no one can predict the specific pattern that it will take over any extended period.

Cumulative processes are distinguished from steady-state processes in that in general they cannot persist indefinitely, but eventually reach a natural termination point. Thus the process of falling must terminate when the falling object hits the ground; the process of floating down-river terminates when the sea is reached, and the process of pouring water from a can terminates when the can is empty. Shanahan (1990) has introduced the term 'autotermination' to describe such processes. Autoterminating processes can very naturally be conceptualized as events—the 'telic' events of Dahl 1981 and Galton 1984.

The distinctions we have drawn amongst different kinds of process are not absolute, but rather grade into one another, there being an infinite variety of possible temporal textures to which one or more of these descriptions can apply. As with the distinction between states and processes, granularity must be taken into account when considering how some process should be described. A description that is apt at one level of granularity must be replaced by other descriptions at other levels. The movement of an individual boat down-river is cumulative in that as more time elapses the net change effected increases; viewed on a larger scale, at which individual boats are not distinguished from one another, the traffic up and down the river looks steady-state. Similarly, if an individual sample of water can be identified notionally as an 'object' then the movement of this down-river and out to sea is also cumulative. But the flow of water in the river as a whole is steady-state: the river looks much the same from one day to the next, even though there is a continuous progression of 'individual' samples of water along it. A periodic process looks homogeneous when viewed on a large enough time scale, and many processes which look steady-state on one time scale may appear cumulative on a larger scale. Consider the flow of the river as it slowly changes course over hundreds or thousands of years. Or consider a person sitting at a desk writing. The pen moves in an almost regular sequence of little gyrations: a periodic process. Ignore the individual gyrations and note the steady progression down the page: a cumulative process. At the end of the page, the writer begins a new one, and this pattern repeats itself, page after page: a periodic process again. But there is a steady build-up of completed pages, so the process is cumulative. And all the while the writer has been sitting there, writing: steady-state.

What of events? The hallmark of an event is that it is a discrete individual which can be clearly distinguished from all other events. Reverting to our sleeping person, we can identify at least two kinds of events related to a person's state of sleep or wakefulness: falling asleep and waking up. Falling asleep and waking up are *transitions* defined in terms of an initial state and a final state. That is, to fall asleep is to undergo a transition from the state of being awake to the state of being asleep, and to wake up is to undergo a transition in the opposite direction. Each such transition occurs at a particular time and can be counted as a unique individual item in a person's history.

Not all events are transitions. If I swim two lengths of a swimming pool, I end up in more or less the same state as I started in. Of course, it is not exactly the same state: I am a bit more tired, I am wet. But the event which is reported in the statement that I swam two lengths cannot be defined as a transition from my being to a certain degree tired to my being to a certain degree more tired, nor as a transition from my being dry to my being wet: these transitions may occur as a by-product of my swimming the lengths, but they are not what my swimming the lengths *are*. So this event is not in itself a transition.

How is my swimming two lengths (an event) different from my swimming (a process)? The main difference concerns how they are considered in relation to the time that they take up. Suppose I take two minutes to swim my two lengths, beginning at 10.45 a.m. and ending at 10.47 a.m. That interval, from 10.45 to 10.47, is the time of the event. I can say that over that interval I swam two lengths. This is not true of any shorter interval contained within the longer one. It is not true that I swam two lengths over the one-minute interval from 10.45 to 10.46. Thus the event must be considered to occupy the longer interval *as a whole*; if you consider a subinterval of the interval, then you must consider a subevent of the event: for example, I swam *one* length over the interval from 10.45 to 10.46.

In the scenario just considered, it is also true that *I swam* throughout the two-minute interval from 10.45 to 10.47. This is a process, and it differs from an event (*I swam two lengths*) in that we *can* ascribe the same process to subintervals of the longer interval. If I swam throughout the interval from 10.45 to 10.47, then I swam throughout both the interval from 10.45 to 10.46 and the interval from 10.46 to 10.47; and I swam on subintervals of these intervals, and so on. Thus the process of swimming does not take up the interval 10.45 to 10.47 as a single whole, but is rather distributed over it in a more homogeneous way.

How homogeneous? How short a subinterval can we consider and still say that I swam during it? Opinions vary on this, but the standard view (Dowty 1979; Taylor 1985) is that there is a lower limit set by the intrinsic granularity (or heterogeneity) of the process. Swimming is a cyclical process involving a sequence of strokes. Arguably, one cannot be said to swim over any interval shorter than the time it takes to do one stroke.[5]

On the other hand, if instead of concentrating on the statement 'I swim' we consider the statement 'I am swimming', then here it would appear that we have complete homogeneity. That is, if I am swimming throughout a certain interval, then it must be the case that I am swimming throughout every subinterval of that interval; and even that I am swimming at every single instant within that interval. At any time during that interval, if asked what I am doing, I can truthfully say that I am swimming ('What were you doing at exactly 10.46 a.m. today?'—'I was swimming'). Thus while the process denoted by the verb 'to swim' appears to be homogeneous only down to a certain level of granularity, the process 'to be swimming' is completely homogeneous, indeed more like a state than a process. One function of the progressive form of the English verb is precisely to present a process or event as state-like.[6] Note, however, that in many languages there is no exact analogue to the English progressive. In French and German, for example, a single present tense does duty for both the process- and state-like aspects: we can translate 'Je nage' and 'Ich schwimme', depending on context, as either 'I swim' or 'I am swimming'. Perhaps we should not make too much of the fact that in English we can speak of the process of swimming in one way which seems to be completely homogeneous and in another way which is only homogeneous up to a certain point. On the other hand, this contrast is only strengthened when we turn from the progressives of verbs denoting processes to the progressives of verbs denoting events: 'At 10.46 I was swimming a length'. This description of me, 'swimming a length', applies homogeneously throughout the interval over which I swam that length. Indeed, it can even apply if I fail to swim the length, giving up when I am only halfway along. In that case the description refers not so much to a matter of fact but to an intention I had: I was engaged in a bout of swimming which, if completed according to my

[5] Compare Taylor's remarks about fruit-cake quoted on p. 113.

[6] For a sketch of a formal account of the progressive, see below, §5.4.4.

intention, would have led to my having swum a length.[7]

Events may be defined in terms of states or processes in various ways. We have already seen one way: events as transitions between states. Here we must interpret states liberally as including processes viewed as state-like: 'The water started to flow' records an event which is a transition between a state (no flow) and a process (flow)—but this process is being considered here under its most state-like aspect, as quite homogeneous. Another way in which we can specify events in terms of processes is by adding one or more limiting conditions: 'The stone falls' is a process, but 'The stone falls to the ground' is an event; 'John runs' is a process, 'John runs a mile' is an event. The limiting condition might be quite vague: 'John runs for a while' is still an event, since it has a definite beginning (when John starts running) and end (when he stops), which confer exactly that discrete, indivisible character required to define it as an individual.

Notice, incidentally, that when an event arises directly from the intention of some agent, we tend rather to call it an action; likewise an intentional process is called an activity. But however different they may be in other respects, when considered purely in terms of the logic of their temporal incidence, actions and events are all of a piece, as are processes and activities.

We must distinguish between generic *event-types* and individual *event-tokens*. An event-token is a particular piece of history, occurring at a definite and unique time. By its nature it must happen once and only once. An event-type, on the other hand, is a general description under which more, or fewer, than one event-token can fall. The event-tokens falling under a given event-type are its *occurrences*; a given event-type may occur on many different occasions. Suppose that Carol cycled to work on the morning of 24 September 1996, setting off from home at 8.15 a.m. and arriving at her destination fifteen minutes later, the distance covered being two miles. What I have just described is an event-token. This event-token is an occurrence of many different event-types such as

Carol cycles to work
Carol cycles for fifteen minutes
Carol cycles two miles
Carol cycles more than a mile
Carol has a cycle-ride
Carol travels to work

[7] See Galton 1984 for further discussion of the meaning of the progressive in English.

as well as the more general event-types obtained from these by replacing 'Carol' by 'a woman', or 'a teacher' or 'a person'. Each of these event-types may have many other event-tokens as occurrences. On the other hand, we can quite easily specify event-types which have no occurrences, such as 'Sir Isaac Newton cycles to work', and we can also specify event-types which happen to have only one occurrence, such as 'Beethoven writes a violin concerto', or which necessarily only have one occurrence, such as 'Sir Isaac Newton dies'. Note that we must distinguish this last event-type, whose unique occurrence might have occurred in many different ways and at many different times, from the event-token which is the actual death of Sir Isaac Newton in London on 19 March 1727.

I shall call an event-type E a 'subtype' of event-type E' if any occurrence of E has to be an occurrence of E' as well.[8] The phrase 'has to be' is crucial here: it is not enough, for E to be a subtype of E', that all the *actual* occurrences of E are occurrences of E': rather, we require all *possible* occurrences of E to be occurrences of E'. This requires there to be a logical connection between the two event-types, not just a contingent one. To illustrate, it is, we may suppose, a matter of fact the case that every time Mozart travelled between Salzburg and Vienna, he rode in a coach; but this does not mean that 'Mozart travels from Salzburg to Vienna' is a subevent of 'Mozart undertakes a coach journey', since it is not logically necessary that every occurrence of the former type is also an occurrence of the latter—rather, it is a purely contingent fact that this happens to be the case. Thus an event-type is not simply a set of occurrences (it is more like a set of possible occurrences—though this definition, if adopted, has to be handled with care).[9]

We have seen that expressions denoting processes behave in some ways similarly to expressions denoting states. They can also behave like expressions denoting events. For example, because of the unitary character of events, one of their distinguishing characteristics is *countability*: for any event-type, and any interval, one can (at least in principle) count how many times that event-type occurred during the interval, for example 'Mary swam across the lake five times this morning'. But it seems that this is often the case with processes, too; for example, 'Mary swam five times this morning'. Here

[8] In Galton 1984 I used the term 'subevent' for this, but since this term has generally been used by others in a different sense, to refer to the temporal *phases* of an event, I have chosen not to use it in this way here.

[9] McDermott (1982) identifies an event-type with a set of intervals, defining E as the set of intervals i for which there is an occurrence of E taking up exactly the interval i.

the process-verb 'swim' is being used in an event-like way, such that 'Mary swims' is understood to refer to an event which begins when Mary begins swimming, continues for as long as she continues swimming, and then ends when she stops swimming. It is a discrete 'chunk' of swimming, separated from any other such chunks, rather like the event 'John swims for a while' that was mentioned above.

Thus processes seem to have a chameleon-like character, appearing now as states, now as events, depending on the context in which they are considered. For this reason, to characterize something as a process is to describe it at a somewhat lower level of abstraction than to describe it as a state or an event. The fundamental logical distinction, which should be reflected in the formalism we adopt for representing what goes on in time, is that between states and events. Processes can be presented under either a state-like aspect or an event-like aspect; they do not constitute a third category on the same footing as the other two.

Normally we shall take the default aspect of a process to be the state-like one, in effect conceptualizing all processes as if they were steady-state. Later we shall introduce the notion of a *fluent* as the technical correlate of this category which encompasses both the undisputed states (in which no change occurs) as well as processes viewed under a state-like aspect (states of change). We should not be misled by this. It does not mean that we are adopting a very coarse scale of granularity; but rather that the variety of different textures possessed by processes constitute a dimension of analysis orthogonal to the basic state/event division, which has to do with the formal logic of temporal incidence and not with the details of the material constitution of what goes on. From one point of view, *all* that goes on is processes; it is to aid our conceptualization of these processes that we either gloss over their inherent textural heterogeneity by treating them as homogeneous states, or else carve out of them the discrete chunks we call events. The boundaries we draw in order to do this may be more or less artificial.

5.3. Time itself: an absolute temporal framework

As already indicated, I shall base my formal account of time on the idea of an absolute temporal framework that can be specified independently of anything that actually happens. Only once such a framework is established will I consider how to populate it with states, processes, and events.

5.3.1. What is a temporal framework?

What does a temporal framework look like? Time itself seems to flow by in a continuous, uniform, unstructured manner, providing us with no way of identifying different positions in time. In order to do this we need to articulate the unstructured flow by imposing some kind of structure on it; typically this structure is derived from events, in keeping with the relational view that ultimately the concept of time is dependent on that of temporal relations amongst events. The kinds of structures that can be used may be illustrated by means of a number of examples.

Calendars

A calendar imposes structure on time by partitioning the uniform flow into a system of *nested cycles*. In most calendrical systems, as in the Gregorian calendar in common use throughout the westernized world, there is a three-layer structure of years, months, and days. Because these are defined in terms of astronomical cycles which bear complex arithmetical relations to each other, and because the calendar now in use is the product of a complicated history, the details are messy, as everyone knows who has tried to remember which months have thirty days, or which years are leap years. But messy details apart, the underlying idea of a calendar is conceptually simple and may be stated in general terms somewhat as follows: at the top level, time is divided into a discrete sequence of at least approximately equal intervals, which may for convenience be labelled by means of the natural numbers, say

$$\ldots, i_{-4}, i_{-3}, i_{-2}, i_{-1}, i_0, i_1, i_2, i_3, i_4, \ldots$$

(Note that in the Gregorian calendar, if we identify i_1 with the year 1 AD, then i_0 is the year 1 BC, there being no year 0—which is why the year 2000 is really the last year of the twentieth century and not, as popularly supposed, the first year of the twenty-first.) At the next level down, each interval i_n is divided into some number m of subordinate units, giving us a finer-grained partition of time as

$$\ldots, i_{1,1}, i_{1,2}, \ldots, i_{1,m-1}, i_{1,m}, i_{2,1}, i_{2,2}, \ldots$$

This process can be continued recursively as many times as we please.

At the lowest level we have a division of time into intervals which are, from the point of view of the calendar, *atomic*, that is, they cannot be divided into smaller intervals. In our three-level calendar this lowest level consists of days. Since all the other intervals are aggregates of intervals at this lowest level (for example, years comprise particular consecutive sequences of 365 or 366 days), we could in principle make do with just the lowest level, as if we were to specify the date simply by saying that it is day 729977 rather than 11 August 1999. Thus the simplest representation of this kind of temporal framework consists of a single sequence of atomic intervals indexed by the natural numbers:

$$\ldots, a_{-4}, a_{-3}, a_{-2}, a_{-1}, a_0, a_1, a_2, a_3, a_4, \ldots$$

Analogue Clocks

Whereas a calendar divides time up into discrete chunks, setting an intrinsic limit on the fineness of temporal discrimination that it can handle, an analogue clock tracks the passage of time continuously, in principle allowing arbitrarily fine discrimination. (In practice, of course, clocks tend to rely on discrete mechanisms, such as toothed gear-wheels, or vibrating atoms, but this is a conceptually inessential feature, and one that clockmakers would circumvent if they could.) Thus instead of dividing time up into intervals, each of which receives some numerical designation, a clock assigns a number to each instant of time. If the flow of time is truly continuous, the appropriate numerical designations for these instants are real numbers. Between noon and 12.01 p.m., we can pick out such instants as '30 seconds past noon' or 30.465 seconds past noon', and so on, in practice limited only by the accuracy of our chronometry.

In order to use real numbers to label instants, we have to settle on some duration to correspond to one unit, for example a second, and also to single out a particular instant to be labelled with the number 0. That done, every time can be assigned a real number corresponding to the number of seconds separating it from time 0—negative numbers for earlier times and positive numbers for later ones.

Digital Clocks

Unlike analogue clocks, digital clocks do not track the passage of time continuously. Rather, they function more like calendars. Let us assume a digital clock that discriminates seconds but no finer units. Then just as the calendar provides a three-layer structure of years, months, and days, so this digital clock provides a three-layer structure of hours, minutes, and seconds: at the top level, time is partitioned into hours:

$$\ldots, 1, 2, 3, 4, 5, 6, 7, 8, 9, 10, 11, 12, 1, 2, 3, 4, 5, \ldots$$

each of which is itself partitioned into sixty minutes:

$$\ldots, 1.56, 1.57, 1.58, 1.59, 2.00, 2.01, \ldots, 2.58, 2.59, 3.00, 3.01, \ldots$$

and each minute is further partitioned into sixty seconds:

$$\ldots, 1.59.58, 1.59.59, 2.00.00, 2.00.01, 2.00.02, \ldots$$

For this clock, the seconds are the atomic intervals. The instant at which the analogue clock registers exactly 1.59.58, say, marks the *beginning* of the atomic interval with that label in the digital system.

If we could imagine a digital clock with an infinite nest of cycles, subdividing seconds into milliseconds, then microseconds, nanoseconds, and so on without limit, we would then have a device capable of arbitrarily fine temporal discrimination, equivalent to a completely accurate analogue clock. Both such clocks are, of course, fictitious idealizations.

As these examples attest, the basic units of a temporal framework may be either instants or intervals. An interval is a stretch of time, as it were a chunk carved out of the temporal flow; as such it has a length or duration, and this on its own suffices to distinguish intervals from instants, which are durationless. An instant is not a chunk or part of time, but rather marks a point of potential division of time. Although from a psychological point of view the present moment seems to be a little 'smeared out', encroaching a little on both the past and the future (the so-called 'specious present'), it is often conceptualized as an infinitely fine knife-edge separating a past and a future which together account for the totality of temporal duration, and it is this conceptualization of the present moment that forms the model for our idea of an instant. It would be idle to pretend that this notion is unproblematic; but in the historical development of the physical sciences, the standard mathematical model for the flow of time has precisely been that of a *continuum*, which since Dedekind and Cantor we have learnt to see as a structure consisting of indenumerably many points forming a strict total ordering, exactly represented by the ordered set of real numbers. For this reason I begin my account of the temporal framework with the instant-based systems; later I turn to systems based on intervals and examine the relation between them.

5.3.2. Instant-based systems

In mathematical terms, an instant-based temporal framework can be specified as a set of instants together with an ordering relation, called 'succession'. This is denoted $\mathcal{T} = (T, \prec)$, where T is the set of instants, and $\prec$ is a relation on T. I shall use lower-case letters to denote individual instants, that is, members of T. A formula such as $t_1 \prec t_2$ may be read as 't_1 precedes t_2', 't_1 is earlier than t_2', or 't_1 is before t_2', all these readings being interchangeable. The same relation can also be written as $t_2 \succ t_1$, read 't_2 follows t_1', 't_2 is later than t_1', or 't_2 is after t_1'.

By imposing various conditions on the succession relation, we can constrain our model to have different properties as we see fit. Two plausible properties which we might insist on our model having are that the relation of succession should be both irreflexive and transitive, that is, that our model satisfies the following axioms:

- No instant precedes itself:

  $\mathbf{Irref_T}$ $\quad\quad \forall t \neg(t \prec t)$

- Any instant precedes any instant any instant it precedes precedes:[10]

  $\mathbf{Trans_T}$ $\quad\quad \forall t, t', t''(t \prec t' \wedge t' \prec t'' \to t \prec t'')$

It may seem obvious that these properties must be possessed by any model of time, if the notion of temporal succession is to mean what we ordinarily understand by it. It is worth pausing to take note, however, of some possible counter-examples.

Consider, for example, the case of *circular* time, which has been held to be useful for modelling various cyclical phenomena in which the same sequence of events repeats itself over and over again. (Traffic lights and lecture timetables have been suggested.) If time is cyclical, then it cannot be both irreflexive and transitive, for that would rule out the relation '$t_1 \prec t_2$ and $t_2 \prec t_1$' which must hold for *any* pair of instants in a cyclical model. Of course, if we allow this relationship, then every instant precedes every other instant as well as itself, so the succession relation becomes *universal* and so effectively devoid of content. To avoid this we might consider a succession

[10] I must confess to a fondness for this grammatically impeccable but somewhat impenetrable formulation in English of the transitivity property; perhaps it is would be clearer, though marginally less elegant, to put a 'that' between the fifth and sixth words.

relation of this form: t_1 precedes t_2 by at most some fixed duration d. The fixed duration might be chosen, in a cyclical model, to be at most half the length of the repeating period. This relation is now not universal, and it is irreflexive though not transitive. Transitivity fails because if t_1 precedes t_2 by, say $0.6d$, and t_2 precedes t_3 by the same amount, then t_1 precedes t_3 by $1.2d$, and hence this pair are not related by our chosen succession relation.

Even if we do not espouse the notion of circular time, we might still want to make use of this modified form of succession relation, for example to model some device which has a fixed upper limit to the length of time it can 'remember' things for, so that from a point of view internal to the device the state of things at a time longer ago than that can have no direct bearing on the present.

From now on I shall ignore such marginal cases and concentrate on those models in which the temporal succession relation is both transitive and irreflexive. Such an ordering relation is called a strict partial order. In stipulating that the succession relation should be a strict partial order, I am imposing a little structure on the temporal framework, but only a little. Further properties can be specified by means of additional postulates. Here I shall discuss a few of the more important ones. These concern both the global or large-scale structure of time (linearity, unboundedness) and the local or small-scale structure (density, discreteness, or continuity).

Linearity

In visualizing the properties of irreflexivity and transitivity, the reader very possibly assumed, at least initially, that the temporal order is linear, that is, that the instants are strung out in a single ordered succession like the number-line. In fact, there is nothing in these axioms that warrants this assumption: the temporal order may, consistently with the axioms, be modelled as branching, converging, or both, as a single linear succession, or as a set of independent 'parallel' successions. These possibilities are illustrated in Figure 5.1.

If we want our temporal model to have a linear structure, we must lay this requirement down explicitly as another axiom. It will suffice to postulate that

- Of any two distinct instants, one precedes the other:

Lin$_\mathbf{T}$ $$\forall t, t'(t \prec t' \vee t = t' \vee t' \prec t)$$

The disjunction in this definition may be summarized by saying that t is 'linearly related' to t', written $t \sim t'$. The linearity axiom simply says that *all*

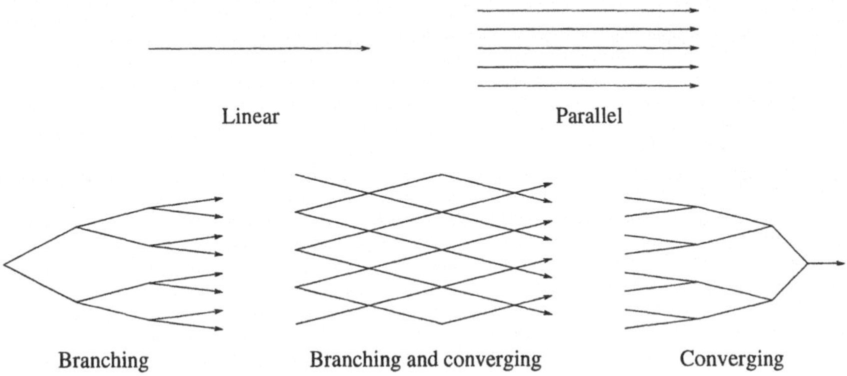

Figure 5.1. Models for temporal succession

pairs of instants must be linearly related.

By incorporating the irreflexivity and transitivity axioms into the linearity axiom, we obtain a stronger version of the latter, which is the form in which it is often cited:

- For any instants t, t', exactly one of the relations (i) $t = t'$, (ii) $t \prec t'$, and (iii) $t' \prec t$ must hold:

$$\mathbf{Lin}'_{\mathbf{T}} \qquad \forall t, t'(t \prec t' \oplus t = t' \oplus t' \prec t)$$

We can straightforwardly derive $\mathbf{Lin}'_{\mathbf{T}}$ from $\mathbf{Lin}_{\mathbf{T}}$, $\mathbf{Irref}_{\mathbf{T}}$, and $\mathbf{Trans}_{\mathbf{T}}$. For suppose (i) and (ii) both hold; then t precedes itself, violating irreflexivity; for the same reason, (i) and (iii) cannot both hold. If (ii) and (iii) hold then again, t precedes itself, by transitivity. Thus no two of the conditions can hold together, and since by $\mathbf{Lin}_{\mathbf{T}}$ at least one of them holds, we infer that exactly one of them holds.

As already indicated by Figure 5.1, there are a number of different ways in which time can fail to be linear. Branching time, for example, is characterized by the following two properties (called 'past-linearity' and 'past-confluence' respectively):

- Any instants t' and t'' preceding a given instant t are linearly related:

$$\mathbf{Lin}^-_{\mathbf{T}} \qquad \forall t, t', t''(t' \prec t \wedge t'' \prec t \rightarrow t' \sim t'')$$

- For any distinct instants t' and t'', neither of which precedes the other, there is an instant t which precedes them both:

$$\mathbf{Confl}_{\mathrm{T}}^{-} \qquad \forall t' t'' \exists t (t \preceq t' \wedge t \preceq t'')$$

Confluence ensures there are no parallel time-lines unrelated to each other, and past-linearity ensures that there is no convergence. Thus branching is allowed, although the axioms do not guarantee that it occurs, since both $\mathbf{Lin}_{\mathrm{T}}^{-}$ and $\mathbf{Confl}_{\mathrm{T}}^{-}$ are satisfied by linear time. The branching could be more prolific than suggested by the diagram: for example, from a given instant there could be infinitely many branches into the immediate future. Branching time has been suggested as a way to model *indeterminism*: in many, indeed most, situations, it is natural to say that more than one course of events is possible in the immediate future. This multiplicity of possibilities may arise either from our ignorance or from a genuine indeterminism in nature.

Unboundedness

An instant which has no other instants preceding it is called an 'initial instant': such an instant marks a beginning of time. Similarly, a 'final instant' is an instant which does not precede any other instant, marking an end of time.[11] If time is linear then there can be at most one initial instant and one final instant. If there is both an initial and a final instant, I shall say that time is 'bounded'. A linear temporal model might also be described as bounded in the past or bounded in the future, and likewise unbounded in the past or unbounded in the future. These combine to give four different possibilities.

With non-linear time the matter is more complicated, since if there are different branches, some of these may be bounded while others are unbounded. We can say that time is 'everywhere bounded in the past' if every instant has some initial instant preceding it, and likewise 'everywhere bounded in the future' if every instant has some final instant following it. The terms 'unbounded in the past/future' are best reserved for cases where there are *no* initial/final instants on any branch.

It is important to appreciate that unboundedness does not necessarily entail infinite duration. It is widely believed that the universe began in a 'Big

[11] An instant which neither precedes nor is preceded by any other instant is both an initial instant and a final instant; such an instant is isolated from the rest of time, and so we may call it an 'isolated instant'. While nothing I have said so far can rule out isolated instants, it is hard to imagine that any sensible model of time will contain them.

Bang' some ten or fifteen thousand million years ago. On this model, there was no time before the Big Bang, but this does not mean that time is bounded in the past. There are two possible versions of the Big Bang model. In one of them, there is an initial instant, the time of the Bang itself, but in the other there is not. In terms of numbers, it is like the difference between the set $\{x \in \mathbb{R} \mid x \geq 0\}$ of all non-negative real numbers, and the set $\{x \in \mathbb{R} \mid x > 0\}$ of all positive real numbers; whether this distinction has any useful role to play in cosmological theory is doubtful, but if one were to attempt to formalize the theory on a rigorously logical basis, a decision would have to be made one way or the other.

From a theoretical point of view it is often simplest to assume that time is unbounded. This is the natural choice to make so long as it does not conflict with any specific features one wishes to build into the model. We can enshrine this assumption in the following pair of axioms, one expressing unboundedness in the past, the other unboundedness in the future:

- For every instant there is an instant which precedes it.

$\mathbf{Unb}_{\mathbf{T}}^{-}$ $\qquad\qquad \forall t \exists t'(t' \prec t)$

- For every instant there is an instant which it precedes.

$\mathbf{Unb}_{\mathbf{T}}^{+}$ $\qquad\qquad \forall t \exists t'(t \prec t')$

I shall write $\mathbf{Unb}_{\mathbf{T}}$ for the conjunction of $\mathbf{Unb}_{\mathbf{T}}^{-}$ and $\mathbf{Unb}_{\mathbf{T}}^{+}$. In conjunction with $\mathbf{Irref}_{\mathbf{T}}$ and $\mathbf{Trans}_{\mathbf{T}}$, unboundedness implies that so long as there is at least one instant there must be infinitely many. For let t_0 be any instant. Applying $\mathbf{Unb}_{\mathbf{T}}$ gives us an instant t_1 such that $t_0 \prec t_1$. Repeated applications of $\mathbf{Unb}_{\mathbf{T}}$ generate an infinite sequence $t_0 \prec t_1 \prec t_2 \prec \cdots$; by $\mathbf{Irref}_{\mathbf{T}}$ and $\mathbf{Trans}_{\mathbf{T}}$ the members of this sequence are all distinct.

Density

In the physical sciences it is customary to model the temporal succession by means of the system of real numbers. This is because we use numbers to measure durations, and we fix positions in the temporal series in terms of the duration of time elapsed from some base time. This is what is involved in designating a particular instant as, say, twenty-one minutes and three seconds past four o'clock in the afternoon of 10 February 1998. This can be converted into a certain number of seconds after the start of the year 1 AD. Each instant of time is allocated a real number, and to each real number it is assumed that there is an instant of time. This is useful for science because it allows us to apply the mathematical theory of the real numbers to the solution of physical problems, with a remarkable degree of success.

Here we are concerned with the qualitative aspects of temporal models, which means that duration measures play no part. It is the properties of the ordering relation on the real numbers that are of immediate concern to us. An important and obvious feature of the ordering of the real numbers is that it is *dense*, which means that between any two real numbers there is a third. In terms of instants, this is equivalent to postulating the axiom

- Whenever one instant precedes another, there is an instant later than the first but earlier than the second:

$$\mathbf{Dens_T} \qquad \forall t, t'(t \prec t' \rightarrow \exists t''(t \prec t'' \wedge t'' \prec t'))$$

In conjunction with $\mathbf{Irref_T}$ and $\mathbf{Trans_T}$, this implies that there are infinitely many instants—so long as there are at least two instants, one of which precedes the other. For suppose $t \prec t'$. Then by $\mathbf{Dens_T}$, t precedes some instant t_1 which precedes t'. By $\mathbf{Dens_T}$ again, t_1 precedes an instant t_2 which precedes t'. In this way we generate an infinite sequence

$$t \prec t_1 \prec t_2 \prec t_3 \prec t_4 \prec \cdots \prec t'.$$

By $\mathbf{Irref_T}$ and $\mathbf{Trans_T}$, these instants are all distinct.

Note that density is the converse of transitivity. Transitivity says that if there is an instant t such that $t_0 \prec t$ and $t \prec t_1$, then $t_0 \prec t_1$. Density says that if $t_0 \prec t_1$ then there is an instant t such that $t_0 \prec t$ and $t \prec t_1$.

Discreteness

At a given instant, does it make sense to speak of the 'next' instant, or the 'previous' instant? In some models we want to say exactly that. For example, consider an abstract representation of a game of chess. Initially the game is in a certain state, consisting of the pieces in their initial positions on the board. This is the first 'instant' of the game. After White has made the first move, say 'P to K4', the state of the game is different, with the white king's pawn advanced two squares. In the physical realization of the game, a multitude of distinct world states intervene between these two states of the game; in these intermediate states, the white king's pawn is not on any square but being moved through the space above the board between its initial position and its new position. From the point of view of our abstract representation, these states have no meaning. All that is meaningful is that initially the pawn is on K2 and then 'immediately' afterwards it is on K4. The model of time presupposed by this kind of representation is one in which it *does* make sense to speak of the 'next' instant after a given one, at least for all instants apart from that at which the game ends. Such *discrete* models of time are useful for modelling many different processes, for example, the execution of a computer program, and the teaching timetable of a university where the minimum unit of tuition is a one-hour slot.

An adequate formulation of discreteness requires some care. It is not enough just to say that for each instant there is a 'next' instant; we must say what is meant by 'next'. I shall use the following definition:

> An instant t' is an *immediate successor* of instant t so long as t precedes t' but does not precede any instant which precedes t', that is: $(t \prec t') \wedge \neg\exists t''(t \prec t'' \prec t')$. In this case I shall also say that t is an *immediate predecessor* of t'.

In linear time the condition for t' to be an immediate successor of t is equivalent to the somewhat neater formulation

$$\forall t''(t \prec t'' \leftrightarrow t' \preceq t''),$$

from which it follows that an instant can have at most one immediate successor.

The axioms I shall give for discreteness only do the job intended for them on the assumption that time is linear. (Rather as, for example, the density axiom only gives true density, in the sense of there being infinitely many instants between a given pair, on the assumption that time is irreflexive and transitive.) The axioms distinguish discreteness in the past from discreteness in the future, as follows:

- If an instant has a predecessor, then it has an immediate predecessor:

Disc$_{\mathbf{T}}^{-}$ $\quad \forall t, t'(t' \prec t \rightarrow \exists t''(t'' \prec t \wedge \neg\exists t'''(t'' \prec t''' \prec t)))$

- If an instant has a successor, then it has an immediate successor:

Disc$_{\mathbf{T}}^{+}$ $\quad \forall t, t'(t \prec t' \rightarrow \exists t''(t \prec t'' \wedge \neg\exists t'''(t \prec t''' \prec t'')))$

I shall write $\mathbf{Disc_T}$ to refer to the conjunction of $\mathbf{Disc_T^-}$ and $\mathbf{Disc_T^+}$. Since these two conjuncts are independent it is possible to have either holding without the other. For example, suppose the instants of our model are identified with *finite bit-strings* such as 1110001 and 100110, and let the succession relation be defined in terms of the ordinary alphabetical order on these strings (with 0 preceding 1). Then each string has an immediate successor (for example, the successor of 1110001 is 11100010), but only strings ending in 0 have immediate predecessors: for example the string 11 is preceded by the infinite sequence

$$1, 10, 100, 1000, 10000, 100000, 1000000, \ldots ,$$

none of whose members is an immediate predecessor of 11. Thus this ordering is 'discrete in the future' but not 'discrete in the past'.

The discreteness axiom does not secure discreteness in, say, branching time, since it will be satisfied so long as there is at least one branch in the future of each instant along which that instant has an immediate successor; other branches need not have this property.[12]

Finite Dissection

The paradigmatic model for a discrete ordering is the set of integers with the usual 'less than' relation, but this is by no means the only example of a discrete ordering. An example of a discrete linear ordering that is interestingly different from that of the integers is the ordering of their *reciprocals*:

$$-1 < -\tfrac{1}{2} < -\tfrac{1}{3} < -\tfrac{1}{4} < -\tfrac{1}{5} < \cdots < \tfrac{1}{5} < \tfrac{1}{4} < \tfrac{1}{3} < \tfrac{1}{2} < 1.$$

Each element with a successor has an immediate successor, and each element with a predecessor has an immediate predecessor. Thus the ordering is

[12] So long as there are at most finitely many branches into the immediate past or future from a given instant, the following definition of discreteness will do:

$\mathbf{Disc'_T}$ If t precedes t', but is not an immediate predecessor of it, then t has an immediate successor which precedes t', and t' has an immediate predecessor which t precedes.

In linear time this reduces to $\mathbf{Disc_T}$. The definition $\mathbf{Disc'_T}$ does not work in cases where there are infinitely many branches into the immediate future from some instant. For example, consider the following curious model. The set of instants consists of (i) rational numbers, with their usual ordering, and (ii) for each pair $q_1 < q_2$ of rational numbers an extra instant $t_{[q_1,q_2]}$ which is stipulated to precede q_2 and to follow q_1, with no other instances of the precedence relation than what are implied by this stipulation. Since this model contains the dense system of rational numbers, we should hardly wish to classify it as discrete; but it does satisfy $\mathbf{Disc'_T}$!

discrete; however, between any negative term and any positive term (for example, between $-\frac{1}{5}$ and $\frac{1}{5}$) there are infinitely many other terms. Between all the negative terms and all the positive terms there is a 'gap', or *singularity*, not marked by any term of the sequence, around which infinitely many terms accumulate.

This particular example is rather unsatisfactory as a model of time, because it does not treat all instants equally: the negative terms are distinguished from the positive terms by their position relative to the singularity. A more homogeneous example is a model suggested by Van Benthem (1991), obtained from the rational numbers by replacing each element by a complete copy of the integers. Each instant in this model can be represented by a pair (q, n), where $q \in \mathbb{Q}$ and $n \in \mathbb{Z}$, with the ordering defined by the rule

> (q, n) precedes (q', n') if and only if either (i) q precedes q', or (ii) $q = q'$ and n precedes n'.

This is discrete but has infinitely many singularities which are densely distributed. It is homogeneous in the sense that any *instant* can be transformed into any other by an order-preserving bijective mapping. On the other hand, there are infinitely many distinct classes of *pairs* of instants: for each natural number n, there are pairs of instants between which exactly n instants fall; and there are also pairs of instants between which infinitely many instants fall. As Van Benthem says,

> It is a pleasing speculation whether such temporal structures, dense at a 'macro-level', but discrete at their 'micro-level', could be models for physical theories of quantized time.
>
> (Van Benthem 1991: 22)

Pleasing though these speculations may be, there does seem to be something rather problematic about a model in which the interval between a pair of instants can contain infinitely many atomic intervals as parts. It is to outlaw this possibility that the 'finite dissection' property is introduced.

The finite dissection condition differs from the preceding conditions in being *second-order*, meaning that it generalizes not just over instants or intervals (the basic temporal entities) but also over *sets* of instants or intervals. This means that the condition has a logical form which cannot be expressed as a formula of first-order predicate calculus. The condition may be expressed as follows:

- Between any two instants there are at most finitely many others:

$$\mathbf{FD_T} \qquad \forall t, t' \exists P(Finite(P) \wedge \forall t''(t \prec t'' \prec t' \leftrightarrow P(t'')))$$

Note the quantification over predicates, making this condition second-order. The second-order predicate $Finite$ used here can be defined in second-order logic (and care must be taken here to define it in such a way that the empty set counts as finite too).

Continuity

I discuss continuity in some detail in Chapter 7, where in particular I consider whether or not the standard mathematical treatment of continuity does justice to the nature of continuity as we sometimes feel we discern it in the physical world. Here I avoid such contentious matters and present an account of continuity in keeping with the standard mathematical picture.

The essence of continuity can be summed up in the phrase 'no gaps'. Assume time is linear, and consider the different ways of dividing all of time into two parts such that the earlier part (which I shall call P for 'past') wholly precedes the later part (which I call F for 'future'); using instants,

$$T = P \cup F, \text{ where every element of } T \text{ precedes every element of } F.$$

This is known as a 'Dedekind cut'. Concentrating on what happens as we pass from P into F, we find there are four possibilities:

(1) P has a last instant, and F has a first instant;
(2) P has a last instant, but F has no first instant;
(3) P has no last instant, but F has a first instant;
(4) P has no last instant, and F has no first instant.

In a discrete order with the finite dissection property, only possibility (1) can occur. In the integers, for example, wherever we choose the division between past and future, we will end up with sets of the form

$$(1)\ \ P = \{\ldots, n-3, n-2, n-1\},\ F = \{n, n+1, n+2, \ldots\},$$

where $n-1$ is the last element of P and n the first element of F. In a dense order, possibility (1) cannot occur, but all of the other three can. For example, in $\mathbb{Q}$, the set of all rational numbers, we can have

(2) $P = \{x \in \mathbb{Q} \mid x \leq 0\}$, $F = \{x \in \mathbb{Q} \mid 0 < x\}$
(3) $P = \{x \in \mathbb{Q} \mid x < 0\}$, $F = \{x \in \mathbb{Q} \mid 0 \leq x\}$
(4) $P = \{x \in \mathbb{Q} \mid x^2 < 2\}$, $F = \{x \in \mathbb{Q} \mid 2 < x^2\}$

The reason why (4) is an exhaustive division (that is, $P \cup F = \mathbb{Q}$) is that there is no rational number x such that $x^2 = 2$.

One can regard the impossible condition $x^2 = 2$ as representing a kind of 'gap' in the ordering of the rational numbers, and for this reason the rational numbers are not thought to provide a good model for continuity, even though it is, of course, densely ordered. Although this is the particular 'gap' that is usually cited in this connection, it is very far from being the only one: in fact the gaps themselves form a dense ordering, and it can be shown that the set of gaps in the rational numbers in an important sense vastly outnumber the rational numbers themselves.

The real numbers are obtained by 'plugging' the gaps in the rationals with new numbers called irrationals. We can define an irrational number by means of a pair (P, F) such that (*a*) $P \cup F = \mathbb{Q}$, (*b*) P has no greatest element, and (*c*) F has no least element. The set of real numbers is the union of the set of all rational numbers with the set of all irrational numbers. The rational numbers are ordered as usual, and each irrational (P, F) is stipulated to follow every element of P and to precede every element of F. In real number time, (2) and (3) are the only possibilities for the division into P and F; this means that any division of reals into a past and future must occur *at* a particular point, which is either the latest element of the past or the earliest of the future. It is this property that is conventionally regarded as the correct mathematical analogue of our intuitive notion of continuity.

The precise expression of continuity is second-order, since it refers to arbitrary sets of instants, as follows:

- For any non-empty sets T_1, T_2 of instants, if every element of T_1 precedes every element of T_2 then there is an instant t which is no earlier than any instant in T_1 and no later than any instant in T_2:

Cont$_{\mathbf{T}}$ $$\begin{aligned}&\forall T_1, T_2 \subset T[T_1 \neq \emptyset \wedge T_2 \neq \emptyset \wedge \\ &\quad \forall t_1 \in T_1, t_2 \in T_2(t_1 \prec t_2) \rightarrow \\ &\qquad \exists t \forall t_1 \in T_1, t_2 \in T_2(t_1 \preceq t \preceq t_2)]\end{aligned}$$

Note that any discrete order satisfies **Cont$_{\mathbf{T}}$**; it is only in conjunction with density that this axiom establishes true continuity.

5.3.3. Introducing intervals

There are two different ways we can set about constructing a temporal framework based on intervals. One way is to take the instant-based framework, as introduced in the previous section, define intervals in terms of instants, and then derive the properties of intervals from the axioms governing the relations between instants. The other way is to introduce intervals from the start as primitive elements, just as I introduced instants in the previous section, and constrain the properties of the framework by means of axioms governing the relations between intervals. I shall discuss these two approaches in turn; first, however, I review some general considerations to be taken into account by either approach.

Temporal succession for intervals

Let us begin by establishing a notion of temporal succession for intervals. Whereas with instants t and t' there are only four possibilities, namely $t \prec t'$, $t = t'$, $t' \prec t$, or t and t' are unrelated, with intervals it is much more complicated. This is clearly seen even in the relatively restricted context of linear time, in which there are no fewer than thirteen qualitatively distinct mutually exclusive possible relations between intervals i and i'; these are listed below in Table 5.1, where it will be seen that all the relations can be defined in terms of a single primitive, which we can take to be the analogue for intervals of the relation $\prec$ on instants. This primitive relation is 'immediate precedence'; it is the relation in which interval i stands to interval i' if the end of i is simultaneous with the beginning of i'. This is expressed by saying that i 'meets' i', and is denoted $i \| i'$. If we adopt the first approach, defining intervals in terms of instants, then we shall need to define the relation $\|$ in terms of the relation $\prec$. If on the other hand we introduce intervals as primitive elements in their own right, not defined in terms of instants, then we must postulate $\|$ as a primitive relation on the set of intervals. Either way, we will end up by characterizing our temporal framework as a structure $\mathcal{I} = (I, \|)$, where I is the set of intervals.

The calculus of instants and intervals

A good deal of attention has been paid in recent years to the *relations* between intervals. As was pointed out by Hamblin (1971) and Allen (1984), if time is totally ordered then there are thirteen exhaustive and mutually exclusive basic relations in which one interval can stand to another, which are listed in

Table 5.1. Here we assume the relation 'meets' as primitive: interval i meets interval j just if i's ending is simultaneous with j's beginning. We write this in symbols as $i\|j$.

In terms of 'meets', we can define other relations between the beginnings and endings of two intervals i and j as follows:

- i and j begin together if there is an interval which meets both of them: $\exists k(k\|i \wedge k\|j)$.
- i and j end together if there is an interval which they both meet: $\exists k(i\|k \wedge j\|k)$.
- i begins before j begins if i begins together with an interval which meets j: $\exists k, l(k\|i \wedge k\|l \wedge l\|j)$.
- i ends before j ends if i meets an interval which ends together with j: $\exists k, l(i\|k \wedge k\|l \wedge j\|l)$.
- i begins before j ends if i begins together with an interval which ends together with j: $\exists k, l, m(k\|i \wedge k\|l \wedge l\|m \wedge j\|m)$.

In terms of these relations we can classify subintervals into three types, as follows:

- j is an initial subinterval of i if i and j begin together, but j ends before i ends.
- k is a final subinterval of i if i begins before k begins, but i and k end together.
- l is an internal subinterval of i if i begins before l begins, and ends after l ends.

It is in terms of this classification that I shall define the thirteen relations between intervals, as shown in Table 5.1. The notations in the second column are those of Allen 1983 and Allen 1984.

As already mentioned in connection with the RCC calculus for spatial regions (§2.5.2), Allen introduced the idea of a composition table for the thirteen relations on intervals. It includes such entries as:

- If i meets j and j meets k then i is before k.
- If i overlaps j and j finishes k then i may be during k, start k, or overlap k.

Table 5.1. *The thirteen qualitative relations between intervals*

Relation	Notation	Definition
i is before j	$i < j$	i meets an interval which meets j
i meets j	i m j	(primitive)
i overlaps j	i o j	Some final subinterval of i is an initial subinterval of j
i starts j	i s j	i is an initial subinterval of j
i is during j	i d j	i is an internal subinterval of j
i finishes j	i f j	i is a final subinterval of j
i equals j	$i = j$	$i = j$
i is finished by j	i fi j	j is a final subinterval of i
i has j during it	i di j	j is an internal subinterval of i
i is started by j	i si j	j is an initial subinterval of i
i is overlapped by j	i oi j	Some final subinterval of j is an initial subinterval of i
i is met by j	i mi j	j meets i
i is after j	$i > j$	j meets an interval which meets i

- If i is before j and j is after k, then the relation between i and k can be any of the thirteen.

Amongst the possible disjunctions of interval relations, the following will be particularly useful to us:

- If i is before or meets j, then i precedes j (notation: $i \lhd j$).
- If j precedes i, then i follows j (notation: $i \rhd j$).
- If i precedes or follows j, then i is disjoint from j (notation: $i \bowtie j$).
- If i starts, is during, or finishes j, then i is a proper subinterval of j (notation: $i \sqsubset j$).
- If i is either a proper subinterval of, or equal to, j, then i is a subinterval of j (notation: $i \sqsubseteq j$).
- If any of the thirteen basic relations holds, then i is linearly related to j (notation: $i \sim j$).

Some of these are defined by Allen and Ferguson (1994) (but note that their definition of 'subinterval' differs from mine in that it omits the disjuncts with 'starts' and 'finishes').

Freksa (1992*a*) introduced the important idea of conceptual neighbourhood, which again I have already discussed in connection with the RCC theory (§2.5.2). To illustrate this in our present context, if i meets j, then by making i finish a little sooner we would have i before j whereas by making it last a little longer we would make it overlap j. Thus the relations 'before' and 'overlaps' are conceptual neighbours of 'meets'. On the other hand 'meets' is not a conceptual neighbour of 'equals': if we wish to convert a situation in which i meets j into one in which i equals j, by continuously deforming one or both of the intervals, then we cannot avoid passing through a situation in which i overlaps j. Freksa found that the clusters of relations appearing as disjunctions in Allen's composition table all form connected sets under the conceptual neighbourhood relation. The complete conceptual neighbourhood diagram for Allen's intervals is shown in Figure 5.2. Later on I shall make use of this diagram as a tool for temporal reasoning (see §6.3.1); I shall derive the diagram from explicit continuity considerations in §8.6.2.

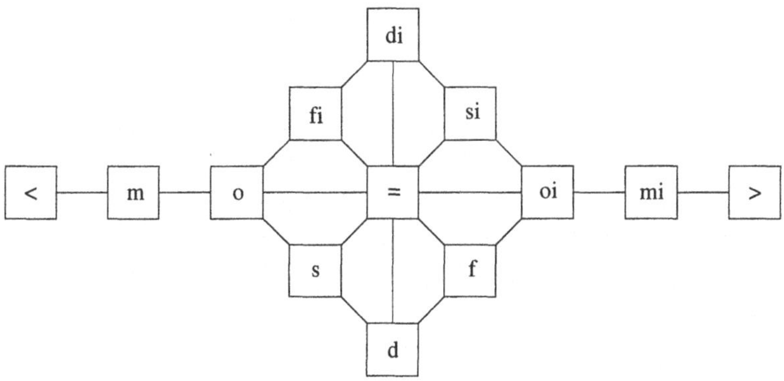

Figure 5.2. Conceptual neighbourhoods for the interval–interval relations

Also important are the relations between intervals and instants, of which, in linear time, there are just five. In order to define them, I first introduce the notions of the beginning and end of an interval; these are instants, which I denote $\mathrm{beg}(i)$ and $\mathrm{end}(i)$ respectively. The beginning and end of an interval are called its limits. For any interval i we must have $\mathrm{beg}(i) \prec \mathrm{end}(i)$, and if $i \| j$ then $\mathrm{end}(i) = \mathrm{beg}(j)$. The position of an instant in relation to an interval can be specified by giving its position in relation to the limits of the interval. If the instant follows the beginning of the interval but precedes its end, then it falls within, or divides, the interval, for which I introduce the special notation

$$t \sqsubset i \triangleq \text{beg}(i) \prec t \prec \text{end}(i).$$

The five relations between instants and intervals are listed in Table 5.2. If time is not linear, there is a sixth possibility: that an instant and interval are temporally unrelated. Notice that I use the same terms 'before' and 'after' to relate both instants to intervals and intervals to intervals. If t either begins or ends i, I shall say that t bounds i, and write $bounds(t, i)$.

Table 5.2. *The five qualitative relations between instants and intervals*

Relation	Definition
t is before i	$t \prec \text{beg}(i)$
t begins i	$t = \text{beg}(i)$
t divides i	$t \sqsubset i$
t ends i	$t = \text{end}(i)$
t is after i	$\text{end}(i) \prec t$

Like the interval–interval relations, the instant–interval relations can be presented in a conceptual neighbourhood diagram; they can be given a composition table, too, though in this case the table shows the composition of the instant–interval relations with the inverses of these, the interval–instant relations. For the set of three instant–instant relations ('precedes', 'equals', and 'follows'), both the conceptual neighbourhood diagram and the composition table are essentially trivial.

Finally, I draw attention to a useful three-place relation involving two intervals and an instant. I shall sometimes want to say that interval i meets interval j at instant t. This is understood to mean that $\text{end}(i) = t = \text{beg}(j)$; I shall notate this compactly as $i \|_t j$.

Intervals defined in terms of instants

There are two main ways in which intervals might be defined in terms of instants. To specify an interval fully we must determine, for each instant, how it is related to the interval. As we have seen, there are six possible relationships between a given instant and a given interval. Any definition of an interval in terms of instants must determine for every instant which of these possibilities holds. This cannot be done arbitrarily, since there are certain constraints that must be obeyed. For example, for a given interval i, if the relation between t

and i comes earlier in the sequence precedes/begins/within/ends/follows than the relation between t' and i, then t must precede t'. Again, if t precedes or begins i, then any instant which precedes t must precede i, and if t_1 and t_2 both fall within i, then any instant which lies between t_1 and t_2 must also fall within i.

Here I consider in turn the two main ways of defining intervals in terms of instants: I shall refer to these as the 'ordered-pair construction' and the 'convex-set construction'. For simplicity, I shall assume that time is linear; although various kinds of intervals and interval-like structures can be defined in non-linear time, the matter is complicated, and I shall not pursue it here—but see Hajnicz 1995 for some preliminary investigations.

1. *The construction of intervals as ordered pairs of instants.* We define an interval as the time between two instants. It is specified by the instants which mark its beginning and its end. An interval is specified as an ordered pair of instants $\langle t_1, t_2 \rangle$, where $t_1 \prec t_2$. Then we have

   - t precedes $\langle t_1, t_2 \rangle$ iff $t \prec t_1$.
   - t begins $\langle t_1, t_2 \rangle$ iff $t = t_1$.
   - t is within $\langle t_1, t_2 \rangle$ iff $t_1 \prec t \prec t_2$.
   - t ends $\langle t_1, t_2 \rangle$ iff $t = t_2$.
   - t follows $\langle t_1, t_2 \rangle$ iff $t_2 \prec t$.

   In this construction, the 'meets' relation is defined as follows:

   $$\langle t_1, t_2 \rangle \| \langle t_3, t_4 \rangle \text{ if and only if } t_2 = t_3.$$

2. *The construction of intervals as convex sets of instants.* We define an interval by specifying which instants fall within it. It is usual to assume that if t_1 and t_2 are both within i, and $t_1 \prec t \prec t_2$, then t is also within i. Given this assumption, it follows that we can specify an interval as a *convex set of instants*, that is a set $i \subseteq T$ such that if $t_1 \in i$ and $t_2 \in i$ then for any instant t, if $t_1 \prec t \prec t_2$ then $t \in i$. More precisely, we require an interval to contain at least two instants, making it a convex set of instants with at least two members.

   For an instant t we then have

   - t precedes i if t precedes some instant which precedes or equals every element of i.

- t begins i if t precedes or equals every element of i, but does not precede any other instant with this property.
- t is within i if $t \in i$.
- t ends i if t equals or follows every element of i, but does not follow any other instant with this property.
- t follows i if t follows some instant which equals or follows every element of i.

In general these five relations are not mutually exclusive. For example, if time is represented by the real numbers, and i is defined as a closed set such as $\{t \mid 0 \leq t \leq 1\}$ (which is indeed convex), then the instant 0 both begins and is within i, while 1 both ends and is within i.

In the convex-set construction, the 'meets' relation is defined by the rule

$$i_1 \| i_2 \quad \text{if and only if} \quad \begin{array}{l} \text{whenever } t_1 \in i_1 \text{ and } t_2 \in i_2 \text{, then} \\ \qquad \text{(i) } t_1 \prec t_2 \text{, and} \\ \qquad \text{(ii) if } t_1 \prec t \prec t_2 \text{ then } t \in i_1 \cup i_2. \end{array}$$

The convex-set construction gives rise to a larger class of intervals than the ordered-pair construction. As already noted, in dense time we can distinguish between open and closed (as well as two kinds of half-open) intervals: a closed interval includes its beginning and end points, an open interval does not. This distinction cannot be made with the ordered-pair construction. In addition, convex sets need not be bounded; hence we can construct intervals which have no beginning, or intervals with no end, which again is not possible in the ordered-pair model. As already indicated, I shall not discuss here the further complications which ensue when time is not linear.

Having defined the structure $(I, \|)$ in terms of $(T, \prec)$, I shall now derive properties of the former structure from the axioms laid down for the latter. I shall work with the ordered-pair construction; the development would proceed somewhat differently if I used the convex-set construction instead. The strategy will be to consider each of the instant axioms in turn, and express its content as nearly as possible in the language of intervals.

Note that in the ordered-pair construction, we have

$$\begin{aligned} \text{beg}(\langle t_1, t_2 \rangle) &= t_1 \\ \text{end}(\langle t_1, t_2 \rangle) &= t_2 \end{aligned}$$

Also, since beg(i) precedes end(i), there is an interval $\langle \text{beg}(i), \text{end}(i) \rangle$, and by the uniqueness of the interval spanning two given instants—which is implicit in the definition of an interval *as* an ordered pair of instants—this must be none other than i itself.

Typically, an instant will be both the beginning of some intervals and the end of others. An exception is when t is either an initial instant or a final instant. An initial instant is not the end, nor a final instant the beginning, of any interval. For an interval i, if beg(i) is an initial instant then I shall call i an initial interval. Likewise, if end(i) is a final instant, then i is a final interval. The 'meets' relation possesses two important properties purely by virtue of its definition in the ordered-pair construction for intervals. First:

- If some interval which meets k also meets l, then *any* interval which meets k meets l:

Int1 $$\forall i, j, k, l(i||k \wedge i||l \wedge j||k \rightarrow j||l)$$

Thus, suppose $i||k$, $i||l$, and $j||k$; then it follows that $j||l$. To prove this, we need merely note that

$$\text{end}(j) = \text{beg}(k) = \text{end}(i) = \text{beg}(l),$$

the three equalities being justified by j meeting k, i meeting k, and i meeting l, respectively.

The second property which follows immediately from the definition of 'meets' is

- If j meets l and is met by i then j is the *only* interval which both meets l and is met by i:

Int2 $$\forall i, j, k, l(i||j||l \wedge i||k||l \rightarrow j = k)$$

Thus if we know both that $i||j||l$ and $i||k||l$, then j and k must be the same interval. This is because both of them could be expressed as $\langle \text{end}(i), \text{beg}(l) \rangle$, which we know specifies a unique interval.

It will be useful to characterize the interval–interval relations in terms of the ordered-pair construction for intervals. It is easy to verify that in this model the relations can be defined as listed in Table 5.3, with the inverse relations defined similarly.

Table 5.3. *Relations between intervals defined as ordered pairs of instants*

Relation	Definition
$i < j$	$\mathrm{end}(i) \prec \mathrm{beg}(j)$
$i \| j$	$\mathrm{end}(i) = \mathrm{beg}(j)$
$i \mathrm{~o~} j$	$\mathrm{beg}(i) \prec \mathrm{beg}(j) \prec \mathrm{end}(i) \prec \mathrm{end}(j)$
$i \mathrm{~s~} j$	$\mathrm{beg}(i) = \mathrm{beg}(j) \wedge \mathrm{end}(i) \prec \mathrm{end}(j)$
$i \mathrm{~d~} j$	$\mathrm{beg}(j) \prec \mathrm{beg}(i) \wedge \mathrm{end}(i) \prec \mathrm{end}(j)$
$i \mathrm{~f~} j$	$\mathrm{beg}(j) \prec \mathrm{beg}(i) \wedge \mathrm{end}(i) = \mathrm{end}(j)$
$i = j$	$\mathrm{beg}(i) = \mathrm{beg}(j) \wedge \mathrm{end}(i) = \mathrm{end}(j)$
$i \lhd j$	$\mathrm{end}(i) \preceq \mathrm{beg}(j)$
$i \sqsubseteq j$	$\mathrm{beg}(j) \preceq \mathrm{beg}(i) \wedge \mathrm{end}(i) \preceq \mathrm{end}(j)$
$i \sim j$	$\mathrm{beg}(i) \sim \mathrm{beg}(j) \wedge \mathrm{beg}(i) \sim \mathrm{end}(j) \wedge$ $\mathrm{end}(i) \sim \mathrm{beg}(j) \wedge \mathrm{end}(i) \sim \mathrm{end}(j)$

Let us now take each of the instant-based axioms in turn and examine its consequence for intervals:

Irreflexivity. From $\mathbf{Irref_T}$ we prove that

- No interval meets itself:

$$\mathbf{Irref_I} \qquad \forall i \neg (i \| i)$$

For suppose $i \| i$. By the definition of 'meets', this means that $\mathrm{end}(i) = \mathrm{beg}(i)$; since we already know that $\mathrm{beg}(i) \prec \mathrm{end}(i)$, this means that $\mathrm{beg}(i) \prec \mathrm{beg}(i)$, which contradicts $\mathbf{Irref_T}$. Thus i cannot meet itself.

Transitivity. From $\mathbf{Trans_T}$ we prove that

- If i meets j, there is an interval which meets all and only those intervals which j meets, and is met by all and only those intervals which meet i:

$$\mathbf{Trans_I} \qquad \forall i, j (i \| j \rightarrow \exists k \forall l ((k \| l \leftrightarrow j \| l) \wedge (l \| k \leftrightarrow l \| i)))$$

The interval we require is $\langle \mathrm{beg}(i), \mathrm{end}(j) \rangle$. Why do we need transitivity here? Without it, we cannot guarantee that this interval exists! The condition for its existence is that $\mathrm{beg}(i) \prec \mathrm{end}(j)$. This follows, using transitivity of $\prec$, from the facts that $\mathrm{beg}(i) \prec \mathrm{end}(i)$, that $\mathrm{end}(i) = \mathrm{beg}(j)$ (since $i \| j$), and that $\mathrm{beg}(j) \prec \mathrm{end}(j)$.

The result I have just proved guarantees the *existence* of an interval which meets whatever j meets and is met by whatever meets i. The earlier result **Int2** guarantees that this interval is *unique*, and because of this we can express it as a function of i and j, using the notation $i^\frown j$ to denote it. Thus if i meets j, then $i^\frown j$ is the unique interval which begins when i begins and ends when j ends, thereby covering exactly the stretch of time covered by i and j together: $i^\frown j = \langle \text{beg}(i), \text{end}(j) \rangle$. The interval $i^\frown j$ is called the concatenation of the intervals i and j. It follows immediately from the definition that concatenation is an associative operation, $i^\frown(j^\frown k) = (i^\frown j)^\frown k$, justifying us in writing $i^\frown j^\frown k$ here.

This notation makes possible a more elegant characterization of the three kinds of subinterval:

- j is an initial subinterval of i if there is an interval k such that $i = j^\frown k$.
- k is a final subinterval of i if there is an interval j such that $i = j^\frown k$.
- l is an internal subinterval of i if there are intervals j and k such that $i = j^\frown l^\frown k$.

Linearity. Assume $\mathbf{Lin_T}$ holds, and consider two intervals i and j. In particular, consider the two pairs of instants $\text{beg}(i)$, $\text{end}(i)$, and $\text{beg}(j)$, $\text{end}(j)$. By $\mathbf{Lin_T}$ we know that each of the first pair is linearly related to each of the second pair:

$$(\text{beg}(i) \sim \text{beg}(j)) \wedge (\text{beg}(i) \sim \text{end}(j)) \wedge \\ (\text{end}(i) \sim \text{beg}(j)) \wedge (\text{end}(i) \sim \text{end}(j)).$$

But this just means that i is linearly related to j, that is, $i \sim j$. The interval formulation of linearity is thus simply:

- Any pair of intervals i and j are related by a relation in the set {<,||,o,s,d,f,=,fi,di,si,oi,mi,>}:

$\mathbf{Lin_I}$ $\qquad \forall i, j (i \sim j)$

The interval version of $\mathbf{Lin'_T}$ is

- Any pair of intervals i and j are related by exactly one relation in the set {<,||,o,s,d,f,=,fi,di,si,oi,mi,>}:

$\mathbf{Lin'_I}$ $\qquad \forall i, j (i < j \oplus i \| j \oplus i \text{ o } j \oplus i \text{ s } j \oplus \cdots \oplus i > j)$

Unboundedness. Consider an interval i. By $\mathbf{Unb_T}$, there is an instant t which precedes $\text{beg}(i)$ and an instant t' which follows $\text{end}(i)$. These give us an interval $\langle t, \text{beg}(i)\rangle$ which meets i and an interval $\langle \text{end}(i), u\rangle$ which i meets. Thus we are led to the formulation of the unboundedness axiom in terms of intervals as:

- For every interval there is an interval which meets it and an interval which it meets:

$$\mathbf{Unb_I} \qquad \forall i \exists j, k(j \| i \| k)$$

which can be split into past and future versions in the obvious way. Equivalently, $\mathbf{Unb_I}$ says that *there are no initial or final intervals.*

If time is assumed to be unbounded, we can reformulate the linearity axiom as follows. Concentrating on just the pair of instants $\text{end}(i)$ and $\text{beg}(j)$, the axiom $\mathbf{Lin_T}$ implies that we have either (i) $\text{end}(i) \prec \text{beg}(j)$, so $i < j$, (ii) $\text{end}(i) = \text{beg}(j)$, so $i \| j$, or (iii) $\text{beg}(j) \prec \text{end}(i)$, in which case there is an interval $\langle \text{beg}(j), \text{end}(i)\rangle$ which begins when j begins and ends when i ends. In this third case, if time in unbounded, we know that i must meet some interval i', and j must be met by some interval j'. In that case, we have $j' \| \langle \text{beg}(j), \text{end}(i)\rangle \| i'$. This leads directly to the linearity axiom given by Allen and Hayes (1985), namely

- If i meets i' and j' meets j, then exactly one of the following must hold: i meets j, i meets an interval which meets j, or j' meets an interval which meets i':

$$\mathbf{Lin''_I} \qquad \forall i, i', j, j'(i \| i' \wedge j' \| j \rightarrow i \| j \oplus i < j \oplus j' < i')$$

This is strictly weaker than $\mathbf{Lin'_I}$, but the difference only shows up in the presence of either initial or final intervals. If j is an initial interval, or i a final interval, then $\mathbf{Lin''_I}$ has nothing to say about them, whereas $\mathbf{Lin'_I}$ says that they must be linearly related. If time is unbounded, the linearity axiom of Allen and Hayes will do just as well as the one I have given.

Density. Assume that $\mathbf{Dens_T}$ holds in some temporal model, and consider an interval i in the model. Then $\text{beg}(i) \prec \text{end}(i)$, so by $\mathbf{Dens_T}$ there is an instant t such that $\text{beg}(i) \prec t \prec \text{end}(i)$. If we put $j = \langle \text{beg}(i), t\rangle$ and $k = \langle t, \text{end}(i)\rangle$ then $i = j \frown k$. The interval-based formulation of density can thus be given as follows:

- Any interval can be expressed as the concatenation of two subintervals:

Dens$_{\mathbf{I}}$ $\quad \forall i \in I\, \exists j, k \in I(i = j^\frown k)$

This can, if preferred, be reformulated without explicit reference to the operation of concatenation:

> For any interval i there are intervals j and k, such that (i) j meets k, (ii) j is an initial subinterval of i, and (iii) k is a final subinterval of i.

Discreteness. To prepare for the interval-based formulation of discreteness, I first introduce the notion of an *atomic interval*:

> An interval i is atomic if end(i) is an immediate successor of beg(i):

$$Atom(i) \triangleq \neg\exists t(\text{beg}(i) \prec t \prec \text{end}(i)).$$

Thus if i is atomic, beg(i) does not precede any instant which precedes end(i). But the existence of such an instant is precisely the condition for i to be the concatenation of two subintervals. This enables us to define an atomic interval without explicit reference to instants:

- An interval is atomic if and only if it is not the concatenation of two subintervals:

$$Atom(i) \triangleq \neg\exists j, k(i = j^\frown k).$$

This in turn provides a simple paraphrase of **Dens$_{\mathbf{I}}$**, namely

> No interval is atomic: $\neg\exists i Atom(i)$.

Now consider an interval i in discrete linear time. Since beg(i) precedes end(i), it follows from the definition of discreteness that beg(i) has an immediate successor, and end(i) has an immediate predecessor. There are two possibilities to consider:

1. beg(i) is the immediate predecessor of end(i), and end(i) is the immediate successor of beg(i). In this case i is atomic.

2. The immediate successor t of $\text{beg}(i)$ precedes $\text{end}(i)$, and $\text{beg}(i)$ precedes the immediate predecessor t' of $\text{end}(i)$. In this case we can put

$$i = \langle \text{beg}(i), t\rangle^\frown\langle t, \text{end}(i)\rangle = \langle \text{beg}(i), t'\rangle^\frown\langle t', \text{end}(i)\rangle,$$

where $\langle \text{beg}(i), t\rangle$ and $\langle t', \text{end}(i)\rangle$ are both atomic.

If an interval i is the concatenation of an atomic interval j with some other interval, I shall call j an 'initial atom' of i. In linear time an interval can have at most one initial atom. Similarly, if i is the concatenation of some interval with an atomic interval k, then I shall call k a 'final atom' of i. In linear time an interval has at most one final atom. I shall also generalize these terms so that an atomic interval is its own initial and final atom. This enables us to formulate the interval version of discreteness as

- Every interval has both an initial and a final atom:

Disc$_\mathbf{I}$ $$\forall i \exists j, k(Atom(j) \wedge Atom(k) \wedge (i = j = k \vee \exists j', k'(i = j^\frown j' = k'^\frown k))).$$

Finite Dissection. The interval version of finite dissection is obtained as follows. Let $i = \langle t, u\rangle$ be any interval. By **FD$_\mathbf{T}$**, either there are no instants between t and u, in which case i is atomic, or the instants between t and u are exhausted by a finite sequence $v_1, v_2, \ldots, v_{n-1}$. In this case we can express i as the sum of n atomic intervals

$$i = \langle t, v_1\rangle^\frown\langle v_1, v_2\rangle^\frown \cdots ^\frown \langle v_{n-1}, u\rangle.$$

Any subinterval of i must be the sum of some subsequence of these atomic subintervals, and hence i has only finitely many subintervals altogether. The interval version of finite dissection is thus

- Any interval contains finitely many subintervals:

FD$_\mathbf{I}$ $$\forall i(Atom(i) \vee \exists i_1, \ldots, i_n(Atom(i_1) \wedge \cdots \wedge Atom(i_n) \wedge i = i_1^\frown i_2^\frown \cdots ^\frown i_n))$$

Note that, like **FD$_\mathbf{T}$**, this is second-order because of the variable number of existential quantifiers, indicating that we are really quantifying over the domain of sets of intervals.

Continuity. The interval version of continuity is:

- For any non-empty sets of intervals I_1, I_2, if every element of I_1 is before every element of I_2 then there is a pair of intervals i, j such that i meets j, every element of I_1 is before or meets j, and i is before or meets every element of I_2:

$$\textbf{Cont}_\mathbf{I} \qquad \forall I_1, I_2 \subset I[\forall i_1 \in I_1, i_2 \in I_2(i_1 < i_2) \rightarrow \exists i, j \in I(i \| j \wedge \forall i_1 \in I_1(i_1 \lhd j) \wedge \forall i_2 \in I_2(i \lhd i_2))]$$

To prove that this follows from $\mathbf{Cont_T}$, suppose I_1 and I_2 are as specified above, and let $T_1 = \{\text{end}(i) \mid i \in I_1\}$, $T_2 = \{\text{beg}(i) \mid i \in I_2\}$. Then for any $t_1 \in T_1, t_2 \in T_2$, we have $t_1 \prec t_2$. By $\mathbf{Cont_T}$, there is an instant t such that for all $t_1 \in T_1$ and $t_2 \in T_2$, $t_1 \preceq t \preceq t_2$. Now let i and j be intervals with $\text{end}(i) = t = \text{beg}(j)$. Then (1) i meets j, (2) if $i_1 \in I_1$ then $\text{end}(i_1) \in T_1$ so $\text{end}(i_1) \preceq t = \text{beg}(j)$, which means that $i_1 \lhd j$, and (3) if $i_2 \in I_2$ then $\text{beg}(i_2) \in T_2$ so $\text{end}(i) = t \preceq \text{beg}(i_2)$, which means that $i \lhd i_2$. Hence $\mathbf{Cont_I}$ holds.

Intervals as primitive elements

The alternative approach to introducing intervals, as championed by Allen (1984) and other authors, is to regard intervals as primitive elements, temporally ordered by means of the relation 'meets'. The properties we want this ordering to have can then be conferred on them by selecting as axioms appropriate formulae (designated with subscript **I**) from those enumerated on pp. 237–43. At the very least we shall need **Int1**, **Int2**, $\mathbf{Irref_I}$, and $\mathbf{Trans_I}$. Others may be chosen as desired.

Even if we take this approach, we may very well want to introduce instants as well. After all, instants and intervals must stand or fall together. An interval can only be defined to the extent that its beginning and end are defined: if its bounding instants are ill defined, then so is the interval, and to just the same extent. If the interval is precisely defined, so are the instants which bound it. This applies whether the interval is defined in terms of clock-times or in terms of events. If an interval is specified as 'the period during which John was running' then its bounding instants are defined as 'the instant at which John started running' and 'the instant at which John stopped running', and if someone objects that these starting and stopping events can only be identified to within, say, a tenth of a second, and therefore are not true instants, the appropriate reply is that by the same token the running event cannot be reckoned a true interval either, since it has ill-defined boundaries. The running event is only as well defined as are the starting and stopping events. Two possible sources of ill-definedness are *imprecision* and *vagueness*, which I shall discuss briefly in turn.

Imprecision, though important as an issue in its own right, is not relevant here. Imprecision brings with it its own notion of an interval, which must be clearly distinguished from the time-intervals which are our primary concern here. Any measurement is subject to experimental error, and hence the onset and cessation of running, say, can only be recorded in the form $t \pm \delta t$. It is customary to refer to the span $(t - \delta t, t + \delta t)$ as an interval, but in this context what this means is that we are uncertain exactly where in that span the instant of onset is located, not that the onset itself spanned an interval of duration $2\delta t$. The interval $\langle t_1, t_2 \rangle$ itself inherits the uncertainty associated with its boundaries: we should properly notate it as $\langle t_1 \pm \delta t_1, t_2 \pm \delta t_2 \rangle$. One might decide to write this as $\langle (t_1 - \delta t_1, t_1 + \delta t_1), (t_2 - \delta t_2, t_2 + \delta t_2) \rangle$, but this could be misleading because of the presence of two different interval-like notations, '(x, y)' to denote an interval of uncertainty, and '$\langle x, y \rangle$' to denote a temporal interval.

The preceding considerations assumed that running is well defined in the sense that at any moment the runner either is or is not running, any uncertainty here being a result of our inability to measure times sufficiently accurately. An alternative objection is that running itself is not sharply defined: even if one knows the precise timing of every bodily movement involved in the transition from standing still to fully fledged running, one still cannot pinpoint the onset of running exactly because there are no clear criteria as to which of the intermediate states between standing still and full running should count as states of running.[13] This is an issue of vagueness rather than uncertainty, and again is an important issue in its own right. But it too is irrelevant to the instant/interval question, because, as already stated, it affects intervals to just the same extent as it affects instants.

We may thus conclude that if it is reasonable to define intervals within an instant-based formalism, then it must be equally reasonable to define instants within an interval-based formalism. As with the previous task, there is more than one approach possible here. We can distinguish two kinds of approach, roughly corresponding to the ordered-pair and convex-set constructions of intervals from instants. I shall describe a representative example of each approach.

[13] Compare my earlier remarks about the homogeneity of 'swimming' (p. 212).

The construction of instants as ordered pairs of intervals

An instant may be characterized as the meeting-point of two contiguous intervals. Thus if i meets j, we may speak of the instant *at which* i meets j. This suggests that we represent the instant as an ordered pair of meeting intervals, which may be denoted $i][j$. The problem with this is that we want some distinct pairs of intervals to represent the same instant: if i and j both meet k and l, then all four pairs $i][k$, $i][l$, $j][k$, and $j][l$ should represent the same instant. The problem is analogous to what happens when we try to define a rational number as a pair of integers; we find distinct pairs, such as $1/2$, $2/4$, $3/6$, all representing the same rational number. The solution to this problem is to use *sets* of pairs rather than individual pairs; specifically, we represent a rational number as a set of pairs of integers, and an instant as a set of pairs of intervals. The sets we require are equivalence classes, that is, maximal sets of equivalent pairs, where the appropriate notion of equivalence has to be defined. For the integers it is that a/b represents the same rational number as c/d if and only if $ad = bc$; for intervals it is that $i][j$ represents the same instant as $k][l$ if and only if i meets l (and hence also, by **Int1**, k meets j).

With this characterization of an instant, we can define temporal succession for instants by the rule

$$i][j \prec k][l \;\triangleq\; i < l.$$

It is necessary, but straightforward, to check that the succession relation so defined treats equivalent pairs in the same way, so that it gives us a relation on the instants themselves rather than just on pairs.

One consequence of adopting this ordered-pair construction for instants is that an initial interval i cannot have a beginning, since there is no interval j such that $j \| i$, and hence there is no instant $j][i$ to identify with $\text{beg}(i)$; and likewise a final interval cannot have an end. This means that if we start with a set of intervals, construct instants as (sets of) pairs of intervals, and then try to recreate the intervals as pairs of instants, we may not get back all the intervals we started with, since the initial and final intervals lack beginnings and ends respectively. This complication does not arise, of course, if our temporal framework is unbounded.

The construction of instants as sets of intervals

An alternative approach is to define an instant as a set of intervals. This is reminiscent of the construction of points as sets of regions discussed in §2.4.3.

There are several different ways of doing this. Allen and Hayes (1989) make use of the idea of a 'nest' of intervals, a maximal set of intervals having a common part. A set of intervals N is a nest if the following conditions hold:

- If any subinterval of i is in N, then i itself is in N.
- If i and j are both in N, then they have a common subinterval, which is also in N.
- If i is not in N, then there is a j in N disjoint from i.

If time is discrete, a nest consists of all intervals containing a given atomic interval as a subinterval, and hence does not succeed in picking out an instant. But if time is dense, the elements of a nest do not have a common subinterval, and in this case we can identify the nest with an instant. The relation of temporal succession on nests can be defined by the rule:

$$N_1 \prec N_2 \quad \triangleq \quad \exists i_1 \in N_1, i_2 \in N_2(i < j).$$

In the remainder of this section, I shall work with the ordered-pair construction of instants. I show that if we take as our axioms for intervals the formulae (with subscript **I**) listed on pp. 237–43, and introduce instants using the ordered-pair construction, then we can prove all the instant formulae (with subscript **T**) introduced in §5.3.2 as theorems.

Irreflexivity. Suppose $i][j \prec i][j$. This means there is an interval m such that $i\|m\|j$. Since $i\|j$, we have $i\|j \wedge i\|m \wedge m\|j$ so by **Int1** we deduce that $m\|m$. But this contradicts $\mathbf{Irref_I}$, and hence we have $\neg(i][j \prec i][j)$, effectively establishing $\mathbf{Irref_T}$.

Transitivity. Suppose $(i_1][i_2 \prec j_1][j_2) \wedge (j_1][j_2 \prec k_1][k_2)$. Then there are intervals $l, m \in I$ such that $i_1\|l\|j_2$ and $j_1\|m\|k_2$. Since $j_1\|j_2$, $l\|j_2$, and $j_1\|m$, we infer from **Int1** that $l\|m$. By $\mathbf{Trans_I}$, therefore, there is an interval $l^\frown m$ such that $i_1\|(l^\frown m)\|k_2$. Hence we have $i_1][i_2 \prec k_1][k_2$ as required.

Linearity. Given any two instants $i_1][i_2$ and $j_1][j_2$, we must have $i_1\|i_2$ and $j_1\|j_2$, so $\mathbf{Lin''_I}$ applies; the consequent can be written in instant notation as

$$(i_1][i_2 = j_1][j_2) \oplus (i_1][i_2 \prec j_1][j_2) \oplus (j_1][j_2 \prec i_1][i_2),$$

exactly as required for $\mathbf{Lin'_T}$.

Unboundedness. Assume $\mathbf{Unb}_\mathbf{I}^-$ holds, and consider an arbitrary instant $i][j$. Applying $\mathbf{Unb}_\mathbf{I}^-$ to i gives k such that $k \| i$. Then we have $k][i \prec i][j$ as required by $\mathbf{Unb}_\mathbf{T}^-$. An exactly analogous argument derives $\mathbf{Unb}_\mathbf{T}^+$ from $\mathbf{Unb}_\mathbf{I}^+$.

Density. Suppose $i_1][i_2 \prec j_1][j_2$, so there is an interval k such that $i_1 \| k \| j_2$. By $\mathbf{Dens_I}$, we can write $k = k_1 \frown k_2$, so $i_1 \| k_1 \| k_2 \| j_2$, giving

$$i_1][i_2 \prec k_1][k_2 \prec j_1][j_2,$$

as required for $\mathbf{Dens_T}$.

Discreteness. Suppose $t \prec u$, where $t = i][j$ and $u = k][l$. Since $t \prec u$, there is an interval m such that $t = i][m$ and $u = m][l$. By $\mathbf{Disc}_\mathbf{I}^+$, either m is an atom, or it can be decomposed as $m = n_1 \frown m_1 = m_2 \frown n_2$, where n_1 and n_2 are atoms. If m is an atom, then u is the immediate successor of t, and t is the immediate predecessor of u. Otherwise, $n_1][m_1$ is the immediate successor of t, and $m_2][n_2$ the immediate predecessor of u. In either case, $\mathbf{Disc_T}$ is verified.

Finite Dissection. Suppose $j][k \prec l][m$. Then there is an interval i such that $j \| i \| m$. By $\mathbf{FD_I}$, we can write $i = i_1 \frown i_2 \frown \cdots \frown i_n$, where the i_r are atomic intervals. Now suppose $j][k \prec t \prec l][m$. The instant t cannot fall within one of the atomic intervals i_r, hence must be one of the meeting-points $i_r][i_{r+1}$. But there are only finitely many such meeting-points, so there are only finitely many instants t such that $j][k \prec t \prec l][m$, as required for $\mathbf{FD_T}$. (Note: this argument assumes linear time; and indeed, for non-linear time, $\mathbf{FD_I}$ and $\mathbf{FD_T}$ are not equivalent.)

Continuity. Suppose $T_1, T_2 \subset T$ are such that every member of T_1 precedes every member of T_2. Define sets of intervals as follows:

$$\begin{aligned} I_1 = & \; \{i_1 \in I \mid \exists j \in I(i_1][j \in T_1)\} \\ I_2 = & \; \{i_2 \in I \mid \exists j \in I(j][i_2 \in T_2)\} \end{aligned}$$

Then for any $i_1 \in I_1, i_2 \in I_2$ we have $i_1 < i_2$, so by $\mathbf{Cont_I}$ there are intervals i and j such that i meets j, and for any $i_1 \in I_1, i_2 \in I_2$ we have $i_1 \lhd j$ and $i \lhd i_2$. Let $t = i][j$. Suppose $t_1 \in T_1$; then $t_1 = i_1][i_1'$ where $i_1 \in I_1$; since $i_1 \lhd j$, it follows that $t_1 \preceq t$. Similarly, if $t_2 \in T_2$, we have $t \preceq t_2$. Thus $\mathbf{Cont_T}$ is satisfied.

I have shown that starting from an instant model with axioms selected from those given in §5.3.2, we can use the ordered-pair construction of intervals to define an interval model which satisfies the corresponding axioms for such models listed above, and conversely, starting from the latter model, we can use the ordered-pair construction of instants to define an instant model satisfying the former axioms.

What is the relationship between the two instant models, the one we started with, and the one constructed from it by applying first the ordered-pair construction for intervals and then the ordered-pair construction for instants? We know they satisfy the same axioms, but that does not make them equivalent. However, so long as the original model is unbounded, satisfying $\mathbf{Unb_T}$, then the two models are isomorphic: that is, they have exactly the same structure, differing only in the labelling of the elements. To show this we must first establish a one-to-one correspondence between the elements of the two models, and then show that this correspondence respects the temporal orderings of the two models.[14]

Let the original model be $(T, \prec)$, and the derived model $(T^*, \prec^*)$. We can define a bijective mapping $\phi : T^* \rightarrow T$ which preserves the temporal ordering as follows. Any $t^* \in T^*$ is defined as a class of interval pairs $i][i'$, where $i\|i'$; these intervals are themselves defined as instant pairs, say $i = \langle t_1, t_2 \rangle$ and $i' = \langle t_3, t_4 \rangle$. Since $i\|i'$, we have $t_2 = t_3$, and we define $\phi(t^*)$ to be this instant. Note that the intervals i, i' here are not in general unique, but for a given t^* the value of $\phi(t^*)$ is unique.

The mapping ϕ is injective, since if $\phi(t_1^*) = \phi(t_2^*) = t$, say, then

$$i][i' \in t_1^* \iff \mathrm{end}(i) = \mathrm{beg}(i') = t \iff i][i' \in t_2^*,$$

so $t_1^* = t_2^*$. And so long as $(T, \prec)$ satisfies $\mathbf{Unb_T}$, the mapping ϕ is also surjective. For consider any $t \in T$. By $\mathbf{Unb_T}$, we have $t^- \prec t \prec t^+$. Then $t = \phi(t^*)$, where $t^* = \langle t^-, t \rangle][\langle t, t^+ \rangle$. Since ϕ is both injective and surjective it is bijective, as required.

All that is now needed to establish isomorphism is to check that the temporal order is preserved by the mapping ϕ. If we put $t_1^* = \langle t_1^-, t_1 \rangle][\langle t_1, t_1^+ \rangle$ and $t_2^* = \langle t_2^-, t_2 \rangle][\langle t_2, t_2^+ \rangle$, then $t_1^* \prec^* t_2^*$ if and only if there is an interval i such that $\langle t_1^-, t_1 \rangle \| i \| \langle t_2, t_2^+ \rangle$, which in turn is the case if and only if $t_1 \prec t_2$. But $t_1 = \phi(t_1^*)$ and $t_2 = \phi(t_2^*)$, so we have

$$t_1^* \prec^* t_2^* \iff \phi(t_1^*) \prec \phi(t_2^*).$$

[14] The argument is a little technical and may be omitted by less mathematically inclined readers.

We can similarly show that if we start with an interval model $(I, \|)$, define instants using the ordered-pair construction, and then from this instant model define a new interval model using the ordered-pair construction of intervals, then so long as the original model satisfies $\mathbf{Unb_I}$, the two interval models are isomorphic. The overall conclusion is that for unbounded time, instant and interval models are fully interconvertible using the ordered-pair constructions.

5.3.4. Duration

Although we are chiefly concerned with the qualitative aspects of time, some words must be said here about quantitative matters. Numbers can be assigned to times in two ways: as measures of duration, and as measures of position. These are interdefinable. Given a measure of duration and a fixed base-time, the positions of other times can be specified by a number giving the duration that has elapsed since the base time (or, in the case of times earlier than the base-time, the duration that elapsed between the time and the base-time). Equally, if each time is assigned a number as its marker of position, subject only to the condition that later times receive larger numbers, then the duration of any interval can be computed as the difference between the numbers marking the positions of its end-points.

I have already mentioned the integers, the rational numbers, and the real numbers in connection with the temporal ordering. These structures serve as canonical exemplars of their respective order-types. A total order is an order satisfying $\mathbf{Irref_T}$, $\mathbf{Trans_T}$, and $\mathbf{Lin_T}$; then

- Any total order satisfying $\mathbf{Unb_T}$, $\mathbf{Disc_T}$, and $\mathbf{FD_T}$ is equivalent to the integers.
- Any denumerable total order satisfying $\mathbf{Unb_T}$ and $\mathbf{Dens_T}$ is equivalent to the rational numbers.
- Any total order satisfying $\mathbf{Unb_T}$, $\mathbf{Dens_T}$, and $\mathbf{Cont_T}$ is either equivalent to the real numbers or contains a subset equivalent to the real numbers.

By 'equivalent' here is meant order-isomorphic, which means that equivalent orderings can be mapped on to one another in an order-preserving one-to-one correspondence.

In the discrete case we can characterize the precedence relation $\lhd$ on atomic intervals in terms of the precedence relation $\prec$ on instants by means of the following equivalence:[15]

$$\langle t, t+1\rangle \lhd \langle t', t'+1\rangle \leftrightarrow t \prec t'.$$

The precedence order on the atomic intervals is equivalent to the precedence order on the instants: each atomic interval can be mapped to its beginning point. Thus the atomic intervals also have the order-type of the integers. As already noted, we exploit this correspondence when we assign numbers to calendrical units such as days and years. At the level of years, for example, we abstract away all the internal temporal structure of a year, effectively regarding each year as an atomic interval; we then number them off using integers; and analogously with the other units.

If time is truly discrete and ordered like the integers, then the fine structure of time provides a natural measure of duration. In general, duration can be regarded as a mapping d from intervals to the real numbers such that

$$d(i) > 0 \text{ for every } i \in I$$
$$d(i^\frown j) = d(i) + d(j)$$

In discrete time with the finite dissection property the natural measure of duration is defined by assigning duration 1 to each atomic interval, thus reducing duration to 'atom count'.

In a discrete ordering without the finite dissection property, atom count is not viable as a measure of duration, since some intervals contain infinitely many atoms. And of course in a dense ordering, atom count is out of the question since there are no atoms. In such orderings there is no way of defining duration on the basis of order-type alone. For example, suppose we have a continuous ordering equivalent to the real numbers, so that there is a bijective function $f : T \to \mathbb{R}$ such that $f(t) < f(t')$ whenever $t \prec t'$. Then the function $d : I \to \mathbb{R}^+$ defined by

$$d(\langle t, t'\rangle) = f(t') - f(t)$$

is a suitable duration function. The trouble is, though, that the function f is only one amongst infinitely many possible ways of setting up an equivalence between T and $\mathbb{R}$, and different equivalences will in general result in different duration functions.

For any but a discrete ordering with the finite dissection property, then, the measure of duration has to come from external considerations. In practice

[15] Note that 'precedes' on intervals is the transitive closure of 'meets'.

we use reliably periodic phenomena such as the rotation of the earth about its axis, the revolution of the earth about the sun, the swings of a pendulum, or, more recently, the vibrations of an atom. It is fortunate for us that such phenomena measure out time at the same rate, that is, that the conversion factors between the period-counts for different phenomena are all linear, if not exactly then at least very nearly.

By using a nested sequence of periodic phenomena, each successive phenomenon in the sequence having a shorter period than its predecessors, we can assign durations to almost indefinitely short intervals. In practice, there is at any time a lower limit to the chronometric resolving power available to us, since we can only ever exploit finitely many distinct periodic phenomena for this purpose. Thus the durations which in practice we are able to assign to intervals are limited to some discrete subset of the real numbers, which means that all our current chronometrical practices can be accounted for in a discrete model of time. However, as far as we know there is no one discrete model which will always suffice however far our ability to resolve ever smaller durations increases, and it is for this reason that the continuous model, equivalent to the real numbers, seems a natural choice to make. None the less, it is an empirical question whether in the final analysis time is continuous or discrete or of some other order type.

5.4. Fluents and events

I return now to the issue of what goes on in time: within our systematic approach, how can we populate the temporal framework of instants and intervals with states, events, and processes? As we saw in §5.2, at the level of processes, the temporal phenomena of the world present a bewildering variety of different forms and textures. Only by means of some clear-cut idealization will it be possible to handle these phenomena formally. The view I shall take is that *all* temporal phenomena can be conceptualized in one (or both) of two primitive forms: as *fluents* or as *events*. I shall begin by discussing each of these in turn.

5.4.1. Fluents

As originally introduced by McCarthy and Hayes in the context of their Situation Calculus (McCarthy and Hayes 1969), a fluent was a function from situations (or world-states) to values. The idea was that different states of the world could be characterized in terms of the different values taken by the time-variable features of the world, so that for example a particular world-state might be specified by, amongst other things, the facts that the temperature in Paris is 23°C (a real-valued fluent), that the population of Greater London is 11,665,000 (an integer-valued fluent), that the Prime Minister of the UK is Tony Blair (a person-valued fluent), and that my computer is switched on (a Boolean-valued fluent). In the Situation Calculus, one reasons about the effects of actions in the world by characterizing them as mappings from situations to situations; their effects show up as changes in the values of various fluents. Time enters the picture in a somewhat incidental way; one can define a *history* as a mapping from times to situations, but this mapping does not play an essential role in the reasoning about actions.

In what follows I shall simplify this picture somewhat by omitting all explicit reference to situations, defining a fluent as a function from *times* to values. In continuous time it is natural to assume that the times are instants; in discrete time, we would use atomic intervals. Thus instead of using a world-history to map time t to situation s, and then using a fluent to map s to value v, our fluents will directly map t to v. Thus, for example, $f : T \rightarrow \mathbb{R}$ might be the fluent which gives the temperature at the front of Buckingham Palace at each time.

A particularly important class of fluents are the Boolean fluents or (partial) states, whose values are taken from the set $\{true, false\}$. Each Boolean fluent divides the set of times into those at which the fluent is *true* and those at which it is *false*. An example is 'It is raining in Exeter', which discriminates the set of all times at which it is raining in Exeter from the set of all times at which it is not raining in Exeter. Any fluent gives rise to Boolean fluents; if f takes values from some set V, then for each $v \in V$ there is a Boolean fluent which assigns *true* to those times at which f has the value v, and *false* to the rest. Again, for each subset $U \subset V$, there is a Boolean fluent which assigns *true* to those times at which the value of f is a member of U, and *false* to the rest. If f is the fluent 'The temperature in Paris', then this enables us to treat statements such as 'The temperature in Paris is 18°C' and 'The temperature in Paris is lower than 18°C' as Boolean fluents. Boolean fluents correspond to the states discussed in §5.2, and also to processes in so far as they are viewed under a state-like aspect. I shall often continue to use the word 'state' in reference to them.

It is customary to say that a Boolean fluent *holds* at those times for which it is assigned the value *true*. For a fluent $s : T \rightarrow \{true, false\}$, instead of writing $s(t) = \mathit{true}$, we write $Holds(s,t)$. Since we have different models of time to consider, we must take care how we use the $Holds$ predicate. It is useful to distinguish the case in which a state holds at an instant from the case in which it holds over an interval. I shall use $Holds_{\mathbf{T}}$ for the former and $Holds_{\mathbf{I}}$ for the latter.

In discrete time, we have little use for $Holds_{\mathbf{T}}$; the history of the world is completely determined by assigning a value to each fluent on each atomic interval. If the Boolean fluent s holds on all the atomic subintervals of an interval i, we can say that s holds on i. We can thus extend the definition of $Holds_{\mathbf{I}}$ so that it applies to all intervals, not just the atomic ones, as follows:

$$Holds_{\mathbf{I}}(s,i) \triangleq \forall a \in I(Atom(a) \wedge a \sqsubseteq i \rightarrow Holds_{\mathbf{I}}(s,a)).$$

In dense time, the approach we take depends critically on whether or not we are prepared to entertain the idea of *the state of the world at an instant.* If we are, then we can simply say that a fluent holds over an interval if and only if it holds at every instant that falls within the interval, giving us the definition

$$Holds_{\mathbf{I}}(s,i) \triangleq \forall t \in T(t \sqsubseteq i \rightarrow Holds_{\mathbf{T}}(s,t)).$$

An immediate consequence of this definition is that states are *homogeneous* over intervals,[16] that is, if a state holds over an interval then it must also hold over all subintervals of that interval:

HOM $\quad \forall i,j(j \sqsubseteq i \rightarrow (Holds_{\mathbf{I}}(s,i) \rightarrow Holds_{\mathbf{I}}(s,j)))$

Although this might seem very straightforward, many people have been unhappy about $Holds_{\mathbf{T}}$. Hamblin (1971) adduced a number of reasons why we might want to disallow talk about states holding at instants, and Allen (1984), in setting up his Interval Calculus, was motivated by a desire to do away with instants altogether. As we have seen, and as both Hamblin and Allen acknowledged, this is not entirely possible since instants can be defined in terms of intervals. However this does not mean we are forced to take instants seriously, and in particular, when we come to assigning states

[16] The use of the term 'homogeneous' in this context is due, I believe, to Taylor (1977), who uses homogeneity as a criterion to distinguish what he calls 'E-predicates' (corresponding to states), which have it, from what he calls 'K-predicates' (corresponding to events), which do not. See §5.2 for a more general discussion of homogeneity in relation to states, events, and processes.

to times, both Hamblin and Allen were adamant that we do not have to assign them to instants. We shall look at three possible approaches that are in conformity with this line of thinking.

Call an interval *quasi-atomic* if the world does not change during it, so that for every fluent f we have $\forall t, t' \sqsubseteq i(f(t) = f(t'))$. One possibility is that every instant either falls within a quasi-atomic interval or forms the meeting-point of two contiguous quasi-atomic intervals. If this is the case, then the history of the world proceeds in little jerks: the world-state is w for a while, then it switches to w', then w'', and so on. In a world like this there is no necessity ever to refer to the state of the world at an instant. We can handle this kind of world perfectly well by treating the quasi-atoms as atoms, in other words by adopting a discrete-time model. Any internal temporal structure in an individual quasi-atom is idle, since it cannot be reflected in any way at the level of world-states.

A second possibility is to retain instants as the domain of the fluents, thereby leaving open the possibility of not having quasi-atoms, but to restrict our Boolean fluents to those which necessarily persist over intervals. An example of such a fluent is 'The car is moving'; it is not possible for this fluent to hold at an instant without also holding over an interval within which that instant falls. If all our fluents are like this, then even though the underlying model might require instants (since for example a world-state may obtain at an instant without persisting over any interval containing that instant), this micro-structure is invisible at the level of the Boolean fluents we are dealing with. Technically, we would still define $Holds_{\mathbf{I}}$ in terms of $Holds_{\mathbf{T}}$, but in addition we would lay down an axiom

$$\forall s, t(Holds_{\mathbf{T}}(s,t) \rightarrow \exists i(bounds(t,i) \wedge Holds_{\mathbf{I}}(s,i))).$$

which states that the only way a fluent can hold at an instant is by holding throughout some interval bounded by that instant.[17]

To anyone with an ontological aversion to instants, this solution is far from satisfactory. The third approach remedies this by cutting out instants altogether, instead introducing $Holds_{\mathbf{I}}$ as a primitive relation between Boolean fluents and intervals. The rule **HOM**, which makes no reference to instants, is retained, but now enjoys the status of an axiom rather than a theorem. This is essentially the solution adopted by Hamblin and Allen.

[17] This includes the possibility that it holds throughout an interval within which the instant falls: if $t \sqsubseteq i$ and $Holds_{\mathbf{I}}(s,i)$, then by **HOM** we also have $Holds_{\mathbf{I}}(s, \langle \mathrm{beg}(i), t\rangle)$, where of course t bounds $\langle \mathrm{beg}(i), t\rangle$.

In many ways Boolean fluents (states) resemble propositions, the main difference being that whereas philosophers have generally regarded propositions as timelessly true or false, states may hold at some times but not at others. This resemblance to propositions prompts the natural question whether states can be combined in the same way as propositions, for example by using the truth-functional connectives of the Propositional Calculus. This proves to be straightforward so long as we are considering the assignment of statements to instants, but becomes more complicated when intervals are brought in as well. I shall consider in turn the cases of conjunction, which is unproblematic, and negation, which needs more care.

The conjunction $s.s'$ is the state which holds whenever the two states s and s' both hold. In an instant-based framework, this is expressed by the rule

$$Holds_{\mathbf{T}}(s.s', t) \leftrightarrow Holds_{\mathbf{T}}(s, t) \wedge Holds_{\mathbf{T}}(s', t),$$

from which it follows that for an interval i,

$$Holds_{\mathbf{I}}(s.s', i) \leftrightarrow Holds_{\mathbf{I}}(s, i) \wedge Holds_{\mathbf{I}}(s', i),$$

which is the rule one would adopt in an interval-based framework.

The negation $-s$ is the state which holds whenever the state s does not hold. In an instant-based framework, this is expressed by the rule

$$Holds_{\mathbf{T}}(-s, t) \leftrightarrow \neg Holds_{\mathbf{T}}(s, t).$$

For an interval i this gives us

$$Holds_{\mathbf{I}}(-s, i) \leftrightarrow \neg\exists t \sqsubseteq i (Holds_{\mathbf{T}}(s, t)),$$

which is called the 'strong' or 'internal' negation, as opposed to the 'weak' or 'external' negation seen in

$$\neg Holds_{\mathbf{I}}(s, i) \leftrightarrow \exists t \sqsubseteq i \neg Holds_{\mathbf{T}}(s, t).$$

The strong negation says that there are *no* instants in i at which s holds, whereas the weak negation merely says that there is at least one instant in i at which s does *not* hold.

Hamblin and Allen, having rejected the assignment of states to instants, had to come up with a different definition of $-s$ suitable for a pure interval-based framework, and what they both chose was

$$Holds_{\mathbf{I}}(-s, i) \quad \triangleq \quad \forall j \sqsubseteq i \neg Holds_{\mathbf{I}}(s, j),$$

which states that s does not hold over any subinterval of i. In general, this is not equivalent to the definition of $-s$ in terms of instants; the condition for the two definitions to be equivalent is that a state cannot hold at an isolated instant; more exactly, a state can only hold at an instant if it holds over some interval bounded by that instant, exactly as suggested above. For more details on this and related matters, see Galton 1990 and Galton 1996.

5.4.2. Events

As discussed in §5.2, events, as individual chunks of history, relate to time rather differently from how states do. I shall use the following notations for events:

- Event-types are represented by upper-case letters such as E.
- Event-tokens are represented by lower-case letters such as e.
- $type(e, E)$ means that e is an occurrence of type E.
- $sub(E, E')$ means that E is a subtype of E'.

If E is a subtype of E', then every occurrence of the former must also be an occurrence of the latter, which means that the predicates $type$ and sub are related by the formula

$$sub(E, E') \rightarrow \forall e(type(e, E) \rightarrow type(e, E')).$$

The formalism suggested here enables facts about events to be expressed in first-order logic, which is technically convenient even if not necessarily philosophically correct. If preferred, expressions of the form $type(e, E) \wedge \Phi(e)$ could be expressed using second-order logic as $\Phi'(E)$, where Φ' is a second-order analogue of the first-order predicate Φ, and E is now a proposition (cf. Galton 1991).

Since an event-token has a definite time of occurrence, we can designate this time by means of a function notation as $time(e)$. This might be either an interval or an instant; we must allow both possibilities, but for any particular event-token it will be definitively one or the other. I introduce two predicates $Occurs_{\mathbf{T}}$ and $Occurs_{\mathbf{I}}$, relating event-types to times. To say that type E has an occurrence which takes up exactly the interval i, I write $Occurs_{\mathbf{I}}(E, i)$. To say that it has an instantaneous occurrence at the instant t, I write $Occurs_{\mathbf{T}}(E, t)$. The definitions of these notations are

$$Occurs_{\mathbf{I}}(E, i) \triangleq \exists e(type(e, E) \wedge time(e) = i)$$

$$Occurs_{\mathbf{T}}(E, t) \triangleq \exists e(type(e, E) \wedge time(e) = t)$$

Notice that the only difference between these two cases is that in the former, the time involved is an interval, whereas in the latter it is an instant. An event-type E is 'durative' if $time(e)$ is an interval for every occurrence e of E; it is 'instantaneous' if all its occurrences are instantaneous.

Historically, the logic of event-types and the logic of event-tokens have generally been studied separately rather than together. The systems of McDermott (1982), Galton (1984), and Allen (1984), albeit in rather different ways, use formalisms involving reference to event-types but not to event-tokens.[18] By contrast, the influential work of Davidson (1967) on the logical analysis of 'action sentences', as well as the more recent Event Calculus of Kowalski and Sergot (1986), involve reference to event-tokens but not (or only derivatively) to event-types. The system we use here is less parsimonious in this respect, but has the advantage of making explicit the relationships amongst the other systems mentioned.

The authors of systems based on event-types have generally sought to impose restrictions on the extension of the predicate $Occurs_{\mathbf{I}}$, thereby in effect restricting the class of admissible event-types. For example, in Galton 1984: 92, an event-type is not allowed to have occurrences which overlap in time. If we assume that the times of the occurrences in question are intervals, this can be expressed by the rule

$$\forall e, e'(type(e, E) \wedge type(e', E) \rightarrow e = e' \vee time(e) \bowtie time(e')),$$

which says that if e and e' are distinct occurrences of the same type then they must occupy disjoint times. (We cannot avoid reference to event-tokens here; at the level of event-types the best we can do is

$$\forall i, j(Occurs_{\mathbf{I}}(E, i) \wedge Occurs_{\mathbf{I}}(E, j) \rightarrow i = j \vee i \bowtie j),$$

which does not rule out the possibility of simultaneous occurrences of E.) The motivation for this rule was that it made the logic cleaner, but in retro-

[18] In fact McDermott does refer to tokens of events, but for him these are in the first instance *intervals*, since he identifies an event-type with a set of intervals; on the other hand he goes on to acknowledge that this does not yield a satisfactory criterion of identity for occurrences, noting that

> A given interval is a token of many different events, which happened to occur at that point. So event tokens must be represented as ordered pairs of events and intervals, or something equivalent.
>
> (McDermott 1982: 111)

A consequence of such a representation would be that it is impossible to have distinct but simultaneous occurrences of the same event-type.

spect it would appear to be a rather artificial limitation. It means, for example, that the system could not handle an event-type such as 'John writes a book', since there is in reality nothing to prevent John from starting one book while he is in the middle of writing another. If we regard the time of each occurrence to be an interval then the times of writing the two books must overlap. If we want to insist that the occurrences are non-overlapping and still be able to handle the example of the two books, then we must allow the time of an occurrence to be a more complex temporal object comprising two or more disjoint intervals as parts. Such 'non-convex' intervals have been studied by Ladkin (1986). The two-book example can now be handled by *interleaving* the periods on which John works on the two books. There is some merit in this suggestion, but it makes for greater complexity (both in an informal sense and in the computational sense) in the logic. For this reason, I shall here allow occurrences of an event-type to overlap.

A similar restriction was imposed by Allen (1984), who proposed the rule

$$\forall i, j \in I(Occurs_{\mathbf{I}}(E, i) \wedge j \sqsubset i \rightarrow \neg Occurs_{\mathbf{I}}(E, j)),$$

which says that if an event occurs on an interval then no event of the same type can occur on any proper subinterval of that interval.[19] The motivation for this was to provide an appropriate counterpart to the formula **HOM**, expressing the homogeneity property of states, which Allen (1984) gives as his Axiom H1. The idea is that if, say, John runs a mile over a certain interval i, then there is no proper subinterval of i over which he runs a mile, since over any subinterval of i he must run a lesser distance. Thus events, in contrast to states, have a unitary character; they cannot be divided into parts of the same type. This is correct, but the correct way to express it is in terms of event-tokens, not event-types. For example, there is no reason why I should not write two books, with one book being started later, and finished earlier, than the other. But if we consider the interval over which I wrote one of the books, it is clear that there is no proper subinterval over which I wrote *that* book in its entirety. The relation between Allen's rule and the non-overlapping rule becomes apparent if we rephrase the former in terms of event-tokens, obtaining

$$\forall e, e'(type(e, E) \wedge type(e', E) \rightarrow \neg(time(e) \sqsubset time(e'))).$$

[19] Effectively the same rule appears in a rather different context as the *ontological assumption* of Pratt and Francez 2000.

This says that occurrences of a single event-type cannot be nested. It is a special case of the non-overlapping requirement, but the arguments levelled against the latter apply equally to the former.

In the system used here the unitary character of event-tokens is secured not by means of a temporal restriction on event-types, but by the decision to relate event-tokens to times by means of a function, $time(e) = i$, rather than, say, a predicate $time(e, i)$. That way, there is one and only one time associated with a given event-token.

5.4.3. Constructing events from states

Although states and events are treated quite differently in the system outlined here, there are close relations between them. Many events can be defined in terms of states, for example as a transition from one state to another, and some states can be defined in terms of events, for example as a state resulting from some event, or contributing to its progress. I now turn to the business of articulating these relations systematically. In this section I concentrate in particular on the various ways in which events can be defined in terms of states; in the next section I consider the reverse direction. The material presented here is very much in the spirit of Galton 1984: ch. 5, although the formal framework is different.

Many event-types can be specified in terms of changes of state. Here I examine a number of ways in which this can be done. In general, to define a new event-type we must specify its *occurrence conditions*, which are necessary and sufficient conditions for there to be an occurrence of that type on a given interval or at a given instant. Event-types that are defined in this way are called derived event-types.

Ingressions

The simplest kind of derived event-type is an 'ingression'. This is the event of some state's beginning to hold. Assuming that, for a given state s, the formula $Holds_{\mathbf{T}}(s, t)$ has a determinate truth-value for all instants t, then the event of s beginning to hold must be instantaneous. I shall denote this event $Ingr(s)$. There is an occurrence of this event-type at the instant dividing an interval over which s holds from an interval over which $-s$ holds, giving us the definition:

$$Occurs_{\mathbf{T}}(Ingr(s), t) \quad \triangleq \quad \exists i, j(i \|_t j \wedge Holds_{\mathbf{I}}(-s, i) \wedge Holds_{\mathbf{I}}(s, j)).$$

Note that this says nothing about whether the state S holds at the instant t itself; it may or may not do so, depending on the nature of S. This issue is best handled in the context of the thory of dominance discussed in Chapter 8. For each t such that $Occurs_{\mathbf{T}}(Ingr(s), t)$, there is a unique event-token e satisfying $type(e, Ingr(s)) \wedge time(e, t)$. For the type $Ingr(s)$, therefore, an ordered pair (event-type,time) (such as $(Ingr(s), t)$) does uniquely identify a single token of the type, in accordance with the suggestion of McDermott's mentioned above. In general, however, one might wish to allow there to be more than one occurrence of a given event-type E at a given time t.

To say that state s ceases to hold we can use the event-type $Ingr(-s)$.

Transitions

Ingressions are special cases of a more general kind of derived event-type, a 'transition' (Gooday 1994; Gooday and Galton 1997). To specify a transition we must pick an 'initial state' s and a 'final state' s'. A transition from s to s' is an event-type which occurs whenever a time at which s holds is followed, not necessarily immediately, by a time at which s' holds. In order to ensure that some genuine change occurs, we shall require that s and s' do not hold together during the portion of history under consideration.

There are several possibilities here, according to whether the times involved are instants or intervals, and whether the initial and final states are separated by an interval, an instant, or not at all. If they are separated by an interval, then the transition is durative; otherwise it is instantaneous. A durative transition from s to s' will be denoted $Trans_{\mathbf{I}}(s, s')$; an instantaneous one is denoted $Trans_{\mathbf{T}}(s, s')$. We cannot give single occurrence conditions to either of these; instead we must split them up into cases according to the nature of the holding times of the initial and final states. To a considerable extent, these variations are determined in any particular case by the nature of the states involved.

In order to give precise occurrence conditions I shall define eight different transition operators, four durative and four instantaneous.[20] My notation for them uses the basic forms $Trans_{\mathbf{I}}$ and $Trans_{\mathbf{T}}$ decorated with superscripts suggestive of the corresponding pattern of temporal incidence. The eight kinds of transition are illustrated schematically in Figure 5.3, in which the labels correspond to the superscripts in the $Trans$ notation. In the diagrams, a state holding over an interval is indicated by a solid bold line; a dashed bold

[20] This section is derived from similar material in Galton 2000*b*. I am grateful to Carola Eschenbach for pointing out a number of 'bugs' in the definitions given there.

line indicates that the state may indifferently hold or not hold over the interval; absence of either indication means that the negation of the state holds. A solid circle indicates a state holding at an instant; a cross indicates a state failing to hold at an instant; absence of either indication means that the state may indifferently hold or not hold at the instant.

Durative transitions

- $Trans^{iii}_{\mathbf{I}}(s, s')$ occurs when an interval over which s holds is followed, after a delay, by an interval over which s' holds:

$$\begin{aligned} Occurs_{\mathbf{I}}(Trans^{iii}_{\mathbf{I}}(s, s'), i) \ &\triangleq \\ &\exists j, k(j \| i \| k \wedge Holds_{\mathbf{I}}(s, j) \wedge Holds_{\mathbf{I}}(-s, i^\frown k) \\ &\quad \wedge Holds_{\mathbf{I}}(-s', j^\frown i) \wedge Holds_{\mathbf{I}}(s', k)). \end{aligned}$$

- $Trans^{iit}_{\mathbf{I}}(s, s')$ occurs when an interval over which s holds is followed, after a delay, by an instant at which s' holds:

$$\begin{aligned} Occurs_{\mathbf{I}}(Trans^{iit}_{\mathbf{I}}(s, s'), i) \ &\triangleq \\ &\exists j(j \| i \wedge Holds_{\mathbf{I}}(s, j) \wedge \wedge Holds_{\mathbf{I}}(-s, i) \\ &\quad \wedge Holds_{\mathbf{I}}(-s', j^\frown i) \wedge Holds_{\mathbf{T}}(s'.(-s), \mathrm{end}(i))) \end{aligned}$$

- $Trans^{tii}_{\mathbf{I}}(s, s')$ occurs when an instant at which s holds is followed, after a delay, by an interval over which s' holds:

$$\begin{aligned} Occurs_{\mathbf{I}}(Trans^{tii}_{\mathbf{I}}(s, s'), i) \ &\triangleq \\ &\exists k(i \| k \wedge Holds_{\mathbf{T}}(s.(-s'), \mathrm{beg}(i)) \wedge Holds_{\mathbf{I}}(-s, i^\frown k) \\ &\quad \wedge Holds_{\mathbf{I}}(-s', i) \wedge Holds_{\mathbf{I}}(s', k)) \end{aligned}$$

- $Trans^{tit}_{\mathbf{I}}(s, s')$ occurs when an instant at which s is followed, after a delay, by an instant at which s' holds:

$$\begin{aligned} Occurs_{\mathbf{I}}(Trans^{tit}_{\mathbf{I}}(s, s'), i) \ &\triangleq \\ &Holds_{\mathbf{T}}(s.(-s'), \mathrm{beg}(i)) \wedge Holds_{\mathbf{I}}((-s).(-s'), i) \\ &\quad \wedge Holds_{\mathbf{T}}(s'.(-s), \mathrm{end}(i))). \end{aligned}$$

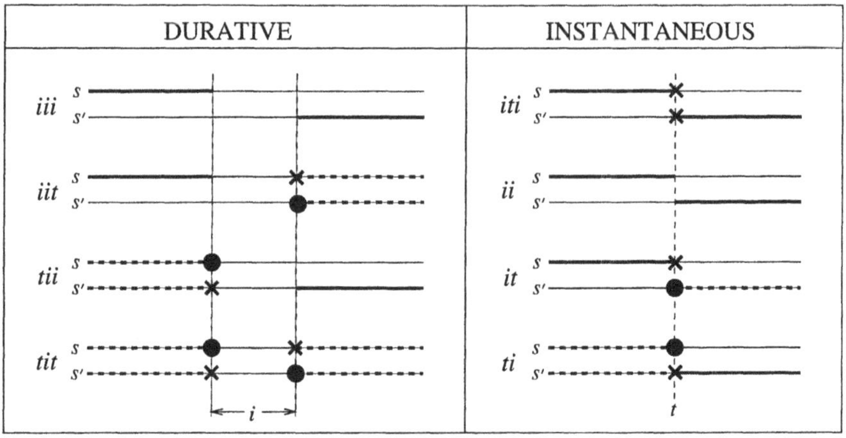

Figure 5.3. Eight kinds of transition from state s to s' (see text for explanation of symbols)

Instantaneous transitions

- $Trans^{iti}_{\mathbf{T}}(s,s')$ occurs when an interval over which s holds meets an interval over which s' holds at an instant at which neither s nor s' holds:

$$Occurs_{\mathbf{T}}(Trans^{iti}_{\mathbf{T}}(s,s'),t) \triangleq \exists i,j(i \|_t j \wedge Holds_{\mathbf{I}}(s.(-s'),i) \wedge Holds_{\mathbf{T}}((-s).(-s'),t) \wedge Holds_{\mathbf{I}}(s'.(-s),j)).$$

- $Trans^{ii}_{\mathbf{T}}(s,s')$ occurs when an interval over which s holds meets an interval over which s' holds:

$$Occurs_{\mathbf{T}}(Trans^{ii}_{\mathbf{T}}(s,s'),t) \triangleq \exists i,j(i \|_t j \wedge Holds_{\mathbf{I}}(s.(-s'),i) \wedge Holds_{\mathbf{I}}(s'.(-s),j)).$$

- $Trans^{it}_{\mathbf{T}}(s,s')$ occurs when an interval over which s holds is ended by an instant at which s' holds:

$$Occurs_{\mathbf{T}}(Trans^{it}_{\mathbf{T}}(s,s'),t) \triangleq \exists i(ends(t,i) \wedge Holds_{\mathbf{I}}(s.(-s'),i) \wedge Holds_{\mathbf{T}}(s'.(-s),t)).$$

- $Trans^{ti}_{\mathbf{T}}(s,s')$ occurs when an instant at which s holds begins an interval over which s' holds:

$$Occurs_{\mathbf{T}}(Trans^{ti}_{\mathbf{T}}(s,s'),t) \triangleq \exists j(begins(t,j) \wedge Holds_{\mathbf{T}}(s.(-s'),t) \wedge Holds_{\mathbf{I}}(s'.(-s),i)).$$

Note that an occurrence of $Trans^{ii}_{\mathbf{T}}(s, s')$ must also be an occurrence of $Trans^{iti}_{\mathbf{T}}(s, s')$, $Trans^{it}_{\mathbf{T}}(s, s')$, or $Trans^{ti}_{\mathbf{T}}(s, s')$; which of these it is will depend on the nature of the states s and s' involved in it. If one wishes to follow Allen and Hamblin in denying that a state can hold at an isolated instant, then the only two transition types available are $Trans^{iii}_{\mathbf{I}}$ and $Trans^{ii}_{\mathbf{T}}$. These could then be unambiguously denoted by $Trans_{\mathbf{I}}$ and $Trans_{\mathbf{T}}$ respectively.

Why is it necessary to distinguish between so many different kinds of transition? The full answer to this will appear later, when I shall be using these different transition operators to model various events of interest, as for example in §6.3.1. For the present I shall just give a simple (perhaps not wholly convincing) example of why we must distinguish between $Trans^{iii}_{\mathbf{I}}$ and $Trans^{tit}_{\mathbf{I}}$; it is easy to construct analogous examples for the other cases. Consider a car which is initially at rest over an interval i ending at t_0, after which it accelerates continuously over an interval j to reach a speed of 100 km/h, which is then maintained over a further interval k, where $i||j||k$. Thus there is an occurrence of the event $Trans^{iii}_{\mathbf{I}}(speed(car) = 0, speed(car) = 100)$ taking up exactly the interval j. By continuity, there must be instants t_1 and t_2 in j at which the speed of the car is 20 km/h and 60 km/h respectively. Assume the increase in speed is strictly monotonic. Then over $\langle t_1, t_2 \rangle$ there is an occurrence of the event $Trans^{tit}_{\mathbf{I}}(speed(car) = 20, speed(car) = 60)$, but no occurrence of $Trans^{iii}_{\mathbf{I}}(speed(car) = 20, speed(car) = 60)$, since neither of these speeds is maintained over any interval, however short, contained in j.

A special case of a transition is when the initial state is the negation of the final state. Such a transition must be of one of the forms $Trans^{it}_{\mathbf{T}}(-s, s)$ or $Trans^{ti}_{\mathbf{T}}(-s, s)$; some, but not all, occurrences of these will also be occurrences of $Trans^{ii}_{\mathbf{T}}(-s, s)$. In any other case the state $s.(-s)$ would have to hold at the time of the transition itself, and this is impossible. The occurrence condition for $Trans^{ii}_{\mathbf{T}}(-s, s)$ reduces to

$$Occurs_{\mathbf{T}}(Trans^{ii}_{\mathbf{T}}(-s, s), t) \leftrightarrow \exists i, j (i ||_t j \wedge Holds_{\mathbf{T}}(-s, i) \wedge Holds_{\mathbf{T}}(s, j)),$$

from which it follows that $Trans^{ii}_{\mathbf{T}}(-s, s) = Ingr(s)$.

If the final state is the *same* as the initial state, then we do not have a transition at all; as defined here, transitions necessarily involve a net change. As noted above, this is ensured by our definitions being framed in such a way as to disallow the initial and final states from co-occurring at any of the times involved in the definition. We may none the less wish to define an event in which some state s holds initially, then stops holding, and finally holds again. Events of this kind are included in the category of 'tenures', which I turn to next.

Tenures

A tenure event consists of the holding of some state for a time, bounded by times at which the state does not hold. It is the sort of event referred to by expressions such as 'John had a swim', 'Carol spent a while reading', or, with duration information added, 'John swam for ten minutes', 'Carol spent half an hour reading'. There are two species of tenure event, depending on whether the constitutive state holds over an interval or over an instant. The examples just given are durative tenures; I shall write $Po_{\mathbf{I}}(s)$ to denote the tenure of state s over an interval (for the notation Po, see Galton 1984: p.81). An example of an instantaneous tenure is the case of a ball, thrown vertically upwards, reaching the topmost point of its trajectory, at which point the state 'The ball is stationary' holds for just an instant. Instantaneous tenure of state s will be denoted $Po_{\mathbf{T}}(s)$. The occurrence conditions for the two kinds of tenure are as follows:

$$Occurs_{\mathbf{T}}(Po_{\mathbf{T}}(s), t) \triangleq$$
$$\exists i, j \in I(i \|_t j \wedge Holds_{\mathbf{I}}(-s, i) \wedge Holds_{\mathbf{T}}(s, t) \wedge Holds_{\mathbf{I}}(-s, j))$$

$$Occurs_{\mathbf{I}}(Po_{\mathbf{I}}(s), i) \triangleq$$
$$\exists j, k \in I(j \| i \| k \wedge Holds_{\mathbf{I}}(-s, j) \wedge Holds_{\mathbf{I}}(s, i) \wedge Holds_{\mathbf{I}}(-s, k))$$

An observation of McDermott's is relevant here: he distinguishes between on the one hand *fact changes* (the prototypical examples of our transitions) and on the other hand 'events that consist of some aimless thing happening for a while such as a rooster crowing in the morning' (McDermott 1982: p.110) (the prototypical examples of our durative tenures).

The expressions $Ingr$, $Trans_{\mathbf{I}}$, $Trans_{\mathbf{T}}$, $Po_{\mathbf{I}}$, and $Po_{\mathbf{T}}$ all denote functions from the domain of states to events. In the terminology of Galton 1984, they are perfective aspect operators; the connection with the phenomenon of aspect in natural language thereby implied supplied the motivation for introducing these operators in the earlier work, although in the current context this connection has perhaps become more tenuous. In the next section we shall examine some imperfective aspect operators, which act in the reverse direction.

5.4.4. Constructing states from events

Just as we define event-types by giving their occurrence conditions, so we can define new states by giving their holding conditions. These are necessary and sufficient conditions for the truth of a proposition, taking one or other of the forms $Holds_{\mathbf{T}}(s,t)$ and $Holds_{\mathbf{I}}(s,i)$. For the sake of completeness I here briefly review some important ways of constructing states from events, although I shall not, in fact, have much use for these in the remainder of the book.

The progressive

If Jane walks from her house to the station over a certain interval, then at any time during the interval we can say that Jane *is walking* to the station. In the sentence 'Jane is walking to the station' we say that the verb is in the progressive form. We use the progressive form in two senses: a 'broad' sense which holds at any time between Jane's departure from home and her arrival at the station (even when she stops to take a rest so is not actually walking), and a 'narrow' sense which implies that she is actually walking. The broad sense might hold even if it is not the case that Jane is walking throughout the interval: this use of the progressive can gloss over interruptions. (Even while I am asleep, it is still true that I am writing this book.)

I shall represent the two senses of the progressive by means of two operators, $prog^-$ and $prog^+$, called the 'formal' and 'material' progressive operators respectively. These operators are intended to represent the broad and narrow senses of the progressive, respectively. An important difference between them is that whereas we can straightforwardly present holding conditions for the formal progressive (hence its name), the material progressive cannot be defined purely formally: the conditions of its application depend rather sensitively on the events it is applied to (see the last three chapters of Galton 1984 for an extended discussion of this issue).

For any durative event-type E, then, we can define a state $prog^-(E)$ which holds just when there is an occurrence of E in progress. Its holding condition is

$$Holds_{\mathbf{T}}(prog^-(E),t) \triangleq \exists i \in I(t \sqsubseteq i \wedge Occurs_{\mathbf{I}}(E,i)).$$

It is clear from the occurrence condition that $prog^-$ cannot be applied to an instantaneous event. To say that an iii-type transition between states s and s'

is in progress at time t, we can write

$$Holds_{\mathbf{T}}(prog^{-}(Trans^{iii}_{\mathbf{I}}(s, s')), t);$$

by applying the occurrence conditions for $Trans^{iii}_{\mathbf{I}}$ and the holding conditions for $prog^{-}$, we see that this is equivalent to

$$\exists i, j, k \in I(i \| j \| k \wedge t \sqsubseteq j \wedge Holds_{\mathbf{I}}(s, i) \wedge Holds_{\mathbf{I}}(s', k) \\ \wedge Holds_{\mathbf{I}}(-s', j^{\frown}i) \wedge Holds_{\mathbf{I}}(-s, i^{\frown}k)).$$

Similarly we have

$$Holds_{\mathbf{T}}(prog^{-}(Po_{\mathbf{I}}(s)), t) \leftrightarrow \\ \exists i, j, k \in I(i \| j \| k \wedge t \sqsubseteq j \wedge \\ Holds_{\mathbf{I}}(-s, i) \wedge Holds_{\mathbf{I}}(s, j) \wedge Holds_{\mathbf{I}}(-s, k)),$$

from which we can derive the implication

$$Holds_{\mathbf{T}}(prog^{-}(Po_{\mathbf{I}}(s)), t) \rightarrow Holds_{\mathbf{T}}(s, t),$$

which says that if a stretch of s's holding is currently under way, then s now holds (thus if Jane is spending a while in the park, then she is in the park).

For the material progressive, we cannot give complete holding conditions. For $prog^{+}(E)$ to hold at a time it is not enough that the time should fall within the time of some occurrence of E; rather, we require that some state which forms a contributory part to the event should hold then. The notion of 'contributory part' cannot be defined within the current formal framework. Here I merely note that for tenure events, the distinction between the two progressives is nullified:

$$prog^{+}(Po_{\mathbf{I}}(s)) = prog^{-}(Po_{\mathbf{I}}(s)).$$

The reason for this is that tenure events are by definition uninterrupted; and it is only the possibility of interruptions which gives rise to the distinction between $prog^{+}$ and $prog^{-}$. Hence the above implication can also be written as

$$Holds_{\mathbf{T}}(prog^{+}(Po_{\mathbf{I}}(s)), t) \rightarrow Holds_{\mathbf{T}}(s, t).$$

The perfect

We sometimes need to refer to a state which obtains by virtue of some past occurrence of an event; for this we use the perfect tense of the verb, as in 'I have read *Hamlet*'. In practice, the perfect tense is used in a variety of different ways, and in particular it is often used to refer to a state *resulting* from an occurrence of the event. As with the progressive, I shall distinguish between a broad and a narrow sense of the perfect, the former being called the 'formal perfect', denoted $perf^-$, and the latter the 'material perfect', denoted $perf^+$.

The formal perfect, sometimes called the 'bare perfect' or the 'experiential perfect', is readily defined for durative events as follows:

$$Holds_{\mathbf{T}}(perf^-(E), t) \quad \triangleq \quad \exists i \in I(\mathrm{end}(i) \preceq t \wedge Occurs_{\mathbf{I}}(E, i))$$

and for instantaneous events:

$$Holds_{\mathbf{T}}(perf^-(E), t) \quad \triangleq \quad \exists u \in T(u \prec t \wedge Occurs_{\mathbf{T}}(E, u)).$$

An immediate consequence of these holding conditions, proved using the transitivity of $\prec$, is that

$$\forall t, t' \in T(t \prec t' \rightarrow (Holds_{\mathbf{T}}(perf^-(E), t) \rightarrow Holds_{\mathbf{T}}(perf^-(E), t')).$$

This is a formal expression of the idea that 'what's done is done': once something has happened, it can never subsequently come not to have happened. This means that the event-type $Ingr(perf^-(E))$ can have at most one occurrence, which takes place at the instant that the very first occurrence of E comes to completion.

The material perfect, which refers to a state resulting from the event, cannot be given a general formal definition, being, like the material progressive, rather sensitive to context (Moens and Steedman 1988). The only cases where we can give clear holding conditions are those events for which the resulting state forms part of the definition, namely transitions and ingressions. The state resulting from $Trans_{\mathbf{I}}(s, s')$ is s', and the state resulting from $Ingr(s)$ is s. In general, if event-type E has resulting state $res(E)$, then if E is durative we have

$$\begin{aligned} Holds_{\mathbf{T}}(perf^+(E), t) \quad \triangleq \\ \exists i(Occurs_{\mathbf{I}}(E, i) \wedge \\ Holds_{\mathbf{I}}(res(E), \langle \mathrm{end}(i), t\rangle) \wedge Holds_{\mathbf{T}}(res(E), t)), \end{aligned}$$

and if E is instantaneous

$$\begin{aligned}Holds_{\mathbf{T}}(perf^{+}(E),t) &\triangleq \\ &\exists u \prec t(Occurs_{\mathbf{T}}(E,u) \wedge \\ &\quad Holds_{\mathbf{I}}(res(E),\langle u,t\rangle) \wedge Holds_{\mathbf{T}}(res(E),t)).\end{aligned}$$

Given that $res(Ingr(s)) = s$, we have

$$\begin{aligned}Holds_{\mathbf{T}}(perf^{+}(Ingr(s)),t) &\triangleq \\ &\exists u \prec t(Occurs_{\mathbf{T}}(Ingr(s),u) \wedge \\ &\quad Holds_{\mathbf{I}}(s,\langle u,t\rangle) \wedge Holds_{\mathbf{T}}(s,t)).\end{aligned}$$

Note that therefore we have

$$Holds_{\mathbf{T}}(perf^{+}(Ingr(s)),t) \rightarrow Holds_{\mathbf{T}}(s,t),$$

so that the formal relationship between $perf^{+}$ and $Ingr$ is parallel to that between $prog^{+}$ and Po.

5.5. Summary

In this chapter I have investigated absolute temporal frameworks based on instants or intervals, and discussed various properties such as unboundedness, linearity, density, discreteness, and continuity which we may consider it desirable to impose on such frameworks. I have shown that it is possible to define instants and intervals in terms of each other in such a way that the resulting temporal orders correspond exactly, so that in the final instance the choice between instants and intervals as the primitive elements of a temporal framework may be a less momentous issue than it has sometimes been made to appear.

I went on to consider the various kinds of substantial information that can be situated in the temporal framework, dividing these into Boolean fluents (states and processes) and events, which differ from each other with respect to their modes of temporal incidence. In particular, I discussed ways of defining event types in terms of states, and states in terms of event types.

This chapter has laid the formal groundwork for handling all forms of change, including of course our central concern, spatial change, which is the subject of the next chapter.

6

Varieties of Spatial Change

In this chapter the survey of spatial attributes from Chapter 4 is combined with the temporal dimension to generate a survey of varieties of spatial change. The plan for the chapter is straightforward. For each spatial attribute, I shall consider what kinds of change objects can undergo in respect of that attribute. I shall be interested both in collecting illustrative examples, and in classifying these in terms of the nature of the change involved. I shall consider the attributes in the same order as they were presented in Chapter 4.

6.1. Change in dimension

Just as I distinguished in §4.1 between strict and apparent dimension, so we must distinguish between change in strict dimension and change in apparent dimension.

Examples of change in strict dimensionality can be found, though they have a somewhat contrived air. Consider for example the sequence of snapshots in Figure 6.1*a*, in which a cone becomes progressively truncated so that its top changes from a point to an area and then to a line. The changes need involve no sharp discontinuities: they can be accomplished by the gradual and continuous erosion of material from the top of the object. But each change of dimension must be accomplished in an instant since the available dimensions form a discrete space, not a continuum (the issue of fractal dimensions in not relevant here). Of course the entity whose dimension is changing here, the top of the cone, is something of an idealization, but it is well defined independently of any arbitrary human conventions and hence cannot be dismissed as merely conventional—it is in fact a parasitic object.

Changes in apparent dimension are both more frequent and more natural. An example is a river overspreading its banks, thereby changing from a linear feature to an areal one. There can be intermediate stages where part of the river is linear and part of it is areal, the river as a whole then being of 'mixed' apparent dimension (not fractional, of course), as shown in Figure 6.1*b*.

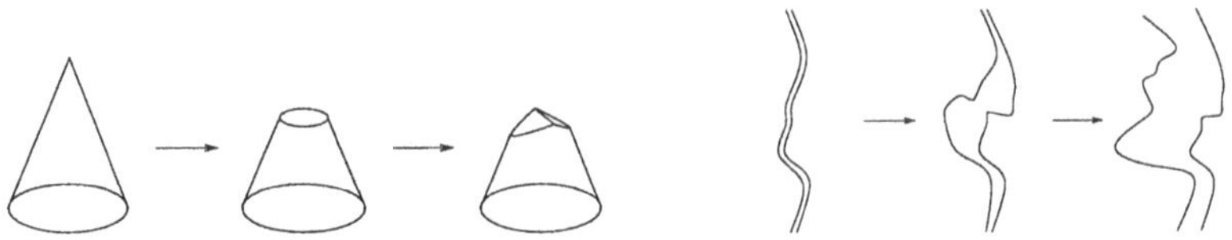

(*a*) Change in strict dimension: the 'top' of the figure changes from point to area to line

(*b*) Change in apparent dimension: a river over-spreads its banks

Figure 6.1. Change in dimension

Another example is an ice-cube, placed on a table and gradually melting to produce a thin film of water spread over the table-top. If we suppose that at one edge of the table there is a small groove, then the water will tend to run off the table-top through this groove, forming a narrow column of water as it descends. Here we have an object whose apparent dimension successively reduces from three to two to one.

In these cases of changing apparent dimension the times at which the changes occur are only as well defined as the demarcation between the different apparent dimensions. In practice one would be hard put to define exactly how broad a river must become for it to be regarded as areal rather than linear; there is an ineradicable element of imprecision here.

Change in dimension would appear to be of marginal interest; it is hard to think of any practical application which might be usefully served by a systematic analysis of it.

6.2. Change in connectivity

Change in connectivity must always involve either the making or breaking of connections. As we saw in §4.2, we can distinguish a range of different kinds of connection, varying from mere spatial adjacency to cases where physical forces have an important part to play. Accordingly, we shall have to recognize corresponding varieties of *change* in connectivity.

With your own body you can readily demonstrate any number of changes of adjacency-connectivity. Put your hands together and the topological genus of your free surface increases to that of a torus. Put your feet together as well and the topological genus increases once more. Join hands with someone

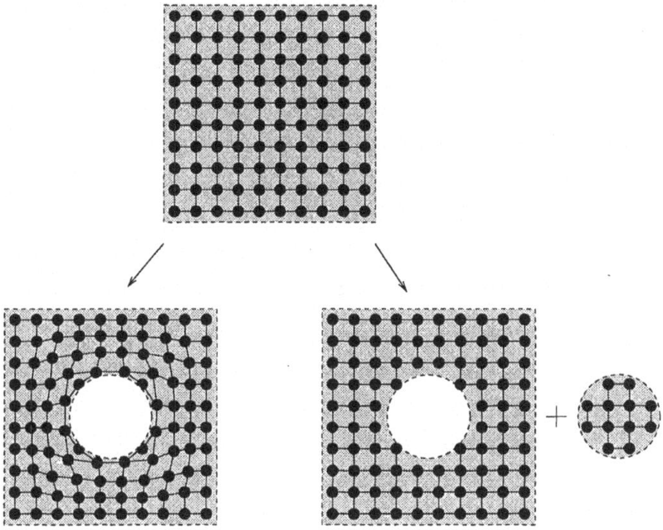

Figure 6.2. Two ways of making a hole (atomistic representation)

else and the previously scattered object consisting of the two of you becomes connected. Since the connections involved here are so readily reversed, it is natural to dismiss these changes as of little significance. Still, for some purposes mere contact can have far-reaching consequences. In an electric circuit, for example, it can make the difference between flow of current and no flow.

I shall consider in some detail various ways in which holes can be made; other forms of connectivity change can be handled similarly. One way of thinking of a hole is as 'missing matter'. So one way of making a hole is by removing matter. This involves severing the connections between the matter to be removed and the rest of the object. This is illustrated using our 'atomistic' representation in the lower right-hand illustration in Figure 6.2. Another way of making a hole is to redistribute the matter in an object, without actually removing any of it. This is shown in the lower left-hand illustration in Figure 6.2.

These two methods of hole-making can be illustrated by different ways of making a hole in a sheet of paper:

1. The crudest way is simply to stick a pencil through it. If you do this, you will obtain a rather jagged-edged hole. None of the material of the paper

is lost, but the internal connections have been severed along a set of lines radiating out from the point where the tip of the pencil first penetrated. Roughly triangular sections of the paper are bent out of the plane of the paper to form a crude crown-shaped 'fence' around the resulting hole. This is illustrated schematically in cross-section in Figure 6.3*a*. As in the left-hand example of Figure 6.2, the hole has arisen as a result of a relocation of the material of the paper; in detail, though, this case looks quite different from that illustration, because the relocated material is moved out of the original plane of the paper.

2. A neater way to make a hole is to cut it out using a knife or a pair of scissors (Figure 6.3*b*). This illustrates hole-making by removal of material. In fact what happens is that once the first cut has been made, the paper already has a hole, but only by virtue of the severing of internal connections; no matter has been removed at this stage. As the cutting out proceeds, the hole (which takes the form of a slit, being apparently one-dimensional) extends round the perimeter of the material to be excised. At the moment that it forms a closed curve, all the connections between the material inside the curve and the rest of the sheet have been severed, and the former material drops away, leaving a 'missing matter' type hole.

3. Another method, also involving the removal of matter, is to use a hole-punch (Figure 6.3*c*). Here the connections between the material to be removed and the rest of the sheet are all severed virtually simultaneously. There is no intermediate stage at which there is only a 'severed-connections' hole.

4. A fourth method, not illustrated here, is to create a hole in the paper by converting some of the paper to some other substance which can then be removed. One way of doing this is to *burn* a hole in it, taking care that the fire is extinguished before it reaches the edge of the sheet; the burnt part then drops away as ash or diffuses as smoke. This hole is formed by chemical rather than mechanical means. A similar method would be to use a strong acid to corrode the paper in the centre of the sheet.

5. Yet another method of using a sheet of paper to make a hole, this time involving the *making* rather than breaking of connections, is to bend the paper round and glue together a pair of opposite edges to form a tube (Figure 6.3*d*). This is a very different kind of hole from the others; it is not a hole *in* the paper in the same sense that they are. Here we are using a single, holeless object to create a holed object by connecting together

two of its parts. This is a common method of hole-formation. It is the principle of padlocks, belts, watch-straps—all of which when viewed at a suitable level of granularity have a holeless and a holed state.

6. A related method is to cut the paper in two, and then glue the two parts together in such a way that a hole is formed. These gluing methods all create holes by adhesion-based connectivity rather than cohesion-based. This illustrates the general idea of making holes by connecting together two or more appropriately configured objects. With two objects you need two separate connecting places in order to obtain one hole.

I shall return to these examples later in the discussion of continuity (see p. 333). For the moment, the main point to be gleaned from them is that a hole may be created either by destroying existing connections or by creating new ones. This point will be taken up further below.

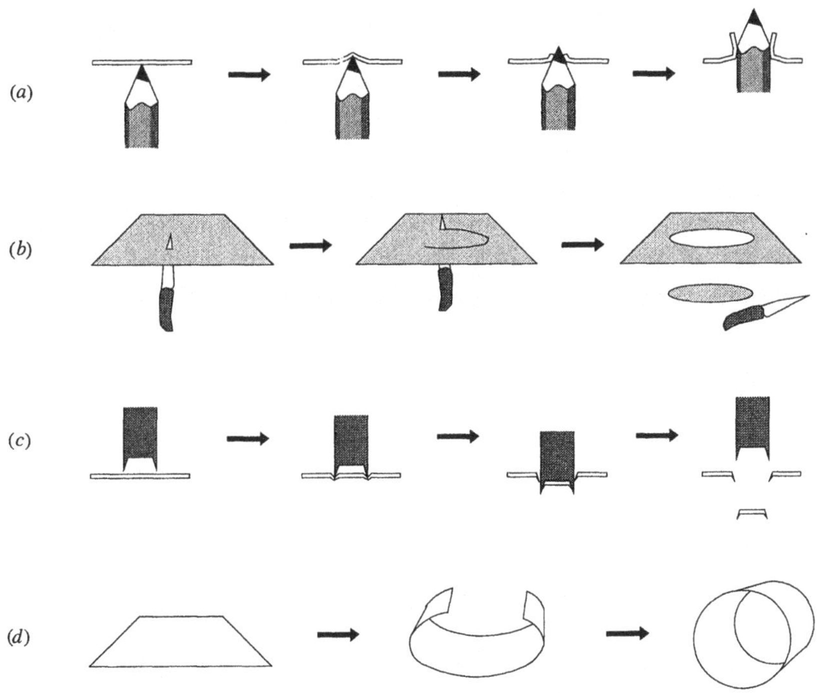

Figure 6.3. Four ways of making a hole in a sheet of paper

The holes I have been considering have all been tunnels. They can be made in three-dimensional objects like blocks of wood as well as in (apparently) two-dimensional objects such as our sheet of paper: any of the methods enumerated above can be used, although we perhaps cannot distinguish clearly between the analogues of methods 2 and 3. In both these cases the connections to be severed lie along a roughly cylindrical surface; they cannot all be severed simultaneously, so the strictest conceivable analogue of method 3 cannot be realized in practice; but the connections can be severed in two different sequences, either by cutting out complete circles at successive positions along the length of the tunnel, or by cutting out complete tunnel-lengths at successive positions around the perimeter. The first method corresponds to using a drill, the second to using a saw. If the latter resembles method 2, the former resembles method 3.

When we turn from making tunnels to making internal cavities, many of the options are closed to us. The connections to be severed are all buried in the interior of the object; one cannot get access to them from outside without first severing some other connections that one actually wants to keep. Hence these have to be restored after the internal work is done. Thus almost all ways of making an internal cavity involve first making a pit and then closing off its entrance in some way. Cavitied objects are extremely useful as containers which function as such at any orientation; the problem is that so long as the cavity remains intact, one has no access to what it contains. Thus we find innumerable devices containing pits which can be converted into cavities and back again—or, as we would say, containers with lids or stoppers.

Although the process of making or destroying a hole may seem quite gradual, there is a sense in which, in idealization at least, it has to be instantaneous. How long does it take to drill a hole through a block of wood? Well, say it takes a minute from when the drill first bites into the top surface to when it first breaks through at the bottom surface. Before that minute, there was no hole; after the minute, there is a hole. But—and this is the crucial point—during the minute there is no hole (in the sense of a tunnel), only a pit. So the creation of the tunnel happens at the end of the minute, at the instant when the tip of the drill-bit first emerges from the wood. The rest of it, the drilling that took place during the minute, was all necessary preparation for the existence of the tunnel, but strictly speaking is not part of the actual coming into being of the tunnel. This instantaneous change in connectivity is referred to as a 'magic moment' by Casati and Varzi (1994: 142)

Assuming that at any moment an object has a well-defined and unambiguous connectivity, and that different connectivities form a discrete space of variation, then changes in connectivity must always be instantaneous. If it should ever appear to the contrary, this must be for one of two reasons. Either the object in question is not sufficiently sharply defined for us to be always able to assign to it a definite connectivity; or our knowledge of the object is insufficiently clear for us to tell what its connectivity is. Thus the instant of change in connectivity can be 'smeared' out into a longer period, for example the hour which Jacquetta Hawkes fancifully assigned to the final severance of Britain from mainland Europe (see p. 128).

Granted that the first appearance of the hole is instantaneous, one can ask a further question, which at least on some views of the structure of time makes good sense: when a hole forms, is there a first moment when the hole exists, or a last moment when it does not? We get different answers depending on our mathematical model. Suppose that at time $t \in \langle 0, 1 \rangle$ the hole is a circle of radius t, remaining with radius 1 for $t \geq 1$. Assume further that there is no hole for $t < 0$. On one possible model, the hole already exists as a point ('a circle of radius zero') at time 0, which therefore, on this model, is the first instant of the hole's existence. If, on the other hand, we do not want to recognize the existence of a hole of size zero, then time 0 is the last moment of there not being a hole; the hole is there at *any* time later than this, and hence has no first moment of existence. It might be hard to see what difference there could be between a hole of size zero and no hole at all; but consider the case of a cylinder of metal with a very thin tunnel through which a fine wire can be threaded. For some purposes an adequate idealization of this might be to treat the tunnel as a linear feature having, like the wire, effectively zero width. This might be the case if we wanted to rule out the possibility of a narrower tunnel. Now if a tunnel can have, under idealization, zero width, then perhaps a hole can have, under a similar idealization, zero diameter. Perhaps; but in what follows I shall assume not. Thus when a cavity is formed in the interior of an object by the breaking of connections, I shall say that there is a last moment when there is no hole, but no first moment when there is a hole.

This follows from the general principle that when a connection is made or broken, the connection exists at the magic moment when the qualitative change occurs. Thus when a connection is made, there is a first moment of its existence but no last moment of its non-existence, and when a connection is broken there is a last moment of its existence but no first moment of its non-existence. This has the following significant consequence: that any number of connections can be made or broken simultaneously, and at the instant that this happens, *all* the connections, both those that are being made and those that are being destroyed, exist, the former being then at the first moment of their existence, the latter at the last moment of theirs.

This is not the end of the story, however. I have been concentrating on the case of holes made by perforation, which involves the severing of connections. But as seen in examples 5 and 6 above, holes can also be made by *creating* connections; Figure 6.3*d* illustrates a case of this. For this kind of hole, which comes into being as soon as the connection is first made, there is a first moment of existence of the hole, but no last moment.

And there are yet further complications. I referred above to the principle that when a connection is made or broken, the connection exists at the moment of qualitative change. In fact this only applies if the connection-making or breaking process is of a particular type: a connection is made or broken by closing or opening up a gap. But there is another way of making a connection: by growing an isthmus from the centre-line outwards. These two methods are illustrated by the two paths in Figure 6.4. In both cases the starting-point is a region consisting of two disconnected rectangles, and the end-point is a connected H-shaped region. The upper path shows an evolution in which the crossbar of the 'H' is formed from two appendages growing out one from each rectangle and finally meeting. The instant at which they meet is the first moment of connection; there is no last moment of disconnection. The lower path shows a different evolution in which the crossbar first appears as a thin isthmus joining the two rectangles, which gradually thickens. There is a last moment of disconnection, call it t_0; for any $\delta t > 0$, at time $t_0 + \delta t$ the crossbar is in existence and its width is, at least until the final stage, proportional to δt. Thus there is no first moment of connection.

It might be objected that the evolution shown in the lower path is not truly continuous; if continuity is jettisoned, almost anything can happen, and we are no longer talking about realistic situations. Whether or not such cases are to count as continuous is an important issue which will be discussed in much greater detail in the next chapter. For now, note that in Figure 6.5 are shown two solid blocks. Imagine each of these to be placed in water, and the water-level *continuously* lowered. If we consider the changes undergone by the islands formed by the parts of the blocks protruding above the water surface, we shall see that these exactly correspond to the two paths in Figure 6.4: the left-hand block to the upper path, the right-hand block to the lower. The two methods of creating a connection are dual to one another in the sense that when a new connection is created in a region by one of the methods, a

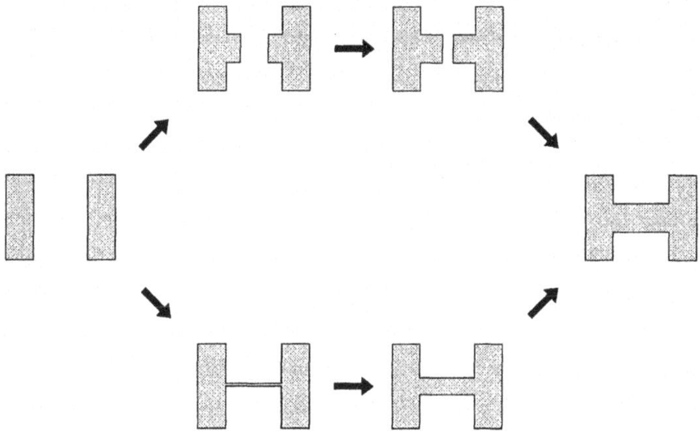

Figure 6.4. Two ways for a disconnected figure to become connected.

connection in the complement of the region is thereby broken by the other method. This shows up in Figure 6.5 as the fact that in each 'island' the part corresponding to the crosspiece of the 'H' resembles the complement of the corresponding part in the other island, but upside down.

In the light of this, we see that there are in fact three distinct types of hole creation. We can create a hole by having a perforation grow from nothing, or we can create a hole by forming a new connection between two parts of a connected object; the latter method occurs in two varieties: we form the new connection either by narrowing the gap between the two parts to be connected, or by widening an isthmus between them. With the first and third methods, there is a last moment of disconnection but no first moment of connection, whereas with the second method there is a first moment of con-

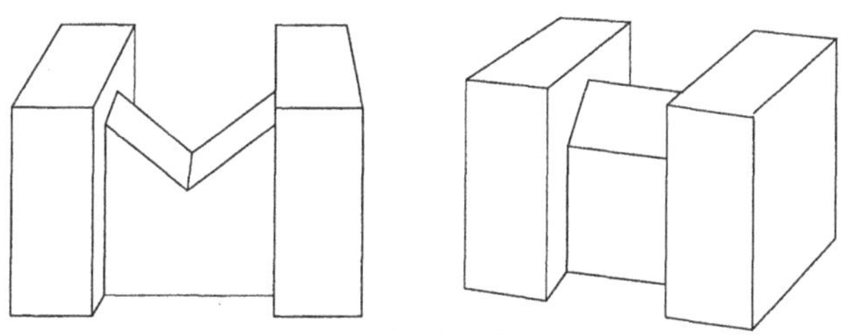

Figure 6.5. Two 'islands', each of which is disconnected when the sea-level is high but becomes connected when the sea-level lowers

nection but no last moment of disconnection. Corresponding statements hold for the time-reversals of these processes, the three kinds of disconnection.

The existence of these various different ways of altering the connectivity of a figure has repercussions for the structure of abstract connectivity space. Recall from §4.2 the notation $(n_1, n_2, \ldots, n_m)$ to denote the planar connectivity type for an object with m components of which the ith component has n_i holes. Can these connectivity types be treated as elements of a mode space structured by the conceptual neighbourhood relation? It seems reasonable to suggest that elements in this space that are linked by a single event of connection or disconnection, or of hole-making or hole-filling, should count as conceptual neighbours. This gives us an initial structuring of the space as portrayed, in part, in Figure 6.6, in which the horizontal or approximately horizontal links correspond to hole creation (moving from left to right), while the vertical or approximately vertical links correspond to splitting. For illustrations of some of these connectivity types, see Figure 4.5.

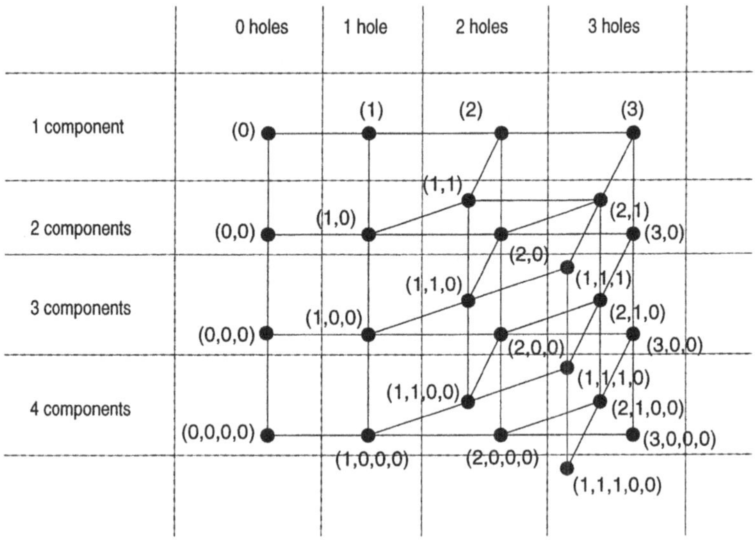

Figure 6.6. Part of connectivity space

The links shown in this figure do not exhaust the possible pathways in connectivity space. Suppose, for example, that starting with a one-piece, unholed region of type (0), a hole is formed simultaneously with the splitting of the region into two components, thus leading us to a configuration of type (1,0). It is natural to suppose that this possibility should be represented by a direct link between (0) and (1,0) in the conceptual neighbourhood diagram. We must be very careful, however. Exactly what path is followed in this envisaged transformation depends on the kinds of disconnection and hole formation we are dealing with. Two possible combinations are shown in Figure 6.7.

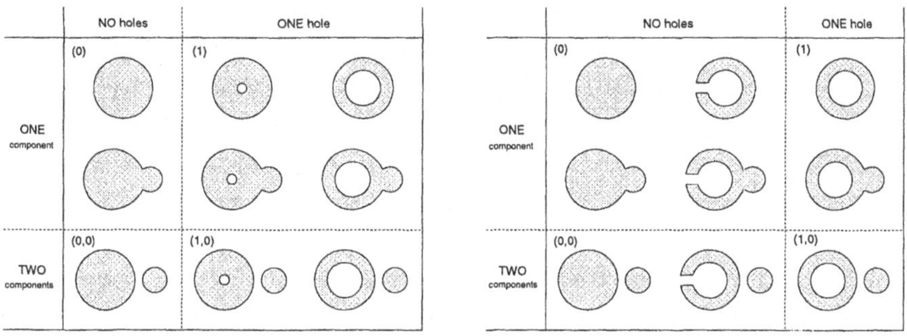

Figure 6.7. A closer look: two maps of a small part of connectivity space

In the left-hand illustration, the hole is produced by perforation (growing from a 'pin-prick'), while the disconnection is accomplished by the reversal of the 'narrowing gap' method of connection shown in the upper path of Figure 6.4. For the hole, there is a last moment of holelessness but no first moment of having a hole; for the disconnection, there is a last moment of being connected but no first moment of being disconnected. If these two moments coincide, then at that moment, there is no hole, and the region is still connected, so the connectivity type is (0). Immediately after this moment, the region has two components, in one of which there is a hole, so the connectivity type is (1,0). Thus there is a direct transition between connectivity types (0) and (1,0), which could be represented by adding a link between these types in Figure 6.6.

In the right-hand illustration of Figure 6.7, the disconnection is achieved in the same way as in the left-hand one, but the hole is produced by making a connection, resulting in the enclosure of an indentation, rather than by perforation. Moreover, the connection by which the hole is formed is made by the 'narrowing gap' method. There is thus now a *first* moment at which the hole exists, and no last moment of its non-existence. If this moment coincides with the last moment at which the region as a whole is connected, then at that moment we have connectivity type (1), and there is thus no direct link, using this method of hole formation and disconnection, between (0) and (1,0). Instead, there is a link across the other diagonal of the square, between (1) and (0,0).

We thus see that the detailed linkages in the conceptual neighbourhood diagram for planar connectivity space depend sensitively on the models adopted for the making and breaking of connections and the creation and destruction of holes. I shall not pursue this further in the present section; issues of this kind can be more satisfactorily handled in the context of the general framework presented in Chapter 8 (see in particular Figure 8.14 and the associated text).

As a final note, consider how changes in connectivity, represented as movements through connectivity space, can be expressed in the formalism of Chapter 5. The conceptual neighbourhood diagram for connectivity space will show a link between any two connectivity types between which a direct transition is possible, that is, between any types CT_1 and CT_2 such that the following pattern of temporal incidence can occur:

$$\begin{aligned}&Holds_{\mathbf{T}}(CT_1, t_1) \wedge Holds_{\mathbf{T}}(CT_2, t_2) \wedge \\ &\qquad \forall t \sqsubseteq \langle t_1, t_2 \rangle (Holds_{\mathbf{T}}(CT_1, t) \vee Holds_{\mathbf{T}}(CT_2, t)).\end{aligned}$$

This enables conceptual neighbourhood diagrams to be translated into axioms of a formal theory.

6.3. Change in location

Change of position in general is motion. Just as position can be resolved into location and orientation, so motion can be divided into change of location, which is displacement or translation, and change of orientation, which is rotation, or turning. I shall handle rotation separately in the next section. Here I consider motions with a displacement component, simply ignoring the question whether there is rotation as well.

Motion of a compact object is along a path. If the end-points of the path are specified (the beginning and end of the motion), then we have a 'motion event' or movement (Galton 1995). Otherwise it is a 'motion process', which is progressive if the path is traversed once but cyclic if it is traversed repeatedly, either in the same direction (for a closed path or orbit—this is circulation or revolution) or alternately in opposite directions (oscillation). These are illustrated in Figure 6.8.

Motion is the prototype of all spatial change (indeed of all change, change in respect of property p being metaphorically represented as motion in 'p-space'). Many of the changes I have discussed can be described at the lower level as *motion of parts*—this is quite explicit in Figure 6.2. What we might dub the 'atomic theory of change' is the thesis that all change can ultimately be explained in terms of the movement of primitive components which individually can partake of no kinds of change apart from motion.

From a formal point of view it is much easier to handle movements, which are specified in terms of their beginning and end positions, than motions in general, whose specification includes the path along which the moving object travels. Using the pos functor, and recalling that $pos(a) = p$ is abbreviated to $at(a,p)$, we can represent the movement of object a from position p to position q by the event-type

$$Trans_{\mathbf{I}}(at(a,p), at(a,q)).$$

This is only useful, of course, if we have a means of specifying the positions p and q exactly. In qualitative descriptions this will not in general be possible; instead, we must describe the position of the moving object in terms of its qualitative relationship to other objects which are treated as features or fixtures.

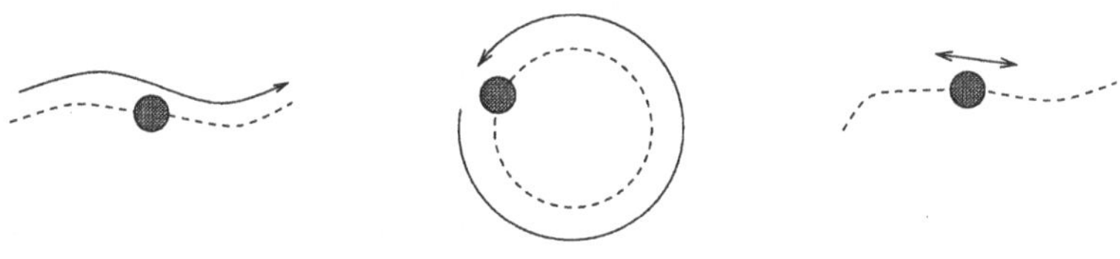

Figure 6.8. Three varieties of displacement: motion along a path, revolution around an orbit, and oscillation back and forth along a path

6.3.1. Change in RCC relations

Using the RCC system, we can readily describe several kinds of movement, giving precise occurrence conditions in each case. I shall illustrate this with a number of examples, adapted from Galton 2000*b*. The moving object is a in each case; its movement may be specified with respect to another object b

or to one or more regions $r, r_1, \ldots$. I shall make use of the various kinds of transitions defined in §5.4.3 (pp. 260 ff.).

Entering a region

I shall describe the event 'a enters region r'. The object a starts off outside r and ends up inside r. Three different ways in which this can happen are illustrated in Figure 6.9. In each case the entering event occurs over the interval $\langle t_1, t_2 \rangle$. It is clearly a transition of some kind, and our task is to determine precisely what transition it is. To this end, I list below all the transitions of which each of the events illustrated is an occurrence.

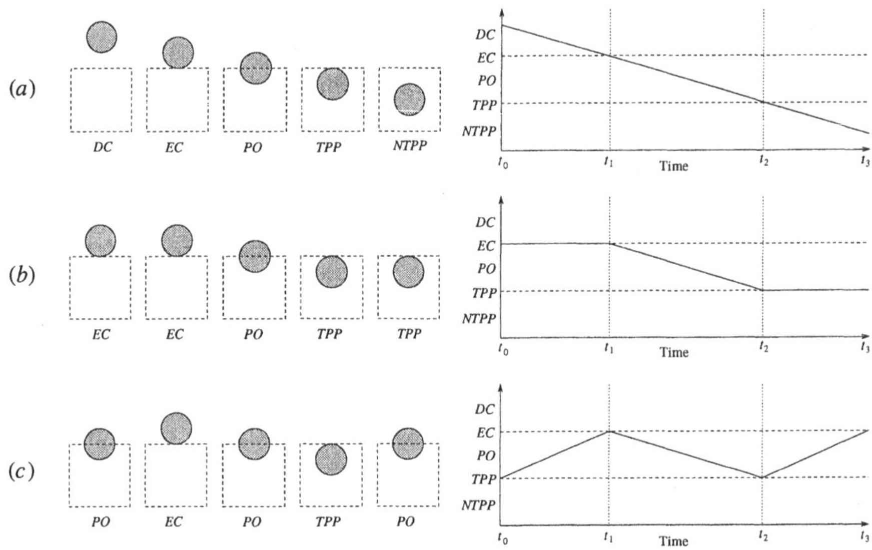

Figure 6.9. Possible sequences of RCC relations when an object (circle) enters a region (square)

In (a), object a moves towards r from a distance, then crosses the boundary of r and moves away from the boundary into the interior of r. One way of describing this is to say that an interval on which $DC(a,r)$ holds meets an interval on which $PO(a,r)$ holds, which meets an interval on which $NTPP(a,r)$ holds. Thus the event is an occurrence of the type

$$Trans_{\mathbf{I}}^{iii}(DC(a,r), NTPP(a,r)).$$

This is not the only way of describing it, however, since we also know from the conceptual neighbourhood diagram (Figure 2.9) that $EC(a,b)$ holds at the instant t_1 which marks the beginning of the interval over which the entering occurs, and $TPP(a,b)$ holds at the instant t_2 which marks its end. Thus the event in this case is also an occurrence of the three types

$$\begin{array}{l} Trans_{\mathbf{I}}^{tit}(EC(a,r), TPP(a,r)), \\ Trans_{\mathbf{I}}^{tii}(EC(a,r), NTPP(a,r)), \\ Trans_{\mathbf{I}}^{iit}(DC(a,r), TPP(a,r)). \end{array}$$

In (*b*), the object rests a while just outside r, touching its boundary, then moves inside r, where it rests a while just inside, again touching the boundary. From considerations of continuity, it is evident that as in the previous case we have $EC(a,b)$ holding at t_1 and $TPP(a,b)$ holding at t_2. Thus the transition is an occurrence of each of the four types

$$\begin{array}{l} Trans_{\mathbf{I}}^{iii}(EC(a,b), TPP(a,b)), \\ Trans_{\mathbf{I}}^{iit}(EC(a,b), TPP(a,b)), \\ Trans_{\mathbf{I}}^{tii}(EC(a,b), TPP(a,b)), \\ Trans_{\mathbf{I}}^{tit}(EC(a,b), TPP(a,b)). \end{array}$$

In (*c*), the object starts off inside the region, then comes outside it, only remaining outside (in the EC position) for an instant before entering it again. Having entered the region it only remains there (in the TPP position) for an instant before once more leaving r. What happens between the instantaneous holding of EC and the instantaneous holding of TPP is quite naturally described as a's entering r; indeed, there seemed nothing unnatural about my so describing it in the previous two sentences. This may not be a paradigmatic case of entering, but it surely *is* a case of entering none the less. It differs from the preceding cases in that there is only one type of transition we can regard it as an occurrence of—namely $Trans_{\mathbf{I}}^{tit}(EC(a,b), TPP(a,b))$.

Reviewing the examples, we see that what they all have in common is that they are occurrences of the type $Trans_{\mathbf{I}}^{tit}(EC(a,b), TPP(a,b))$, and it seems reasonable therefore to take this as our definition of the entering event:

$$Enter(a,r) \quad \triangleq \quad Trans_{\mathbf{I}}^{tit}(EC(a,b), TPP(a,b)).$$

We do not in fact have to specify the type so narrowly. If we were to put

$$Enter(a,r) \quad \triangleq \quad Trans_{\mathbf{I}}(EC(a,b), TPP(a,b))$$

instead, that would do fine, since we can infer from continuity that any occurrence of the latter type must be an occurrence of the former, whatever other types it may also be an occurrence of. However, the advantage of using the more narrowly defined type is that we have an explicit occurrence condition for events of this type, which in this instance, since EC and TPP are mutually exclusive, reduces to

$$\begin{aligned} Occurs_{\mathbf{I}}(Enter(a,r),i) \leftrightarrow\ & Holds_{\mathbf{T}}(EC(a,r),\mathrm{beg}(i)) \wedge \\ & Holds_{\mathbf{T}}(TPP(a,r),\mathrm{end}(i)) \wedge \\ & Holds_{\mathbf{I}}(-EC(a,r).-TPP(a,r),i). \end{aligned}$$

This occurrence condition implies that throughout the time of any occurrence of $Enter(a,r)$, the position of a must be neither EC nor TPP to r. Assuming that the motion of a is continuous, we can read off from the conceptual neighbourhood diagram for the RCC relations (Figure 2.9) that, throughout that period, the position of a must be PO to r, hence

$$Holds_{\mathbf{T}}(prog^{-}(Enter(a,r)),t) \rightarrow Holds_{\mathbf{T}}(PO(a,r),t).$$

This is a welcome conclusion, for it simply says that when an object is in the process of entering a region, part of it must already be inside the region, and part of it must still be outside.

Coming into contact

For two objects to come into contact is for there to be a transition from their being separated in space to their being in contact. There are essentially three ways in which this can happen, as illustrated in Figure 6.10.

In (*a*), the two objects approach one another and when they have made contact they remain in contact over an interval:

$$i \| j \wedge Holds_{\mathbf{I}}(DC(a,b),i) \wedge Holds_{\mathbf{I}}(EC(a,b),j).$$

This is a clear case of $Trans^{ii}_{\mathbf{T}}(DC(a,b),EC(a,b))$, occurring at the instant at which i meets j. Notice, though, that if the motion is continuous, then the final state $EC(a,b)$ must already hold at the instant of the transition, which means that more specifically the transition is an instance of $Trans^{it}_{\mathbf{T}}(DC(a,b),EC(a,b))$.

In (*b*), the objects are only in contact for an instant, moving away again as soon as they have met. We can describe this as follows:

$$\begin{aligned} i \|_t j \wedge Holds_{\mathbf{I}}(DC(a,b),i) \wedge Holds_{\mathbf{T}}(EC(a,b),t) \wedge \\ Holds_{\mathbf{I}}(DC(a,b),j). \end{aligned}$$

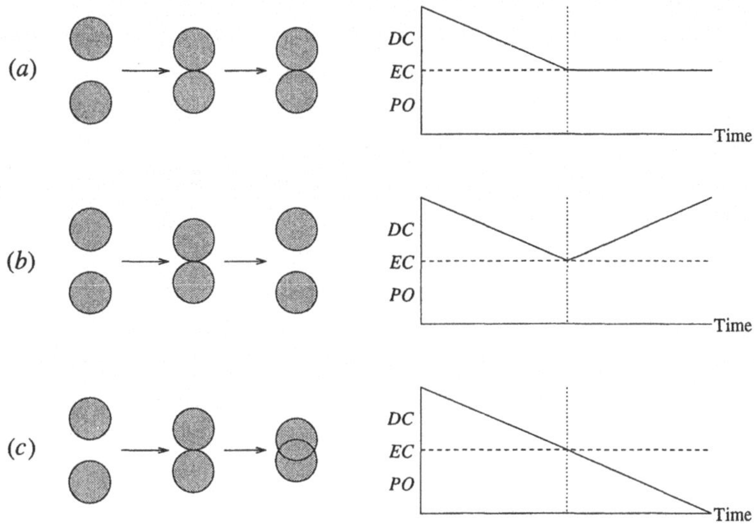

Figure 6.10. Possible sequences of RCC relations when two objects come into contact

This means that we have an occurrence of $Trans^{it}_{\mathbf{T}}(DC(a,b), EC(a,b))$ at t; symmetrically we also have an occurrence of $Trans^{ti}_{\mathbf{T}}(EC(a,b), DC(a,b))$ at the same instant: the bodies make and break contact simultaneously. It is debatable whether such an event could happen in exactly this way in real life, but it provides a useful idealization for describing events such as a pair of billiard balls bouncing off one another, where the moment of contact is effectively an instant at the granularity of everyday observation.

In (*c*) we have interpenetrable objects which begin to overlap as soon as they have made contact. The making contact in this case may be described as either $Trans^{it}_{\mathbf{T}}(DC(a,b), EC(a,b))$ or $Trans^{ii}_{\mathbf{T}}(DC(a,b), PO(a,b))$.

The upshot of these observations is that every case of making contact can be described using the transition type $Trans^{it}_{\mathbf{T}}$, enabling us to define:

$$connect(a,b) \quad \triangleq \quad Trans^{it}_{\mathbf{T}}(DC(a,b), EC(a,b)).$$

Note the implied asymmetry between DC and EC here; one might wonder why we could not instead use $Trans^{ti}_{\mathbf{T}}(DC(a,b), EC(a,b))$, but a little reflection will show that this would violate continuity: if two bodies are separated at t, then an interval must elapse during which the distance of separation smoothly diminishes to zero before they can make contact.

Moving from one region to another

My final example is 'a moves from r_1 to r_2'. This is to be distinguished from the case of a moving from one position to another, which I have already represented as $Trans_{\mathbf{I}}^{tit}(at(a,p), at(a,q))$. In that case p and q were possible positions for a, that is, regions of space geometrically congruent to a. Here I attempt to approach more closely the kind of locutions we use in everyday life, as in 'Ann went from the kitchen to the bathroom', or 'Bob moved from the desk to the wall'. Here Ann is congruent to neither the kitchen nor the bathroom, and Bob is congruent to neither the desk nor the wall.

These two cases are quite different, in that in Ann's case the movement is described as starting inside the first region and ending inside the second, whereas in Bob's case the starting-point is only adjacent to the first region and the end-point is adjacent to the second (taking adjacency as a first approximation to the relation Bob bears to the desk and the wall).

It is unnecessary to labour the details of these cases. Considerations similar to those of the preceding cases lead one to define two kinds of $Move$ event as follows:

$$Move_1(a, r_1, r_2) \triangleq Trans_{\mathbf{I}}^{tit}(TPP(a, r_1), TPP(a, r_2))$$
$$Move_2(a, r_1, r_2) \triangleq Trans_{\mathbf{I}}^{tit}(EC(a, r_1), EC(a, r_2))$$

If need be, one could define two other cases to represent movement from TPP to EC and vice versa.

6.3.2. Change in prepositional relations

The meanings of some prepositions are strongly associated with motion. Examples are 'into', 'on to', 'out of', 'off', 'through', 'along', 'across', 'around', and 'past'. They are typically used for describing the path of a moving body in relation to a fixed body or region.[1] Some of them can, it is true, be used for describing static relationships as well, but arguably these uses are derivative, the primary use being connected with motion. Thus here I take the view that the meaning of a sentence like 'The railway goes into a tunnel', in which no motion is involved, is parasitic on the meaning of sentences like 'The train goes into a tunnel', in which motion plays an essential role.

[1] Lakoff (1987: case study 2), drawing on work of Claudia Brugman, regards the basic meaning of 'over' as combining elements of both 'above' and 'across', and therefore involving a path, as exemplified by 'The plane flew over'. Bennett (1975), however, regards the basic meaning of 'over' to be that exemplified by 'My hand is over the table'.

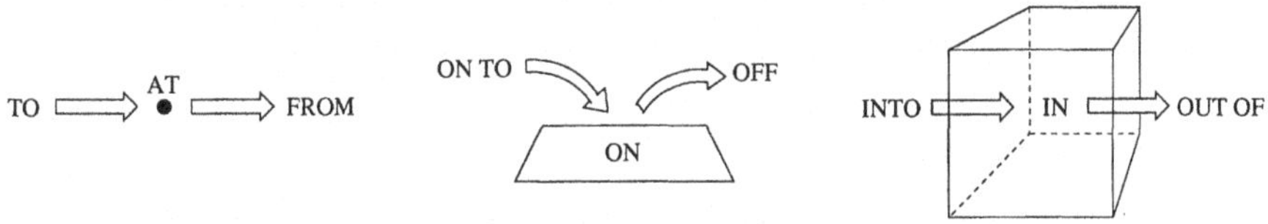

Figure 6.11. Prepositions describing change of topological relationship

The meanings of these motion prepositions can be at least approximately characterized in terms of transitions. Thus 'into the tunnel' represents a transition from a position which is 'not in the tunnel' to a position which is 'in the tunnel'. We can symbolize this as

$$\text{into} \quad = \quad \text{not in} \rightarrow \text{in}.$$

Similarly we have

$$\begin{aligned} \text{on to} \quad &= \quad \text{not on} \rightarrow \text{on} \\ \text{to} \quad &= \quad \text{not at} \rightarrow \text{at} \\ \text{out of} \quad &= \quad \text{in} \rightarrow \text{not in} \\ \text{off} \quad &= \quad \text{on} \rightarrow \text{not on} \\ \text{from} \quad &= \quad \text{at} \rightarrow \text{not at} \end{aligned}$$

These relationships are illustrated in Figure 6.11.

In some languages these relations are expressed in the morphology by means of affixes. Hungarian, for example, has a neat system of nine suffixes, which might be thought of as case endings (although the agglutinative structure of the language might equally allow one to think of them as postpositions), corresponding fairly closely to the nine English prepositions illustrated in Figure 6.11. They are tabulated in Figure 6.12. The variant forms for each suffix are determined by the Hungarian rules of vowel harmony.

The other motion prepositions, 'through', 'along', 'across', 'around', and 'past', all involve the path of the motion, not just its end-points. Although it is not easy to give fully general definitions of these terms, the illustrations in Figure 6.13 give an idea of the basic meanings, organized according to the dimension (actual or apparent) of the reference object. A complication, pointed out by Bennett (1975: 18), is that some locative prepositions, when used in the context of movement, are ambiguous between a 'goal' interpretation and

-ba /-be into	*-ban /-ben* in	*-ból /-ből* out of
-ra /-re on to	*-n /-on /-en /-ön* on	*-ról /-ről* off, from
-hoz /-hez /-höz to, towards	*-nál /-nél* at, by	*-tól /-től* (away) from

Figure 6.12. Locative suffixes in Hungarian

a 'path' interpretation. Thus 'The boy ran behind the tree' can mean that the boy ran to a position behind the tree and stayed there (goal), or that he ran from a position not behind the tree, via a position behind the tree, to another position not behind the tree (path).

The meanings of these motion prepositions are quite closely tied to the meanings of the topological prepositions. For example, when one is in the process of going *through* something, one is *in* it, whereas when one is going *across* something, one is *on* it. Thus, we can be in the garden, and on the lawn, so we walk through the garden, but across the lawn. A mole, on the other hand, can be in the lawn, and hence also can tunnel through the lawn. For humans, the lawn is a two-dimensional surface, but for the mole it is a three-dimensional volume.

The motion prepositions also have secondary uses as locative prepositions, as when we say 'The office is through that entrance there', 'The church is across the river', 'The post office is around the corner', and 'The library is along the high street'. In these cases, something is located at a place which is presented as the end-point of some motion. Similar examples could be given with 'over' and 'past'.

6.3.3. Movement of different kinds of objects

So far I have discussed movement in a general way without making any specific assumptions about the nature of the moving object. However, some interesting differences emerge when this element is introduced. I shall consider in turn several of the different object-types introduced in Chapter 3.

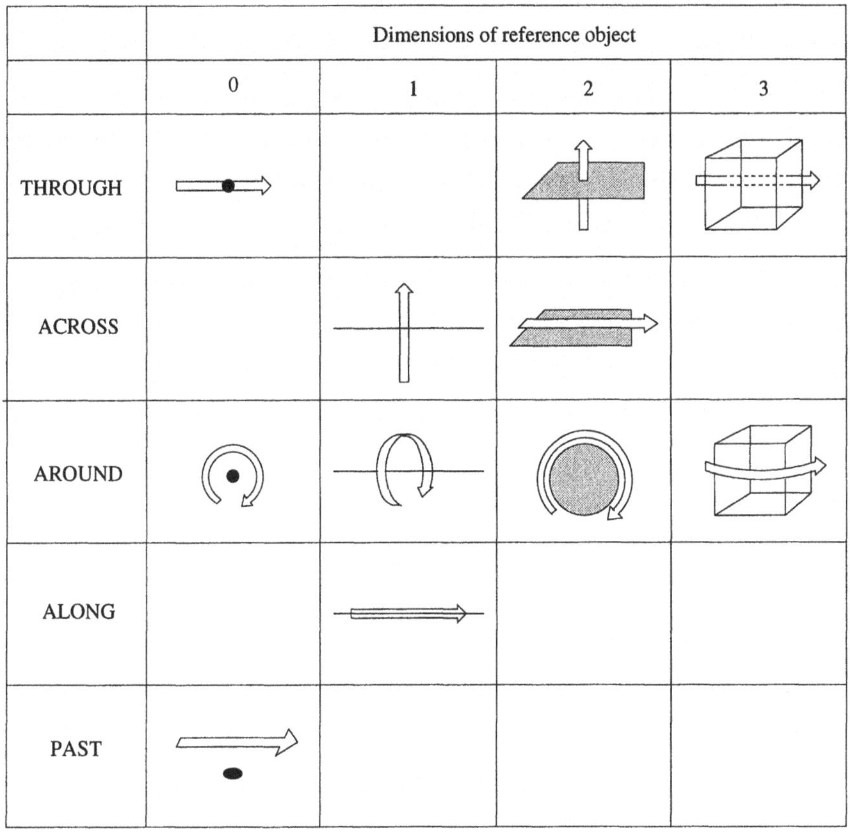

Figure 6.13. Prepositions describing paths

Movement of material objects

The simplest way for a material object to move is for all of its constituent matter to move. If each particle of matter in the object undergoes a certain displacement, then so does the object as a whole, as in Figure 6.14a. It is worth bearing in mind that on a fine scale of granularity the component atoms of a material object are in constant thermal motion all the time, even when the object itself is stationary; thus when the body is moving, the motion of the atoms has to be resolved into two components, the random thermal motion being superposed on the systematic displacement that corresponds to the motion of the body.

As we have seen, the identity of material objects may be more or less tightly coupled with the identity of their constituent matter. In many cases matter can come and go without disrupting the identity of the object itself. Hence another way in which a material object can change its position is by means of a systematic difference between the position of matter that is leaving the object and the position of matter that is entering it, as in Figure 6.14*b*. In general, this will be a much slower process than the bulk motion discussed in the previous paragraph. If trees are removed from the south end of a forest and new trees are planted at the north end, then the forest as a whole can be said to move northward, even though no individual tree moves. Similar phenomena can sometimes be observed with crowds, queues, and other temporary aggregations.

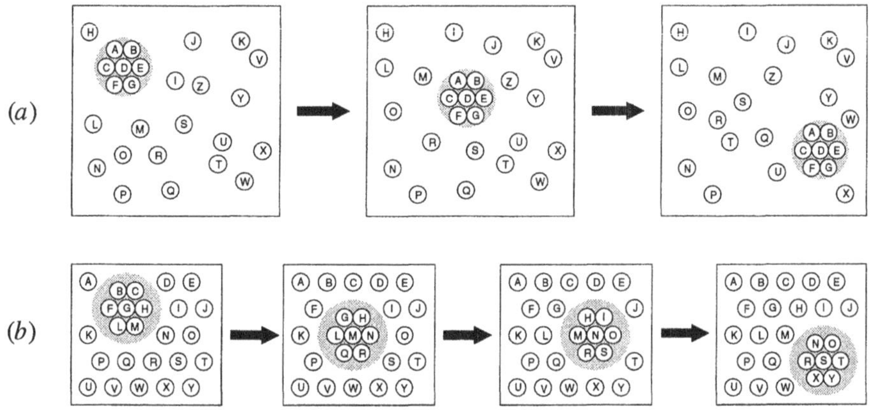

Figure 6.14. Two ways for a material object to move: (*a*) by movement of its constituent matter; (*b*) by replacement of matter

Viewed at a granularity at which the constituent matter of the object is not identifiable, these two kinds of motion are indistinguishable. The distinction between them can be formalized with the help of the two-level ontology discussed in §3.2. I shall use the apparatus of temporal reference introduced in Chapter 5, treating $Comp(x, q)$ and $pos(x)$ as fluents. The case of Figure 6.14*a*, where the object moves solely as a result of its constituent matter moving, can be described using a formula such as

$$Holds_{\mathbf{I}}(Comp(x,q),i) \wedge Holds_{\mathbf{T}}(pos(q)=r, \mathrm{beg}(i)) \wedge \\ Holds_{\mathbf{T}}(pos(q)=r', \mathrm{end}(i)),$$

which says that throughout a certain interval the object x is composed of a fixed quantity of matter q, whose position at the start of the interval is different from its position at the end. An extreme version of the contrasting case shown in Figure 6.14*b* is described by the formula

$$Holds_{\mathbf{T}}(Comp(x,q),\mathrm{beg}(i)) \wedge Holds_{\mathbf{T}}(Comp(x,q'),\mathrm{end}(i)) \wedge \\ Holds_{\mathbf{I}}(pos(q)=r,i) \wedge Holds_{\mathbf{I}}(pos(q')=r',i),$$

in which the motion of an object x over an interval arises solely through its reidentification over that interval with different quantities of matter which do not themselves move. In practice we are usually dealing with a mixture of these extreme cases.

Another possibility, exemplified by lakes, rivers, and more generally the kinds of objects I have called flows (p. 109), has the object staying still while its constituent matter moves through it:

$$Holds_{\mathbf{I}}(pos(x)=r,i) \wedge Holds_{\mathbf{T}}(Comp(x,q).DC(pos(q'),r),\mathrm{beg}(i)) \wedge \\ Holds_{\mathbf{T}}(Comp(x,q').DC(pos(q),r),\mathrm{end}(i)),$$

where it is assumed that $Comp(x,q)$ implies $pos(x)=pos(q)$.

Movement of parasitic objects

The simplest way for a parasitic object to move is for its host to move. You cannot throw a ball into the air without *ipso facto* throwing its surface and its interior cavity. If this sounds odd, it is because there is no way of throwing the surface and the cavity without throwing the ball itself, so we would not normally advert to the former throwings. Of more interest is the question of whether parasites can move independently of their hosts. Regarding holes, for instance, Casati and Varzi note that

> [a]lthough the tie linking a hole to its host in causal interactions is very tight, the movement of the one does not generally imply movement of the other.
>
> (Casati and Varzi 1994: ch. 8)

Thus an object can so to speak *flow past* its own hole, and conversely a hole can move within its host. Since a hole is associated with absence of matter, motion of a hole can be inverse to that of matter. A foreman at a building site instructs one of his workers: ‘Move that hole three feet to the left’. The simplest way of obeying this instruction is to dig a new hole, using the newly excavated earth to fill in the old. A hole-sized quantity of matter has been moved three feet to the right. Has the *hole* moved? The foreman's language implies this, but my (perhaps more careful) description referred to a *new* hole. I shall not pursue this, but refer once again to Casati and Varzi's extensive discussion of the ontology of holes.

One may be a little doubtful about holes moving with respect to their host, but it is well known and uncontroversial that another kind of parasitic entity, waves, routinely display motion independent of that of their hosts. A water wave moves horizontally across the surface of the water, but the water itself moves only in vertical circles. Objects floating on the water ride up and down as the waves pass them by, but are not swept along by the waves. Sound waves likewise travel rapidly through the air, but the individual bits of air only move back and forth by a tiny amount.

I included shadows in the inventory of parasitic objects, though I noted that these obey a more complex logic. The shadow cast by X on Y in the light emanating from Z can move if any of X, Y, or Z moves. In Figure 6.15*a*, the light source has moved to the right, with the result that the shadow has moved to the left. In (*b*), the object casting the shadow has moved to the right, with the result that the shadow has also moved to the right. In (*c*), the surface on which the shadow is cast has moved back, away from the light source, with the result that the shadow has moved back with it (and become larger at the same time). Note that if the surface were moved to the left or right, this would make no difference to the absolute position of the shadow, but relative to the surface itself the shadow would appear to move right or left. In an eclipse of the sun, the relative motion of the earth, moon, and sun result in the moon's shadow sweeping across the earth's surface from west to east at a speed of about 3,000 kilometres per hour; meanwhile the earth's rotation carries a point on the earth's surface from east to west at no more than 1,600 km/h. The resulting motion of the moon's shadow relative to a point on the earth is thus from west to east. This could not be ascertained purely qualitatively: one has to know the numbers involved!

Movement of perspectival objects

Perspectival objects depend on the position and direction of gaze of an observer. They also usually depend on some host object, and sometimes on a source object as well. For a reflection, the host object is the reflective surface in which the reflection is seen, and the source is the object which it is the reflection of. When you look at yourself in a mirror you are both source and

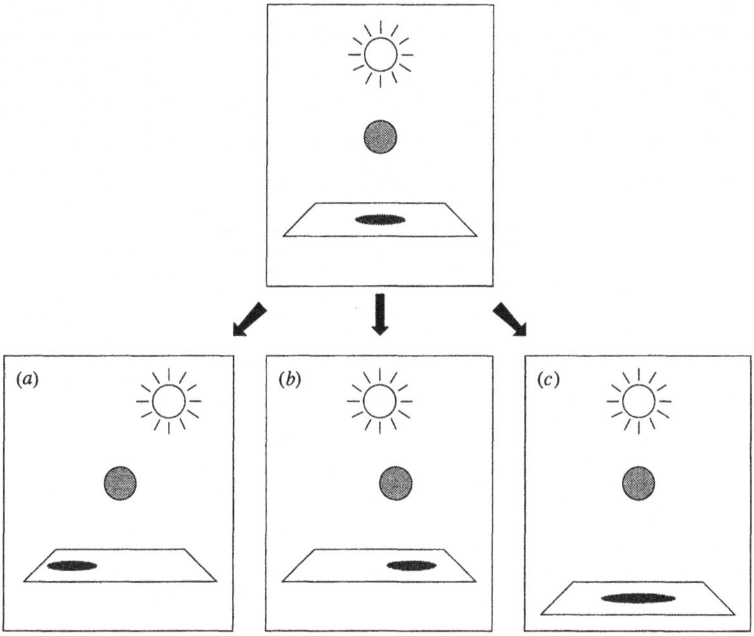

Figure 6.15. Three ways for a shadow to move: (*a*) movement of light source, (*b*) movement of object casting the shadow, (*c*) movement of host surface

observer of the reflection, the mirror being the host. When you look at the reflection of a cloud in a puddle, the cloud is the source, the puddle the host, and you the observer. Do reflections move? Only if we can assign positions to them. The position of the fictitious 'virtual image' of optical theory depends on the positions of both source and host, and will therefore move whenever either of these does. Rainbows and other atmospheric refractive phenomena only possess part of what is required to have a position: they have direction but no distance. The direction of a rainbow changes as the sun moves round.

6.4. Change in orientation

Change in orientation is turning or rotation. In two dimensions, rotation is about a point, whereas in three dimensions it is about an axis. A turn is an individual turning event. It can be specified in two different ways, either as a transition from an initial orientation to a final orientation, or in terms of the angle traversed and the sense of the rotation. The former kind of description is at a somewhat higher level of abstraction than the latter. A drill sergeant might issue the command 'Turn to face me'; given a suitable initial configuration, the effect of this command might equally have been produced by the command 'Turn ninety degrees to your right'. However, if the people the command is being given to are all facing in different directions, no one command of the latter type can have the effect of the former command for all the people involved.

Note that the terms 'clockwise' and 'anticlockwise' (or 'counterclockwise') must be used with care, since they assume a point of view as given. Viewed from the back, the hands of a clock turn anticlockwise. Similarly, we tend to think of the bodies of the solar system (with a few exceptions) as rotating and revolving anticlockwise. This is an example of 'boreal chauvinism': viewed from the south, all these motions are clockwise.

Because the space of possible orientations is completely homogeneous, there is no intrinsic way to carve it up into qualitative modes. Hence rotation of a featureless sphere (in three dimensions) or circle (in two) is qualitatively unremarkable. As soon as the rotating object is endowed with features to break the rotational symmetry, more structure emerges. This can lead to puzzles when the object is immaterial: the elliptical shadow of a rotating ellipsoid itself appears to rotate, but the circular shadow of a rotating sphere does not (Casati and Varzi 1994: figure 8.4). We saw in Figure 4.15 that the orientation space for a cube with individually marked faces has twenty-six modes, naturally defined in terms of the faces visible from a given viewpoint. The changes of orientation for such a cube can be represented as a path through this qualitative mode-space, following the lines shown in the diagram.

The axes of the cube, which I have labelled 'up/down', 'back/front', and 'left/right', are not intrinsically distinct but acquire their labels only from the fact that the cube is thought of as embedded in a space in which we have already distinguished certain directions as 'up', 'right', and so on. In our everyday environment, the earth's gravitational pull provides the 'up/down' direction, but for 'front/back' and 'left/right' we require something extra, as discussed in §4.4.2. In this connection, the language of boats and sailing provides an interesting case of specialized vocabulary for orientation change, namely 'pitch', 'roll', and 'yaw'. If we call the direction in which the boat is heading the x-axis, the horizontal normal to this the y-axis, and the vertical direction the z-axis, then pitch, roll, and yaw are rotations about the y-axis, the x-axis, and the z-axis, respectively. The actual motion of the boat in a rough sea is a complex time-varying superposition of these three components.

6.5. Change in size

Many of the changes already discussed may be accompanied by a change of size. Pure change of size, preserving geometrical similarity, is comparatively unusual. Increase in size is either growth or expansion. The difference between these two, in the case of a material object, is that whereas expansion preserves constitution, the same material being spread over a larger space with a corresponding decrease in density, growth involves the accretion of new material, thus resulting in a change of constitution while the density can (though need not) remain the same. These are illustrated in Figure 6.16.

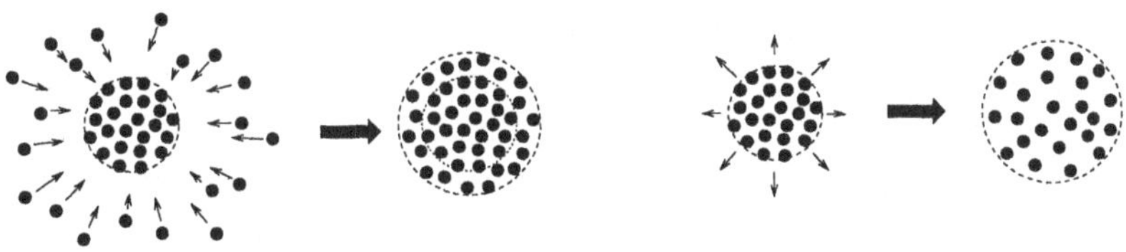

(*a*) Growth (*b*) Expansion

Figure 6.16. Change in size

Whereas expansion necessarily involves movement of parts, growth need not. Thus a forest, whose parts (the trees) are immovable, can increase in size by growth—new trees springing up in previously unwooded area around the forest—but not by expansion (which would require individual trees to move outwards from the centre). None the less we sometimes speak of a forest expanding—so the definitions of these terms given here do not always coincide with ordinary usage. Similarly, when we speak of the expansion of a town, we mean its growth. Note, however, that the contrast between growth and expansion really only applies to material objects; a town *qua* administrative unit—which is a conventional object—can increase in size by having its boundary redrawn, but since there is no material change this cannot be classified as either growth or expansion.

Decrease in size can likewise be divided into two types. The reverse of expansion is contraction or shrinkage: constitution remains the same, density increases as the material draws in towards the centre. For the reverse of growth we have no obvious term: it involves the loss of material, so 'erosion' or 'attrition' are appropriate here, though these terms carry rather specific indications as to the manner in which the material is lost.

Cyclical change in size, which may consist of expansion alternating with contraction, growth alternating with attrition, or a mixture of both processes, can sometimes appropriately be described as pulsation, although there does not seem to be any general term to cover all cases.

6.6. Change in shape

As already noted, shape is the most complex of all spatial attributes, and we can therefore expect change in shape to be the most complex form of spatial change. Many natural objects undergo systematic changes of shape; think of the varying proportions of a human body during an individual's lifetime from birth to old age. We might say that in this case the change in shape is essentially quantitative: individual body parts tend to grow longer and relatively thinner, the head becomes relatively smaller. There is no gross qualitative change such as the growth of extra limbs or the fusing together of previously separate parts. On the other hand, the changes in body shape before birth, from zygote to embryo to foetus to neonate, do involve some striking discontinuities of this kind.

In handling natural phenomena, it is difficult to dissociate shape change from the processes that give rise to it. The action of the sea on the pebbles of the beach has the general effect of smoothing outlines and reducing asymmetry. A typical well-worn pebble has a smooth ellipsoidal shape, with three planes of symmetry. This is the effect of abrasion all round the surface. Rather different are the shaping processes envisaged by Leyton (see §4.7.2), which act locally at specific points.

Shape space as a whole is so complex that a systematic investigation of qualitative shape change had better begin either by restricting itself to shapes within some more-or-less narrowly defined geometric class, such as triangles or polygons, or else by focusing on one specific aspect of shape such as symmetry, convexity, or curvature.

I illustrate the former approach by considering again the space of possible triangles shown in Figure 1.2. As three points move about in space, the triangle they form undergoes continuous change in its quantitative specifications (area, angles, side-lengths); at critical points, it will undergo discontinuous changes in its qualitative form. Continuous motion in the phase space of

Figure 1.2*a* or 1.2*b* shows up at the qualitative level as a sequence of discrete transitions between neighbouring modes in Figure 1.2*c*. A particular instance of this is shown in Figure 6.17, where one point is moving with respect to a fixed baseline formed by the other two. Initially, to the left of position 2, we have a scalene obtuse-angled triangle, except at one isolated position, 1, at which it is isosceles obtuse. At 2, it is scalene right-angled; between 2 and 4, it is scalene acute, except at the single position 3 where it is equilateral. At 4 it is isosceles right, and then after that it is scalene obtuse again. The modes realized at 1, 2, 3, and 4 are all landmarks in the sense that they are both cognitively salient and also, in the context of the current sequence of transitions, each realized only for an instant, the other modes occupying the intervals between these instants. This is a very typical pattern. The instants at which the landmarks hold are like the 'magic moments' of Casati and Varzi (see p. 274).

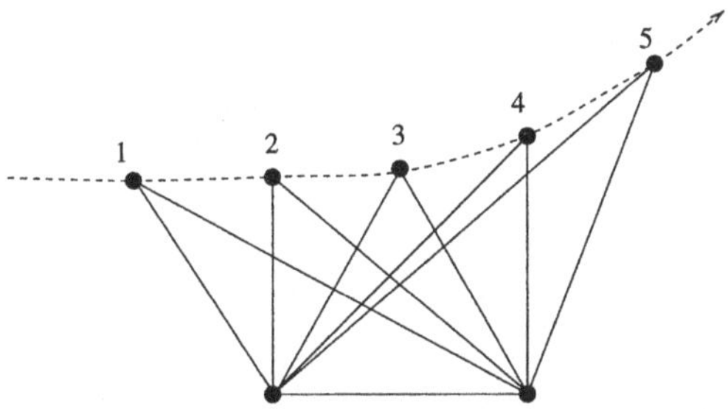

Figure 6.17. Change in triangular region formed by a moving point and two fixed points: 1. isosceles obtuse, 2. scalene right, 3. equilateral, 4. isosceles right, 5. scalene obtuse

Schlieder (1996) has adopted a similar approach to reasoning about configurations of points forming polygons. He too invokes the conceptual neighbourhood structures of qualitative relations arising as equivalence classes of a continuous configuration space. The conceptual neighbourhood structure he gives for the qualitative classification of quadrilaterals is incomplete in the sense that it fails to explicitly acknowledge that when a convex quadrilateral is continuously transformed into a concave one by the inward movement of a vertex, there is a 'magic moment' at which three of the vertices are lined up, making the configuration triangular at just that instant.

To illustrate the second approach to handling qualitative change in shape, I shall consider in turn the cases of change in convexity, change in symmetry, and change in curvature.

In keeping with our usual practice, we need to establish a 'map' of qualitative convexity space, so that changes in convexity can be characterized in terms of paths traced through this map. As already noted in §4.7.1, we can characterize the convexity type of a shape recursively in terms of the convexity type of each of the shape's concavities. It will be helpful to introduce a compact notation for this, called the 'convexity index' of the shape. A convex shape has convexity index V; a shape with n concavities, of convexity indices $I_1, I_2, \ldots, I_n$, has convexity index $c(I_1, I_2, \ldots, I_n)$. The convexity indices of a number of simple shapes is shown in Figure 6.18.

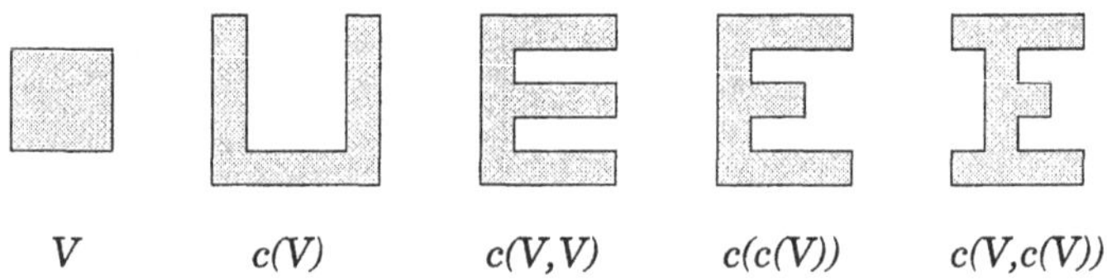

Figure 6.18. Some simple shapes and their convexity indices

We next need to determine how the convexity index can change under continuous deformation of the shape. Figure 6.19 shows three paths by which a rectangle, of index V, can be continuously deformed into an 'E'-shaped figure, of index $c(V, V)$. Along the uppermost path, there first develops a convex concavity, inside which there then develops a peninsula which grows until the concavity is divided into two concavities. The path of development is thus $V \to c(V) \to c(c(V)) \to c(V, V)$. The middle path involves the development of two convex concavities, one after the other, illustrating the sequence $V \to c(V) \to c(V, V)$. In the lowermost path, both concavities develop simultaneously, giving a direct transition $V \to c(V, V)$.

This simple example should already suffice to show that the detailed mapping of convexity space, and the pathways through it, will not be a straightforward matter. I shall not pursue it further here, except to observe that there is a continuity rule for convexity change: at the moment when a concavity first appears, the concavity itself is absent. Thus when a convex shape becomes concave, under continuous deformation, there is a last instant at which it is still convex, but no first instant at which it is concave. This is a simple conse-

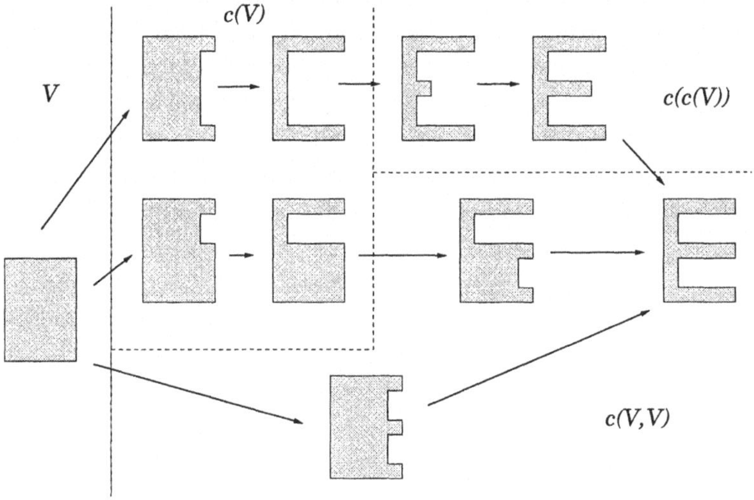

Figure 6.19. Paths through convexity space

quence of the fact that the size of a concavity has to grow continuously from, or shrink continuously to, zero, and, when it is zero, there is no concavity. This applies to concavities at all levels of the recursive decomposition of a complex shape according to the 'c, V' analysis. (A different kind of concavity formation occurs when, for example, an L-shaped object is constructed by bringing together two convex bars. This is not an instance of a convex object becoming concave—the mereological sum of the two components was already concave—and hence the fact that there is a first instant of existence of the L does not contradict the continuity rule given above.)

Turning now to change of symmetry, in Figure 6.20 we see some changes that a four-sided figure may undergo as a result of flexing of the sides. In this figure, each step to the right involves a doubling of the rotational symmetry. A step down adds reflectional symmetry. In addition to these single-step transitions (indicated by the short horizontal and vertical lines, some two-step transitions are possible. The long curved arcs indicate gain of fourfold rotational symmetry without passing through an intermediate stage of twofold symmetry. The diagonal lines indicate simultaneous doubling of rotational symmetry with gain of reflectional symmetry. Why do the diagonals go only one way? This is because of another continuity rule, that when the amount of symmetry is increased or decreased it is the state with greater symmetry which holds at the instant of transition. As far as I am aware, this rule invariably operates. In order to pass from symmetry-type 1′ to symmetry-type 2, it is necessary to double the rotational symmetry and to lose the reflectional symmetry. So long as the former event occurs no later than the latter, then the type 2′ intervenes between the beginning and end states of the transition, whereas if the reflection loss occurs first, we have symmetry-type 1 as an intermediate state; either way there is no direct transition.

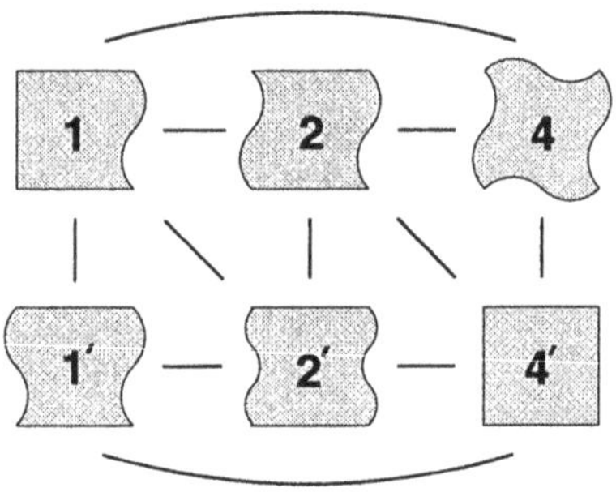

Figure 6.20. Paths through a small portion of symmetry space

As with convexity, I have only scratched the surface of symmetry space. In both these cases the approach I took was first to establish a notation for the elements of the qualitative space and then to use continuity considerations to begin drawing pathways through them. Unlike the position and orientation spaces, for which there exist theories which in their own terms of reference may be regarded as reasonably complete, here we are in largely unknown territory.

The remarks of the previous paragraph apply even more strongly to my third example, change in curvature. Here I shall deal with the qualitative curvature sequences around an outline, as expounded in §4.7.1. Ideally, as in the other cases I have considered, what one would like to have is a 'map' of qualitative curvature space showing which transitions between curvature-type sequences are possible under conditions of continuous deformation. Such a map—a conceptual neighbourhood diagram for the space of qualitative outlines—would, in fact, do less than one might think towards reducing the infinite complexity of quantitative shape space to a tractable representation. The reason is that the map would itself still contain infinite complexity, every curvature-type sequence having infinitely many immediate neighbours, that is, curvature-type sequences into which it can be continuously deformed without passing through any other sequences.

To see why this is, note first that any smooth portion of an outline can, under continuous deformation, become simultaneously 'buckled' in as many places as one wants. The buckling may take the form of a $\supset\subset\supset\subset\supset\subset\cdots$ sequence, or it may involve the interposition of cusps or angles, as shown in Figure 6.21*a*. If the zigzag contour of the edge is characterized by the angle between adjacent segments, then it is obvious that this can evolve continuously away from 180°, so that a profile having any number of kinks in it must count as a neighbour of a straight-line profile in qualitative curvature space. Similarly, a single 'angle out' as shown on the left in Figure 6.21b, can evolve continuously to the jagged profile on the right. To see why this can be continuous, note that the area A above the jagged edge and the dotted lines, representing the difference between the initial and final states in the diagram, can be made as small as we like while preserving the jagged profile, so we can envisage that the left-hand profile exists up to and including time t_0, and for any $\epsilon > 0$ the size of A at time $t + \epsilon$ is proportional to ϵ.

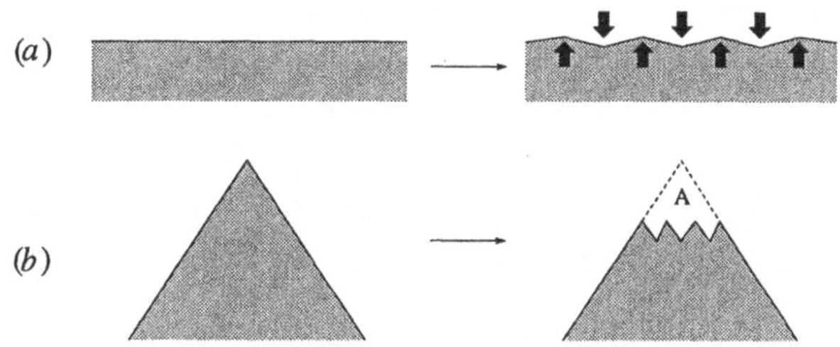

Figure 6.21. Qualitative change in outline under continuous deformation

6.7. Change in posture

Change in posture is reconfiguration. It is most simply described in the case of articulated objects, which consist of a number of individually rigid parts joined together at fixed linkage-points. The bodies of humans and animals approximate to this condition, though the parts are not really rigid: they are mostly relatively rigid in the sense that the shape changes that an individual part undergoes are small compared with the postural changes of the body as a whole. This is the principle underlying the representation of human bodies as assemblages of simple geometrical shapes such as cylinders (Marr and Nishihara 1978). Many machines are articulated, to a very high degree of approximation. Because the various parts are linked together in a particular way, they do not have complete freedom of movement relative to each other; the possible postures of an articulated body are determined by the extent of the possible relative movements of its parts. Each placement of the parts consistent with the given linkages is, in principle, a possible posture. The detailed articulation of the skeleton renders impossible certain reconfigurations such as swivelling the head through a full circle. Dislocation occurs when the normal linkage of the parts is disrupted.

The space of possible postures of a human body is continuous, as is obvious from the fact that all the movements by which we change from one posture to another are continuous. Not all the postures are stable, however. Stable postures form the salient landmarks of posture space, many of which we have special names for: standing, sitting (on a seat or on the ground), squatting, crouching, kneeling, and lying. The postures we adopt in passing from one of these positions to another are mostly rather unstable; if we tried to maintain them for more than a short while we should either fall over or develop muscular pain. In addition, the various forms of bodily locomotion such as walking, running, hopping, skipping, and swimming are characterized by sequences of individually unstable postures. The possible gaits for an n-legged creature can be worked out from first principles using symmetry considerations, as explained by Stewart and Golubitsky (1993); many of these occur in nature (thus for quadrupeds there are walking, pacing, trotting, cantering, galloping, bounding, and pronking). Stewart and Golubitsky report the case of a dog with a missing leg, which was seen to adopt both of the theoretically computed possible gaits for a three-legged creature.

6.8. Change in non-geometrical spatial attributes

As already mentioned, change in density can be the result of expansion or contraction. To have change of density without either, one needs the expansion or contraction to be exactly compensated by loss or gain of material, which is a change in constitution. Change in constitution can occur in the absence of geometrical change: material lost is exactly replaced by material gained. This happens in a lake, for example, with a river flowing in at one end and out at the other. The water-level remains constant, so the geometric attributes of the lake (considered, in the manner of Hayes 1985, as a region of space filled with water) do not vary, yet there is a progressive change in constitution: eventually, perhaps, every single water molecule has been replaced, yet it is the same lake, with the same geometrical character.

Changes of physical state are perhaps not really spatial in character at all, but they do bring about concomitant spatial changes. Recall the example of an ice-cube melting on a table: a cuboid gradually transmutes into a thin layer, so this is not just change of shape but change of (apparent) dimension too. Moreover, change of state provides a classic example of our general picture of a continuous quantitative phase space being partitioned into a discrete qualitative mode space along natural discontinuities. As illustrated for a substance like water in Figure 6.22*a*, the relevant continuous variables are temperature and pressure. The various transitions between neighbouring portions of phase space are melting or liquefaction (solid to liquid), boiling or evaporation (liquid to gas), condensation (gas to liquid), freezing or solidification (liquid to solid), and sublimation (solid to gas or gas to solid). The mode space with these transitions is shown in Figure 6.22*b*.

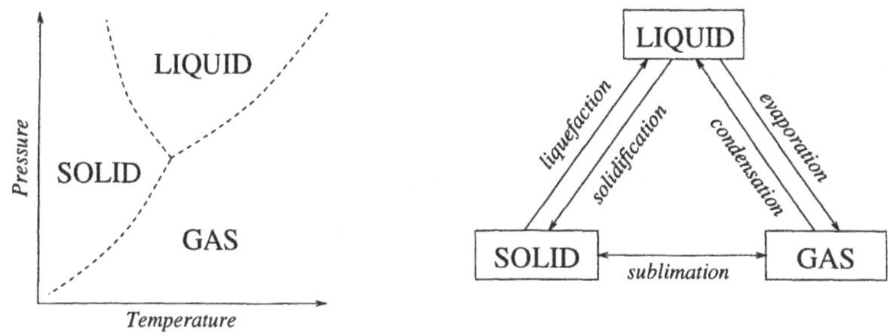

(*a*) The quantitative phase space. (*b*) The qualitative mode space.

Figure 6.22. Change of state

6.9. Summary

In Chapter 4, we saw that many qualitative spatial attributes can be organized into discrete mode spaces which may be viewed as partitions of underlying continuous phase spaces. This picture carries over to the considerations of spatial change in this chapter. An important thread running through the diverse observations I have made concerning different kinds of spatial change is the idea that change with respect to some spatial attribute can be represented in the abstract as the tracing of a path through the phase space formed by the quantitative values of that attribute; qualitative change in that attribute can then be recovered by tracking the concomitant discrete transitions in the qualitative mode-space obtained from the phase space by abstracting away all non-salient distinctions. In particular, I considered three cases in some detail: change in connectivity (§6.2), change in location (§6.3), and change in shape (§6.6).

Although the paths traced in phase space by these changing spatial attributes have largely been continuous, continuity has so far played a rather incidental role in this book. Indeed, the general picture I have built up applies equally well to discontinuous change; a sudden leap from one part of phase space to another will, in general, entail a discrete transition between two modes of the associated mode space. The difference from the continuous case will show up in the fact that the leap need not follow the links of the conceptual neighbourhood diagram for that mode space, which is based on an assumption of continuity. In the absence of continuity, it is tempting to say, 'anything can happen'; at any rate, continuity provides an important constraint which we implicitly appeal to all the time in our reasoning about spatial change, and for that reason it must be regarded as an important aspect of any study of such change.

So much so, indeed, that I shall devote the remaining two chapters of this book to it. In the next chapter I discuss in some detail how to characterize continuity of change associated with spatial regions, since very many forms of spatial change can be described as the successive occupancy, by one or more objects, of different regions in space. This discussion is pursued in a rather quantitative framework: essentially I look at continuity of change in terms of the organization of phase spaces. In the final chapter I turn to the characterization of continuous qualitative change, which involves rather the organization of mode spaces and its relation to the underlying phase spaces.

7
Continuity

I have had cause to refer to continuity on numerous occasions throughout the preceding chapters; it is now time to turn the focus of our attention on to this topic.[1]

7.1. Varieties of continuity

Continuity was an important topic in ancient Greek philosophy, and Aristotle in particular had a great deal to say about it; see White 1992 for a detailed discussion. The distinction between the continuous and the discrete can be applied to any of our key notions of space, time, matter, and change.

Continuous space is indefinitely divisible: any region of space can be regarded as being composed of smaller regions and there is no limit to the number of times this decomposition can be repeated. Continuous space is also seamless: between two separated regions there can be traced a connected path that overlaps them both. In discrete space, by contrast, each region can be decomposed into a set of minimal regions which are atomic in the sense that they cannot be decomposed further. Continuous time can be characterized in the same way as continuous space, but reading 'intervals' instead of 'regions'.

Continuous matter completely fills the space that it occupies, and any sample of such matter can be decomposed into smaller samples of the same stuff which can in turn be decomposed *ad infinitum*. Discrete matter on the other hand is atomic, being an aggregation of minimal atoms which cannot be further decomposed; moreover, it does not fill space, the component atoms being separated from one another by greater or lesser distances.

Continuous change proceeds without instantaneous jumps from one state to a different state: between any two distinct states in the course of the change there must occur states intermediate in character. The amount of change occurring in an interval can be made arbitrarily small by choosing sufficiently short intervals. Discrete change on the other hand does proceed by jumps: a

[1] This chapter is a revised and considerably expanded version of Galton 1997.

state holds over an interval and then over an immediately following interval a different state holds.

It is roughly true that our everyday conception of the world is one in which space, time, matter, and change are all continuous, whereas the scientific conception of the world regards matter, and perhaps change too (in the wake of quantum theory), as discrete.

The distinction between discrete and continuous is very often a matter of granularity. A phenomenon that looks continuous at one level of granularity may very well appear discrete at a different level, either finer or coarser. A good illustration of this, in the case of change, is provided by a cine-film, which looks continuous when viewed at normal speed by a human subject, but which we know in fact consists of a sequence of discrete frames—although when looked at more closely still, the replacement of one frame by the next can be seen to involve only continuous motion. With regard to matter, it is granularity that allows us to reconcile the everyday view of matter as continuous with the scientific view of it as atomic. Some substances, such as sand, salt, and rice, have an intrinsic granularity at just the level where at our human scale we can regard them as either continuous or discrete: we treat them in many ways like homogeneous liquids (pouring or spilling them, measuring them by volume), but at the same time are aware of their nature as collections of individual solid particles ('grains') which could in principle be counted and which are sometimes encountered individually (as when a grain of sand gets in one's eye, at which point the solid, discrete nature of the substance becomes painfully apparent).

As regards the continuity of time and space, the mathematical approach is to regard these continua as aggregates of infinitely many infinitely small components called 'points'. From a mathematical point of view there is an important threefold division into discrete, continuous, and merely dense, the paradigmatic exemplars of these categories being the standard orderings on the integers, the real numbers, and the rational numbers. The technical details underpinning this distinction were presented briefly in §5.3.2. The key feature of the mathematical definition of a continuum is *completeness*, the 'no gaps' property that confers upon the aggregate of points a seamlessness that is at any rate a good simulacrum of the continuity we believe we perceive in space and time themselves. This mathematical approach is strongly opposed to Aristotle's view of a point as either a limit or potential division of a line-segment; on Aristotle's view the continuum comes first, and the points within it have no independent existence, so it would make no sense to regard the continuum as being composed of points. From a technical point of view it matters little whether we regard the continuum as composed of points or conversely regard the points as marking potential divisions within a pre-existing continuum; what matters is that whatever we conceive the 'real nature' of the points to be, their numerosity and relative ordering can be appropriately modelled by representing them as tuples of real numbers. This is a natural extrapolation from our practices of measurement.

Assuming that both time and space can be represented by means of the real numbers, the continuity of spatial change can then be modelled using the mathematical notion of a continuous function, which is defined (in the specific case of functions on the real numbers) as follows:

A function $f : \mathbb{R} \to \mathbb{R}$ is continuous at the point $x_0 \in \mathbb{R}$ if

$$\forall \epsilon > 0\, \exists \delta > 0\, \forall x \in \mathbb{R}\, (|x - x_0| < \delta \to |f(x) - f(x_0)| < \epsilon).$$

In other words, a function is continuous at x_0 if its values at points close to x_0 are close to its value at x_0 itself. More exactly, for any pre-assigned tolerance $\epsilon > 0$, you can always find a number δ such that for all points within distance δ of x_0 the value of the function is within distance ϵ of its value at x_0. This can be generalized in a straightforward obvious way to cover functions whose domain and co-domain are arbitrary metric spaces, that is, spaces in which the notion of 'distance' is defined in an appropriate way.

In order to use mathematical functions to represent change, we assume that the domain of the function, that is, the set of real numbers x to which the function assigns a value $f(x)$, represents instants of time, and that the range of the function, that is, the set of values assigned, represents whatever physical quantity it is whose variation is under consideration.

The position of a point moving along a line can be given as a function $s : \mathbb{R} \to \mathbb{R}$ such that at time $t \in \mathbb{R}$ the point is located at a distance $s(t)$ along the line as measured from some fixed origin (also on the line), to which is assigned the co-ordinate 0. For the motion to be continuous, it is necessary that the function s should be continuous in the mathematical sense. I gave some examples of such functions in §1.4.1. There I noted that the position of a body moving under the influence of a constant force is given by a function of the form

$$s(t) = s_0 + v_0 t + \tfrac{1}{2} a t^2,$$

and that the position of a body moving under the influence of a force that is proportional to the displacement from some fixed equilibrium position, and directed towards that position, is given by a function of the form

$$s(t) = A\sin(kt + e).$$

Both of these functions are continuous at all values of t, as they should be.

The mathematical definition of a function as *any* pairing off of values with arguments allows us to construct some weird 'pathological' functions whose continuity properties are severely counter-intuitive. A well-known example is the function which assigns to each irrational number the value zero, and to the rational number expressed in lowest terms as p/q, the value $1/q$. This function is easily shown to be continuous at all the irrational points and discontinuous at all the rational ones. What is intuitively disturbing about this example is that the points where the function is discontinuous are densely intermingled with those at which it is continuous. Our intuition may be happy with discontinuities, but it prefers them to be isolated. Our intuitive notion of continuity, which is related to the continuity of change that we find, or appear to find, in the physical world, is really a notion of continuity *over an interval* rather than continuity at a point.

In mathematics, one way of defining continuity over an interval is

A real-valued function f is continuous on the interval I if

$$\forall x \in I\, \forall \epsilon > 0\, \exists \delta > 0\, \forall x' \in I \\ (|x - x'| < \delta \rightarrow |f(x) - f(x')| < \epsilon).$$

In other words, f is continuous on an interval just if it is continuous at each point in the interval. Since any interval in the real numbers contains rational numbers, there is no interval over which the pathological example described above is continuous. The interval I in this definition may have any of the forms (x_1, x_2), $(x_1, x_2]$, $[x_1, x_2)$, $[x_1, x_2]$.

We can still construct examples that might be thought to be pathological, however. A famous example is the function $\sin \frac{1}{x}$, which takes the value 1 at the points $\frac{2}{\pi}, \frac{2}{5\pi}, \frac{2}{9\pi}, \frac{2}{13\pi}, \ldots$, and -1 at the points $\frac{2}{3\pi}, \frac{2}{7\pi}, \frac{2}{11\pi}, \frac{2}{15\pi}, \ldots$. It follows that as we approach zero, moving left along the positive x-axis, the value of the function oscillates more and more rapidly. By the time we reach zero, an infinite number of oscillations have occurred. It seems inconceivable that any physical phenomenon can behave in this way: assuming that each oscillation consumes the same amount of energy, the whole process would require an infinite amount of energy to be expended in a finite amount of time. Although the function remains continuous however closely we approach 0, we will find that for a given value of ϵ, we have to choose ever smaller values of δ as the oscillations come ever more rapidly. If we take ϵ as a measure of effect, then the corresponding δ is a measure of the time within which this effect is produced: as we approach 0, the time taken to produce a given effect becomes vanishingly small.

If we wish to exclude this kind of function, we can adopt a more stringent requirement for continuity over an interval, namely, that for a given value of ϵ, some single value of δ will suffice for *all* points in the interval. This is known as 'uniform' continuity, and is defined as follows:

> A real-valued function f is uniformly continuous on the interval I if
>
> $$\forall \epsilon > 0\, \exists \delta > 0\, \forall x, x' \in I \quad (|x - x'| < \delta \rightarrow |f(x) - f(x')| < \epsilon).$$

Uniform continuity accords better with our idea of a process of continuous change in the physical world, since it ensures that there is a built-in upper limit to the amount of change that can occur in a fixed length of time. The function $\sin \frac{1}{x}$ is not uniformly continuous over any interval of the form $(0, x)$.

7.2. Measuring the separation between two regions

Most of the varieties of spatial change I have considered involve change in the geometrical attributes of an object. These are the attributes which belong equally to the body and to its position, the region of space that it occupies. Such changes can be modelled by means of a function pos from time instants to positions. In simple cases, the position of a body can be specified by means of a tuple of real numbers of fixed finite dimension. The position of a rigid spherical body in three-dimensional space can be specified by three real numbers giving the co-ordinates of its centre. If the body is rigid but asymmetrical, another three numbers are required to specify its orientation: thus the position of the body is given by six numbers in all. If it is not rigid, but always maintains the same shape, then a seventh number is needed for specifying its size. In all these cases, where the position of an object is defined by a tuple of real numbers, we can say that the variation in position of the object is (uniformly) continuous over a given interval so long as each of the numbers defining its position is given by a function of time that is (uniformly) continuous over that interval. If the shape of the object is allowed to change, then things become more complicated. If the changes in shape are completely unconstrained, then no fixed finite list of numbers will be sufficient to track all the possible changes. Despite this, we still want to be able to describe such changes as being continuous or not.

What is required is a way of characterizing continuous change of position without having to specify all aspects of an object's position. We want to preserve the essential idea of the mathematical conception of continuity, which is that, in small time intervals, only small changes occur. We need to define what we mean by a small change in position; this requires us to have a measure of the extent to which two regions differ—their 'separation'. As a minimal requirement, we should expect any such measure to define a metric over the space of all possible positions, that is, we require a binary function $\delta(R_1, R_2)$ mapping pairs of regions to real numbers so that[2]

1. $\delta(R_1, R_2) \geq 0$;
2. $\delta(R_1, R_2) = 0$ if and only if $R_1 = R_2$;
3. $\delta(R_1, R_2) = \delta(R_2, R_1)$;
4. $\delta(R_1, R_2) + \delta(R_2, R_3) \geq \delta(R_1, R_3)$.

Any particular choice of metric will give rise to a particular conception of what it is for an object to undergo continuous change of position; different metrics may give rise to different forms of continuity. Our intuitions as to what changes we want to regard as continuous may tally more or less well with the different conceptions of continuity arising from the different metrics. Up to a point, we should be open-minded about this, and be prepared to countenance variant interpretations of separation and continuity.

There are a number of different principles we might adopt in order to furnish a measure of separation between two regions. In what follows I shall treat regions as closed sets; this allows me to handle regions of positive co-dimension—for example, a line segment in the plane—without difficulty. Since regions are to be closed sets, the complement of a region must be defined to be, not its set-theoretic complement (which is open), but rather the closure of this, which I shall denote $cc(X)$.

[2] In the presence of 2, conditions 3 and 4 can be replaced by the single condition $\delta(R_1, R_2) + \delta(R_2, R_3) \geq \delta(R_3, R_1)$.

Thinking of separation as a kind of *distance*, an obvious strategy is to try to define the separation between two regions in terms of the distances between their points. Since we want separation to be a metric, we require that it takes the value zero only when the two regions exactly coincide. In this case, every point of either region coincides with a point of the other. If the regions differ, by however little, there will be at least one point in one region which is separated by a positive distance from the nearest point(s) in the other.[3] An obvious measure of the extent to which the two regions differ is the *largest* distance between any point in one region and the nearest point in the other. This measure is called the *Hausdorff distance* between the two regions, and is defined by the formula

$$\delta_H(X,Y) \triangleq \max\left(\sup_{x\in X}\inf_{y\in Y} d(x,y), \sup_{y\in Y}\inf_{x\in X} d(x,y)\right),$$

where $d(x,y)$ is the distance between points x and y.

Another way of expressing this is as follows. Given a region X and a distance d, define the 'd-neighbourhood' $N_d(X)$ of X to consist of all points at distance d or less from some point of X, i.e., in the notation of §2.3.1,

$$N_d(X) \triangleq \bigcup_{x\in X} cl(B(x,d)).$$

Thus $N_d(X)$ will contain the whole of X itself, together with a 'fringe' of width d around the outside of X. The Hausdorff distance between two regions can now be defined as the smallest value of d for which each region is contained within the other's d-neighbourhood.

Figure 7.1 shows some examples to illustrate the Hausdorff distance between regions in the plane. Illustration (*a*) shows a circle and an ellipse. The solid arrow indicates the greatest distance from any point on the ellipse to the nearest point on the circle. The dashed arrow shows the greatest distance from any point on the circle to the nearest point on the ellipse. The longer of these is the solid arrow, which therefore gives us the Hausdorff distance. Illustration (*b*) shows two mutually tangential circles; their Hausdorff distance is the diameter of the larger circle. In (*c*), we have two rectangles, one of which

[3] This would not be true if one of the regions was a closed set and the other its interior, but I have ruled out this kind of case by stipulating that all regions are to be identified with closed sets.

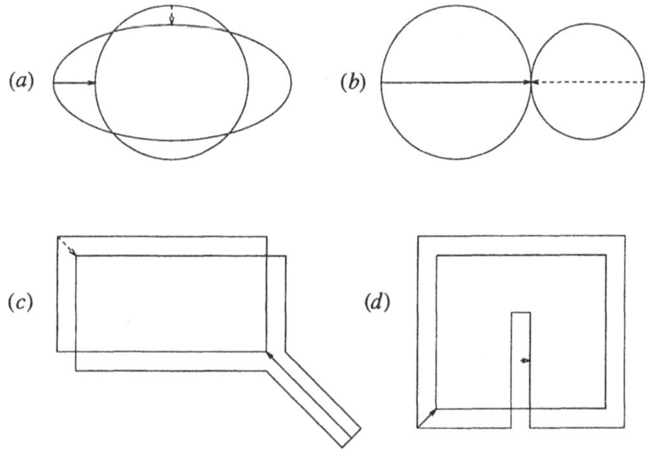

Figure 7.1. Examples to illustrate the Hausdorff distance between regions

has a protuberance extending diagonally from one corner. The Hausdorff distance between the figures is the distance from the end of the protuberance to the nearest corner of the other rectangle. Finally, in (*d*), another two rectangles, but this time instead of a protuberance there is an indentation penetrating deeply into the interior of one of the rectangles. Unlike the protuberance in (*c*), this indentation has no effect on the Hausdorff distance.

The last two examples point to a curious anomaly in the definition of the Hausdorff distance. Two regions may be in most respects very close to one another, but if one of them has a narrow 'spike' protruding well beyond the boundary of the other, this adds substantially to the Hausdorff distance between the two regions, whereas if, instead, it has a thin 'fissure' extending well into its interior, this makes comparatively little difference to the Hausdorff distance. There are a number of ways in which this anomaly might be rectified. The effect of such rectification might be either that both spikes and fissures contribute substantially to the separation, or that neither of them do. Here I consider two ways of the former kind, and one of the latter.

One possibility is to consider, not the Hausdorff distance between the whole regions, but just the Hausdorff distance between their boundaries. I shall call this the *Hausdorff boundary-separation* (or HB-separation for short). It is defined as

$$\delta_{HB}(X,Y) \triangleq \delta_H(\partial X, \partial Y).$$

Hausdorff boundary-separation is used, for example, in Alt *et al.* 1993. If we apply HB-separation to the examples of Figure 7.1, we find no difference in the first three cases, but in the fourth example, we note that the HB-separation of the two regions is a little shorter than the length of the indentation in the outer region, and therefore considerably larger than the Hausdorff distance proper.

Although Hausdorff boundary-separation resolves the anomaly by bringing fissures into line with spikes, both of them now contributing significantly to the separation between the regions, it brings other, more serious problems in its wake. This is illustrated in Figure 7.2, where the two regions in question are a square and a long thin region close to its perimeter. The longer arrow shows the Hausdorff distance between the regions, which is dominated by the size of the square; the shorter arrow shows the HB-separation, for which the size of the square is irrelevant, the important factor being the width of the narrow outer region and its distance from the square. We could make this outer region thinner and thinner, drawing it ever closer to the square; as we do this, the Hausdorff distance between the regions remains large, tending in the limit to half the length of one of the square's sides, but the HB-separation becomes small, tending to zero. Since the two regions are not at all similar, it would seem that the Hausdorff distance here accords better with our intuitive requirements for a measure of separation.

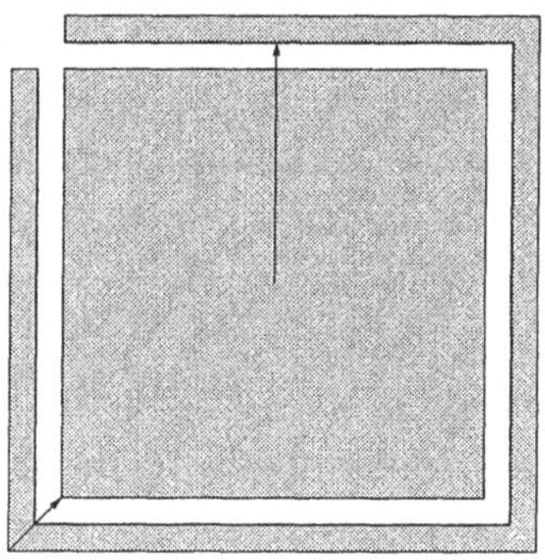

Figure 7.2. An example to illustrate the difference between Hausdorff distance and Hausdorff boundary-separation

Hausdorff boundary-separation proves to be most useful as a measure of the separation of two regions when the regions themselves are *convex*. For convex regions it can be said that closeness of boundaries implies closeness of interiors. In the case of convex regions we therefore do not see anomalies such as that shown in Figure 7.2, in which the boundaries of the regions are close but their interiors are not. Indeed HB-separation is precisely the standard measure of distance used in the study of convex figures (Yaglom and Boltyanskii 1961: 46–9). I show below, in §7.3.2, that for convex regions of co-dimension zero, $\delta_{HB}(X,Y)$ is always equal to $\delta_H(X,Y)$. Note that Hausdorff boundary-separation is not a true metric for all classes of regions; for example, in the plane, any region and its complement have zero HB-separation, and this is also true for regions inscribed on other surfaces. Moreover, if we allow regions of positive co-dimension, we can find infinitely many distinct two-dimensional regions in three-dimensional space with the same boundary (compare §2.3.3). In short, HB-separation can only be a metric for classes of regions which satisfy the constraint that distinct regions must have distinct boundaries.

Perhaps a more satisfactory way of resolving the problem I noted with the Hausdorff distance is to consider the *dual-Hausdorff distance*, introduced by Davis (1995). It is defined to be whichever is the greater out of the Hausdorff distance between the regions themselves and the Hausdorff distance between their complements (that is, the closed complements given by *cc*):

$$\delta_D(X,Y) \quad \triangleq \quad \max\left(\delta_H(X,Y), \delta_H(cc(X), cc(Y))\right).$$

If a region has a spike, its complement has a fissure, so the dual-Hausdorff distance treats both spikes and fissures equally in measuring the separation of regions. Regions differing from one another only by the presence or absence of long thin spikes or fissures will have dual-Hausdorff separation of the order of the (straight-line) length of the longest spike or fissure.

For some purposes, it might be convenient to be able to ignore such spikes and fissures: to regard two regions as close so long as all but a small part of each region is contained in the other. The small part of one region that is not contained in the other may then be indefinitely far from it. A suitable measure in this case is provided by the *area of the symmetric difference* between the two regions, that is, the total area of either region that is not contained in the other. I shall call this the 'area-separation' of the regions, and define it as follows:

$$\delta_A(X,Y) \quad \triangleq \quad \|X \triangle Y\|.$$

For the proof that area-separation is a metric, see Galton 1997. Aspects of area-separation are handled in Alt *et al.* 1996. For two-dimensional regions,

$||X||$ is interpreted as the area of X, whereas in three dimensions it would be the volume, and the measure of separation accordingly called 'volume-separation'.

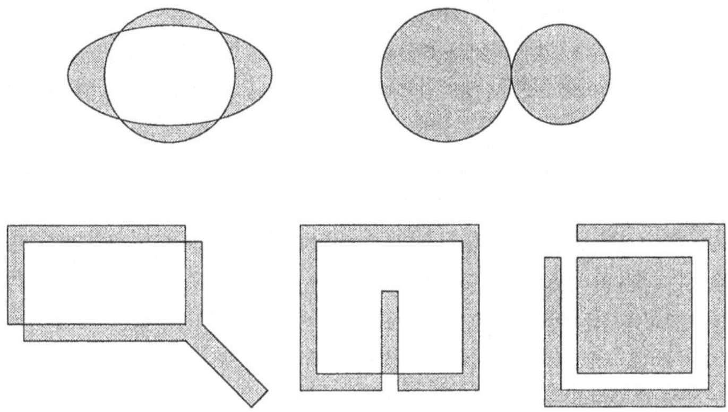

Figure 7.3. Separation of regions as measured by the area of the symmetric difference ('area-separation')

In Figure 7.3, the examples from the previous figures are given again, this time with the area-separation indicated by shading the symmetric difference. Since the area-separation is an area, whereas the measures of separation based on the Hausdorff distance are all distances, they cannot be directly compared. What *can* be compared, however, is the ordering of distances: for example, we might meaningfully say that whereas the Hausdorff distance between X_1 and Y_1 is greater than the Hausdorff distance between X_2 and Y_2, the area-separation of the former pair is less than the area-separation of the latter. In particular, since our main interest in these measures of separation is that they enable us to characterize changes in regions as continuous or discontinuous, we shall be particularly interested in cases of the kind in which, say, the area-separation decreases smoothly to zero whereas the Hausdorff separation jumps discontinuously from some large positive value to zero (I shall provide an example of this phenomenon below).

My final measure of separation is the *Fréchet distance*. Unlike the measures considered up to now, this one is restricted in its application to *paths*, that is, to strictly one-dimensional regions, which can, however, be embedded in space of any dimension. A path in n-dimensional space is defined by means of a continuous one-to-one mapping $f : [0, 1] \rightarrow \mathbb{R}^n$. We may consider the path to be a set of points, in which case we are identifying it with the *range* of the function f, but if we are interested in the *course* of the path then we must also consider the sequence traversed through those points by the function $f(t)$ as t runs from 0 to 1. Given two such paths, one can, of course, consider their Hausdorff distance, which provides some measure of their closeness considered merely as sets of points. Taking in view their path-like nature, however, this measure can be seriously misleading: consider, for example, the two paths AB and CD of Figure 7.4. These have a small Hausdorff distance (equal to half the width of one of the little squares into which the two curves together divide the area they occupy), but if two points were to simultaneously traverse the two curves (from A to B and from C to D, say), then most of the time their distance from one another would greatly exceed this: note, for example, that the distance between an end-point of one curve and an end-point of the other is over seventeen times as great as the Hausdorff distance between the curves. Yet in measuring the closeness of two paths, one should surely expect close paths to have correspondingly close end-points.

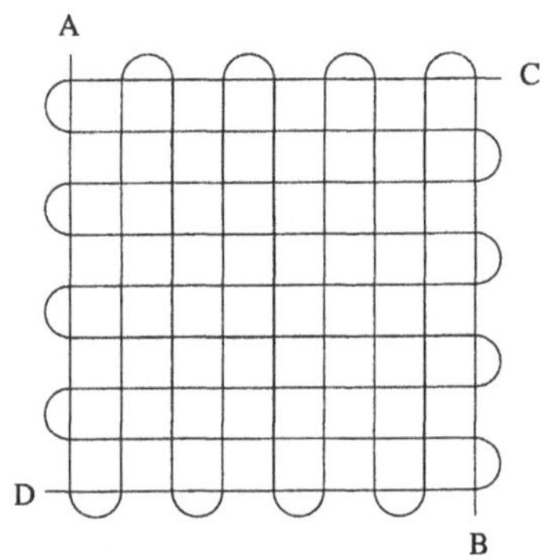

Figure 7.4. Two paths of small Hausdorff distance but large Fréchet distance

The Fréchet distance is designed to take account of the courses of paths rather than merely the locations of their constituent points. As Alt and Godau note,

> In many applications ... the course of curves is important. One example are curves which are input by a digitizer such as handwriting, where a parametrization of the curve is already given by the input device. Also if curves are to be approximated by simpler ones like, for example, curves representing rivers or borderlines in cartography, the

> course is significant and should be reflected by the metric measuring the quality of the approximation.
>
> (Alt and Godau 1995: 76)

They go on to describe the following 'popular illustration' of the intuition behind the definition of Fréchet distance:

> Suppose a man is walking his dog, he is walking on the one curve, the dog on the other. Both are allowed to control their speed but are not allowed to go backwards. Then the Fréchet-distance of the curves is the minimal length of a leash that is necessary.
>
> (76)

The definition of the Fréchet distance is as follows. Here f and g are two paths, given as functions from $[0, 1]$ to $\mathbb{R}^n$ as specified above:

$$\delta_F(f, g) \triangleq \inf_{\phi,\psi \in C} \max_{t \in [0,1]} |f(\phi(t)) - g(\psi(t))|,$$

where C is the set of all continuous non-decreasing functions from [0,1] onto [0,1].

Computing the Fréchet distance between two paths is by no means a straightforward matter, and constitutes the entire subject-matter of the paper by Alt and Godau that I have quoted from above—indeed that paper confines itself to the relatively easy case of polygonal paths or polylines. For the example shown in Figure 7.4 one can 'see' that the Fréchet distance must be a little longer than the length of one of the straight sections of either path (the two paths are congruent), and therefore considerably greater than the Hausdorff distance.

Although, as I have said, the Fréchet distance only applies directly to one-dimensional regions, it would be possible to use it as a separation measure for two-dimensional regions by applying it to their one-dimensional boundaries. That is, we could define the *Fréchet boundary-separation* (or FB-separation) of two two-dimensional regions to be the Fréchet distance between their boundaries:

$$\delta_{FB}(X, Y) \triangleq \delta_F(\partial X, \partial Y).$$

Use of the Fréchet distance here makes for a more discriminating measure of separation than use of the Hausdorff distance. The two regions in Figure 7.5, for example, have small Hausdorff boundary-separation but large Fréchet

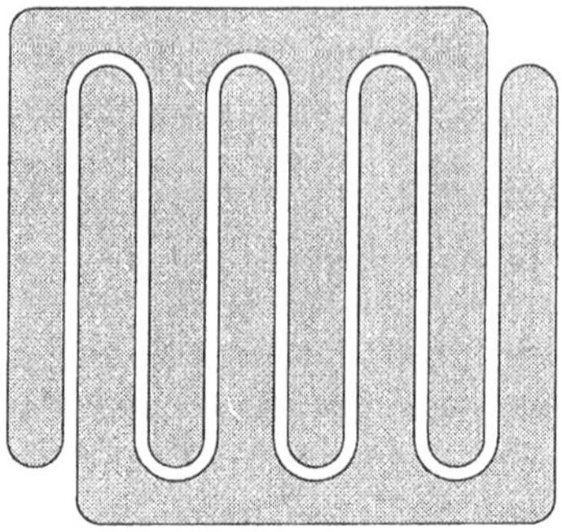

Figure 7.5. Two regions with small Hausdorff boundary-separation but large Fréchet boundary-separation

boundary-separation, the latter according better with our intuition that the regions are not close.

As with HB-separation, FB-separation will fail to provide a metric for classes of regions where distinct regions can have identical boundaries.

7.3. Comparison of the separation measures

In this section I investigate the relations between various pairs of measures taken from those introduced in the previous section. It should be borne in mind throughout that we are dealing with closed regions in Euclidean space.

7.3.1. Hausdorff distance and dual-Hausdorff distance

From the definition of the dual-Hausdorff distance, we can immediately infer that for any regions X and Y, $\delta_H(X,Y) \leq \delta_D(X,Y)$. This means that δ_D is a more discriminating measure than δ_H: the former requires at least as many conditions to be satisfied before it will assign to two regions a measure of closeness as small as that assigned by the latter. Hence, in using these measures to determine which changes are to count as continuous, we may find changes which are continuous from the point of view of Hausdorff distance but not from the point of view of dual-Hausdorff distance.

Note that for regions X and Y of positive co-dimension, the two measures coincide. For in that case both $cc(X)$ and $cc(Y)$ are equal to the whole space, which means that $\delta_H(cc(X), cc(Y)) = 0$.

7.3.2. Hausdorff distance and Hausdorff boundary-separation

For general regions, these two measures are independent in the sense that either can become arbitrarily small while the other becomes arbitrarily large. This is illustrated by Figure 7.1*d*, which has small δ_H but large δ_{HB}, and Figure 7.2, which has large δ_H but small δ_{HB}. For *convex* regions, however, the two measures always coincide:

Theorem. For any convex regions X and Y, $\delta_{HB}(X,Y) = \delta_H(X,Y)$.

Proof. Let X, Y be convex regions with $\delta_H(X,Y) = d$. Let x_0 be a point in X for which the distance to the nearest point in Y is maximum, say d_{XY}, and similarly let y_1 be a point in Y for which the distance to the nearest point in X is maximum, say d_{YX}. Then $d = \max(d_{XY}, d_{YX})$. Let y_0 be a point in Y for which $d(x_0, y_0) = d_{XY}$, and let x_1 be a point in X for which $d(x_1, y_1) = d_{YX}$. Clearly $x_1 \in \partial X$ and $y_0 \in \partial Y$. We shall show that $x_0 \in \partial X$ and $y_1 \in \partial Y$.

Suppose x_0 is in the interior of X. Then in whichever direction you move away from x_0, your path is initially inside X, and hence, by the defining extremal property of x_0, the distance to the nearest point of Y does not increase. But this is impossible if Y is convex, and hence x_0 cannot be in the interior of X. Hence x_0 must be on the boundary of X. By parity of reasoning, y_1 is on the boundary of Y.

We have now shown that for convex regions the points whose distance defines the Hausdorff separation of the regions lie on the boundaries of those regions. This means that they are also the points whose distance defines the Hausdorff boundary-separation (since any point on the boundary is also in the region, the regions being closed), and hence $\delta_H(x,y) = \delta_{HB}(x,y)$. □

7.3.3. Dual-Hausdorff distance and Hausdorff boundary-separation

Figure 7.2 shows that we can have small δ_{HB} with large δ_H, and hence large δ_D. Is the converse relation possible, large δ_{HB} with small δ_D? In fact it is not, as shown by the following theorem:

Theorem. For any regions X and Y, $\delta_{HB}(X,Y) \leq \delta_D(X,Y)$.

Proof. Suppose $\delta_{HB}(X,Y) = d$. Then either there is a point on ∂X from which the distance to the nearest point on ∂Y is *d*, or there is a point on ∂Y from which the distance to the nearest point on ∂X is *d*. Without loss of generality, suppose the former. We thus have $x \in \partial X$ such that the open ball $B(x,d)$ does not intersect ∂Y. Hence either $B(x,d) \subseteq Y$ or $B(x,d) \subseteq$

$cc(Y)$. If the former then x is a point in $cc(X)$ from which the distance to the nearest point in $cc(Y)$ is d, in which case $\delta_H(cc(X), cc(Y)) \geq d$; if the latter then x is a point in X from which the distance to the nearest point in Y is d, in which case $\delta_H(X, Y) \geq d$. In either case we have $\delta_D(X, Y) \geq d$. □

7.3.4. Hausdorff and Fréchet boundary-separations

We need only consider two-dimensional regions, since the Fréchet boundary-separation is only defined for such cases. If the FB-separation of X and Y is d, this means that there is some pair of traversals of their boundaries such that points passed over simultaneously are never more than distance d apart. Thus for each point on either boundary there is a point on the other boundary at a distance no greater than d. It follows that the HB-separation of X and Y cannot be greater than d. Hence we have

$$\delta_{HB}(X, Y) \leq \delta_{FB}(X, Y).$$

For *convex* regions the HB-separation and FB-separation are equal. We can prove this as follows. We consider separately cases in which the interiors of the convex figures overlap and cases in which they do not. An example of the former kind is shown in the left-hand illustration in Figure 7.6. When two convex figures overlap, their symmetric difference appears in the form of a ring of *lunes* surrounding the shared area. The meeting-points of these lunes are the points where the boundaries of the two figures meet (A, B, C, and D in the figure). In some cases there may be a shared stretch of boundary which we may think of as a 'smeared' meeting-point. To determine the FB-separation, I shall construct a suitable 'man-and-dog' traversal of the two boundaries. First, the man and dog will coincide at the points where the boundaries meet (whether at a point or along a segment). In each lune, the walker on the inner boundary segment always stays as near as possible to the walker on the outer segment. The radiating lines in the figure indicate sample positions of the leash. Given that the figures are convex, it is not hard to see that for any point on an outer boundary segment this determines a *unique* position on the inner segment, and that, moreover, the walker on the inner segment never has to retrace his footsteps. The FB-separation is no greater than the greatest distance between the walkers in the traversal I have specified: but this is just the HB-separation.

When the convex figures do not overlap, the situation is, perhaps surprisingly, rather more complex. An example is shown in the right-hand illustration of Figure 7.6. For each region, consider the set of normals running

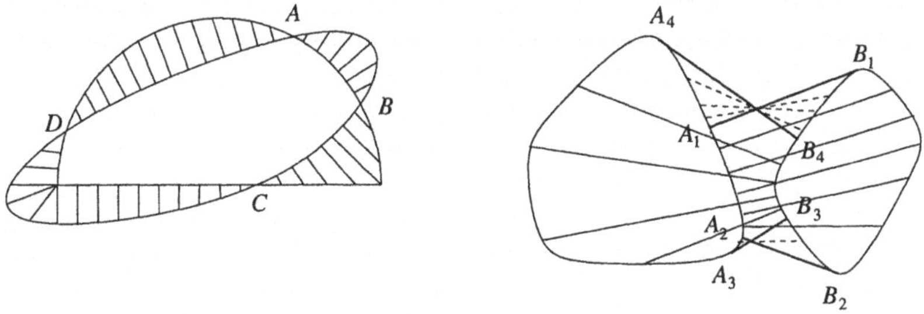

Figure 7.6. For convex figures, Hausdorff and Fréchet boundary-separations are equal

outward from each point on the boundary and intersecting the other region. (At a kink in the boundary, include all lines radiating outward from the kink without intersecting any of the normals from the segments on either side of the kink.) In the figure, these are, for the left-hand region, the lines enclosed between A_1B_1 and A_2B_2, and for the right-hand region the lines enclosed between B_3A_3 and B_4A_4. Since the regions are convex, the eight labelled points are uniquely determined. Our traversal is now defined as follows. Man and dog walk clockwise round the left- and right-hand curves respectively. While the man walks from A_1 to A_2, the dog walks from B_1 to B_2, always staying as near as possible to the man (and hence on the normal from the man's position). Likewise, as the dog walks from B_3 to B_4, the man walks from A_3 to A_4, always staying as near as possible to the dog. For A_2A_3 and B_2B_3, the walkers always adopt positions which will ensure that the taut leash passes through the point of intersection of lines A_2B_2 and A_3B_3, and likewise for the section A_4A_1 and B_4B_1. Once again, it is not hard to see that a leash of length equal to the HB-separation of the two regions will suffice for this traversal, and that therefore this is also the value of the FB-separation.

7.3.5. Hausdorff distance and area-separation

We cannot directly compare these two measures since one is a distance and the other an area. This does not mean we cannot compare the measures indirectly. In Figure 7.7 we see two regions X and Y whose Hausdorff separation is $\delta_H(X, Y) = d$. A margin of width d around X is indicated by the dashed line; region Y falls entirely within the outer limits of this margin. A similar margin could be drawn round Y, within which the whole of X falls. The shaded area represents the region $Y \setminus X$ consisting of those parts of Y which fall outside X. This region lies entirely within the margin outside X. The area of this margin is at most $p_X d + \pi d^2$, where p_X is the length of the perimeter of X. This gives an upper limit on the area of $Y \setminus X$. Likewise, the area of $X \setminus Y$ is at most $p_Y d + \pi d^2$. Hence

$$\delta_A(X,Y) = ||X \setminus Y \cup Y \setminus X|| \leq (p_X + p_Y)d + 2\pi d^2.$$

If we exclude 'pathological' cases in which the perimeters of the regions increase without limit as the Hausdorff distance between them tends to zero, we see that as $\delta_H(X,Y)$ tends to zero, so does $\delta_A(X,Y)$. This means that any change in a region which is uniformly continuous with respect to the Hausdorff distance will also be continuous with respect to the area-separation.

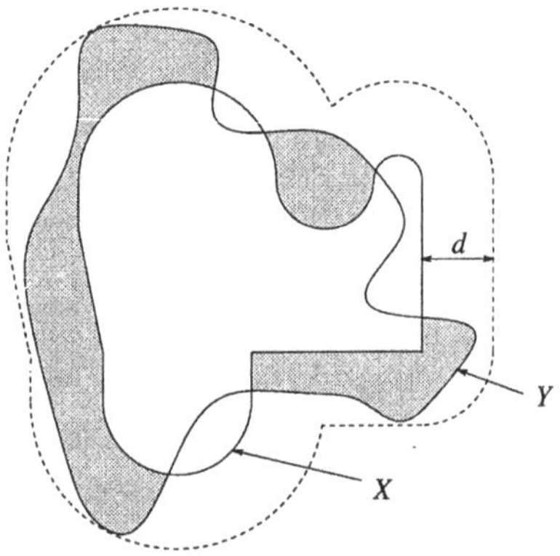

Figure 7.7. Comparing Hausdorff distance and area-separation

The assumption here is that the regions are two-dimensional and of co-dimension zero. It is easy to find counter-examples when these conditions are not met. Consider, for example, a flat disc in continuous motion at right angles to its own plane. Its positions at times t and $t+\delta t$ have small Hausdorff distance (being equal to the distance traversed by the disc over the interval between those times), but the area-separation is twice the area of the whole disc. As δt tends to zero, the Hausdorff distance tends smoothly to zero but the area-separation remains constant. Only as δt finally attains the value zero does the area-separation jump discontinuously to zero. Thus this motion is continuous with respect to Hausdorff distance but discontinuous with respect to area-separation (compare Figure 2.6*b*).

7.4. Different kinds of continuous change

Different separation measures between regions give rise to different criteria for continuity when a region changes. For the change in a region at time t to be continuous, it is necessary that by making δt small enough it is always possible to ensure that the separation between the position of the region at time t and its position at any time in the interval $\langle t - \delta t, t + \delta t\rangle$ is less than any pre-assigned tolerance. This gives a different criterion for continuity for each separation measure. In this section I shall discuss some examples of this; to avoid excessive complication, I shall confine myself to just three of the separation measures, namely Hausdorff distance (δ_H), Hausdorff boundary-separation (δ_{HB}), and area-separation (δ_A). I shall refer to Hausdorff boundary-separation simply as boundary-separation from now on.

Each measure of separation gives rise to a different notion of continuity, according to the general recipe:

> The spatial history of entity E is $\star$*-continuous* at the instant t_0 (where '$\star$' can be any of 'H', 'HB' and 'A') if for any $\epsilon > 0$ there exists a $\delta > 0$ such that for all t, if $|t - t_0| < \delta$ then $\delta_\star(pos(E,t), pos(E,t_0)) < \epsilon$.

By way of illustration,

(*a*) If the thin 'snake' wrapped around the square in Figure 7.2 moves in towards the boundary of the square, getting thinner as it does so, and then just as it disappears assumes the form of the square, then the change at that final moment is HB-continuous but neither H-continuous nor A-continuous.

(*b*) If a circular disc moves through space at right angles to its own plane, then this motion is both HB-continuous and H-continuous, but it is not A-continuous.

(*c*) If a sector opens up in a circular disc, its angle growing continuously from zero (Figure 7.8*a*), then this change in shape is A-continuous and H-continuous, but not HB-continuous.

(*d*) If a circular disc develops a triangular spike of constant length whose apical angle increases continuously from zero (Figure 7.8*b*), then this change of shape is A-continuous, but neither H-continuous nor HB-continuous.

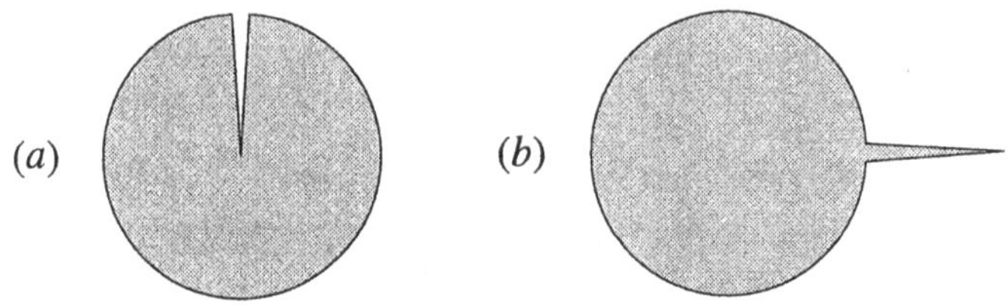

Figure 7.8. A circle with (a) a sector, and (b) a spike

From these examples, it would appear that the intuitively most acceptable definition of continuity is H-continuity. This is because the changes in (b) and (c) seem to us to be continuous, whereas those in (a) and (d) do not. Note that in (d), if the spike were to grow outwards, so that its *length* increases continuously from zero, then we would have a case of H-continuity, and indeed this way of growing a spike does seem to us to be continuous.

Why should H-continuity conform better to our intuitions of continuity than the other two kinds? There is a good reason for this, and it is connected with the idea that in many cases spatial change in an object arises from the motion of small parts. This is implied by what I earlier referred to as the 'atomic theory of spatial change' (see §6.3). It is a natural (though not strictly valid) inference from this that *continuous* spatial change should arise from the continuous motion of small parts. To the extent that this is true in any particular case, we should expect to find H-continuity. Each atomic part effectively occupies a point-like position within the region occupied by the object as a whole. If each part moves continuously, then during a short time interval the displacement of any point in the region will be small; this in turn means that the H-separation between the positions occupied by the whole object over that interval will be small, so the change as a whole is H-continuous.

To find an example of a change that is in some intuitive sense continuous but is not H-continuous, we must look for a case where the spatial change in an object arises not from the continuous motion of parts of the object but from the continuous motion of things outside the object. Think of a new island appearing as the sea-level drops: the rock of which the island is composed does not move at all, what moves is the surrounding water. The object whose change is considered here is not the whole rock but rather that part of it not covered by the sea. I shall describe an idealized version of this in which the islands are circular in form.

Consider an entity E which evolves as follows. For $t \leq 0$, $pos(E,t)$ is a circular region R of radius r, as shown on the left of Figure 7.9. For $t > 0$, E has an extra component, which grows from nothing in a continuous fashion; thus $pos(E,t)$ now consists of the original circular region R together with a second circular region $S(t)$ of radius t, whose centre is located at a distance $d > r$ from the centre of R. This is shown on the right of Figure 7.9. The new component is like an island newly emerging as the sea-level lowers. The change in the position of E can be recorded using our three measures of separation as follows:

$$\left.\begin{array}{r}\delta_H(pos(E,t),R)\\ \delta_B(pos(E,t),R)\end{array}\right\} = \left\{\begin{array}{ll}0 & (t \leq 0)\\ d+t-r & (0 < t \leq d-r)\end{array}\right.$$

$$\delta_A(pos(E,t),R) = \left\{\begin{array}{ll}0 & (t \leq 0)\\ \pi t^2 & (0 < t \leq d-r)\end{array}\right.$$

The change in position of E is continuous as measured by area-separation, but is discontinuous at $t = 0$ as measured by either Hausdorff distance or boundary-separation.

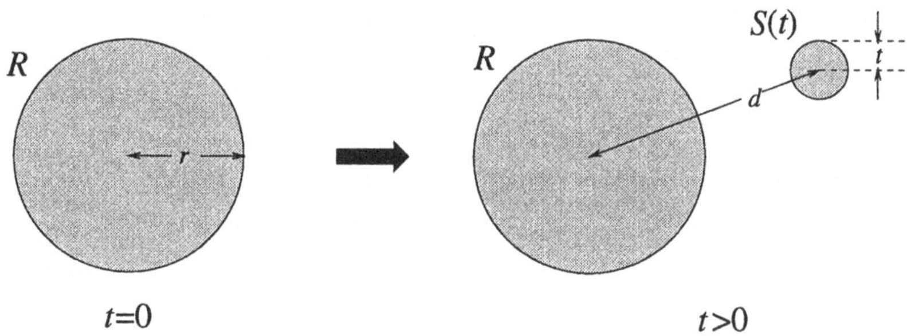

Figure 7.9. A circular entity acquires a remote outlier

The mathematical result here conforms quite well to our intuition that the change involved in this case is not unequivocally either continuous or discontinuous: there is an element of discontinuity, in that the main island suddenly acquires a new outlier, but since this outlier grows in size from nothing in a continuous manner there is an element of continuity as well. If we shift our attention to the sea rather than the island, we see an entity which acquires a new hole, which grows continuously from a point; this change is both H-continuous and A-continuous, though it is not HB-continuous. This again conforms well to our intuition that the change in the sea is more nearly continuous than the change in the land, but still not completely so. It is instructive to contrast this example with the case where the new island has a level top surface, like a table-land, and thus emerges from the sea all at once, with discontinuous growth in area. This change is now discontinuous by all our three measures; as also is the corresponding change in the sea. One might object now that in reality the motion of the water would be continuous in this case too, and therefore we ought to see H-continuity for the change in the sea. The reason why we do not is that the particular idealization we have used is too coarse to track the motion of the water as it runs off from the emerging flat surface.

7.5. Continuity and fusion

I mentioned earlier the idea, associated with the atomic theory of spatial change, that continuous change in position of the parts of an object must result in continuous change in position of the object as a whole. In order for this to be true, it is necessary that the separation measure which provides our criterion of continuity must behave appropriately with respect to fusion: specifically, we require that the separation of two fusions should preserve continuity in the separations of the corresponding parts, in other words that as the values of $\delta(R_1, R'_1)$ and $\delta(R_2, R'_2)$ tend to zero, so does the value of $\delta(R_1 \sqcup R_2, R'_1 \sqcup R'_2)$. (Here we may suppose that R_1 and R_2 represent the positions of two objects at time t, and R'_1 and R'_2 their positions at $t + \delta t$; then $R_1 \sqcup R_2$ and $R'_1 \sqcup R'_2$ are the positions of their fusion at those times.)

I shall show that for area-separation and Hausdorff distance this is indeed the case; but for boundary-separation it is not. For simplicity, I shall confine myself to the case of a region which is expressed as the fusion of just two subregions; the results for this case readily generalize to arbitrary finite fusions.

7.5.1. Area-separation

The area-separation of two fusions does not exceed the sum of the area-separations of corresponding parts:

$$\delta_A(R_1 \sqcup R_2, R'_1 \sqcup R'_2) \leq \delta_A(R_1, R'_1) + \delta_A(R_2, R'_2).$$

Proof. In Figure 7.10 (bearing in mind that some of the fifteen regions may not in fact exist—but this does not vitiate the argument here), $R_1 \triangle R'_1$ is the fusion of the regions marked with a '1', $R_2 \triangle R_2'$ is the fusion of the regions marked with a '2', and $(R_1 \sqcup R_2) \triangle (R_1' \sqcup R_2')$ is the fusion of the regions marked with a '3'. Since every '3' is accompanied by a '1' or a '2' it follows that

$$(R_1 \sqcup R_2) \triangle (R_1' \sqcup R_2') \sqsubseteq (R_1 \triangle R_1') \sqcup (R_2 \triangle R_2').$$

Hence we have

$$\begin{aligned}
\delta_A(R_1 \sqcup R_2, R_1' \sqcup R_2') &= ||(R_1 \sqcup R_2) \triangle (R_1' \sqcup R_2')|| \\
&\leq ||(R_1 \triangle R_1') \sqcup (R_2 \triangle R_2')|| \\
&\leq ||R_1 \triangle R_1'|| + ||R_2 \triangle R_2'|| \\
&= \delta_A(R_1, R_1') + \delta_A(R_2, R_2')
\end{aligned}$$

as required. □

Note that the inequality thus proved becomes equality if the four regions are disjoint in pairs, for then both sides of the inequality are equal to the sum of the sizes of the four regions.

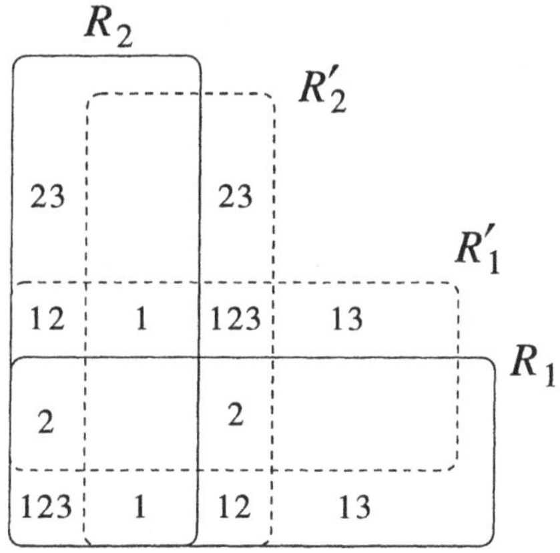

Figure 7.10. The area-separation of a pair of fusions is at most the sum of the area-separations of the corresponding parts

7.5.2. Hausdorff distance

For Hausdorff distance we can prove a stronger result:

Theorem. The Hausdorff distance of two fusions does not exceed the larger of the Hausdorff distances of the corresponding parts:

$$\delta_H(R_1 \sqcup R_2, R_1' \sqcup R_2') \leq \max(\delta_H(R_1, R_1'), \delta_H(R_2, R_2')).$$

Proof. Let p be a point in $R_1 \sqcup R_2$. Then either p is in R_1 or p is in R_2. In the former case, there is a point p' in R_1' within a distance of $\delta(R_1, R_1')$ from p. But then p' is in $R_1' \sqcup R_2'$. Similarly, in the latter case, there is a point in $R_1' \sqcup R_2'$ within a distance of $\delta_H(R_2, R_2')$ from p. Thus in either case there is a point in $R_1' \sqcup R_2'$ within a distance $\max(\delta_H(R_1, R_1'), \delta_H(R_2, R_2'))$ of p. The result follows. □

7.5.3. Boundary-separation

When we turn to boundary-separation, we find that neither of the results proved for the other two measures holds. A counter-example is shown in Figure 7.11. Here R_1 is a circle of radius 1, and R_2 is an annulus of radii 3 and 4; their fusion is of course a disconnected region. R_1' is a circle of radius 2 and R_2' is an annulus of radii 2 and 4; their fusion is a circle of radius 4. All circles and annuli are concentric. The boundary separations are easily seen to be

$$\begin{aligned} \delta_B(R_1, R_1') &= 1 \\ \delta_B(R_2, R_2') &= 1 \\ \delta_B(R_1 \sqcup R_2, R_1' \sqcup R_2') &= 3 \end{aligned}$$

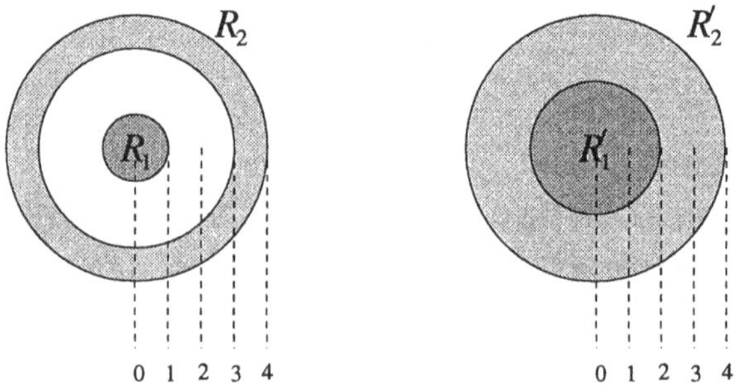

Figure 7.11. Boundary-separation of two fusions can exceed the boundary-separations of either pair of components

so in this case we have

$$\begin{aligned}\delta_B(R_1 \sqcup R_2, R_1' \sqcup R_2') &> \delta_B(R_1, R_1') + \delta_B(R_2, R_2') \\ &> \max(\delta_B(R_1, R_1'), \delta_B(R_2, R_2')).\end{aligned}$$

7.5.4. Continuous change for fusions

These results imply that if two objects change position either H-continuously or A-continuously, then so does their fusion:

Theorem. If the changes in E_1 and E_2 are continuous at time t, as determined by either area- or Hausdorff separation, then so is the change in $E_1 \sqcup E_2$.

Proof. In this proof we shall use $\delta(R_1, R_2)$ to refer indifferently to area- or Hausdorff separation. Let $\epsilon > 0$. Since E_1 and E_2 are both changing continuously at t, there is a $\delta > 0$ such that for any t' for which $|t - t'| < \delta$, we have

$$\begin{aligned}\delta(pos(E_1, t), pos(E_1, t')) &< \tfrac{1}{2}\epsilon \\ \delta(pos(E_2, t), pos(E_2, t')) &< \tfrac{1}{2}\epsilon\end{aligned}$$

Let

$$R_1 = pos(E_1, t),\ R_2 = pos(E_2, t),\ R_1' = pos(E_1, t'),\ R_2' = pos(E_2, t'),$$

so $\delta(R_1, R_1') < \frac{1}{2}\epsilon$ and $\delta(R_2, R_2') < \frac{1}{2}\epsilon$. Note also that

$$pos(E_1 \sqcup E_2, t) = R_1 \sqcup R_2, \quad pos(E_1 \sqcup E_2, t') = R_1' \sqcup R_2',$$

since *pos* preserves the mereological structure.

We now have, using the results just proved,

$$\delta(R_1 \sqcup R_2, R_1' \sqcup R_2') \leq \delta(R_1, R_1') + \delta(R_2, R_2') < \tfrac{1}{2}\epsilon + \tfrac{1}{2}\epsilon = \epsilon,$$

which means that the variation in $E_1 \sqcup E_2$ is continuous at t, as required. (For the Hausdorff separation case, we here used the fact that for non-negative values of x and y, $\max(x, y) \leq x + y$.) □

This theorem does not hold when $\delta(R_1, R_2)$ is interpreted as boundary-separation. A counter-example is shown in Figure 7.12. At time $t > 0$, the position of E_1 is a circular region of radius t. The position of E_2 is fixed:

an annulus concentric with E_1, with radii 1 and $1+d$ for some $d > 0$. Note that for $t < 1$ the fusion $E_1 \sqcup E_2$ has three boundary components, but for $t \geq 1$ only the outermost component remains, the others having been as it were absorbed into the interior when the expanding boundary of E_1 made contact with the inner boundary of E_2. Adopting the position at $t = 1$ as a reference point, the history of the boundary-separations over the interval (0,1) is as follows:

$$\begin{aligned} \delta_B(pos(E_1,t), pos(E_1,1)) &= 1-t \\ \delta_B(pos(E_2,t), pos(E_2,1)) &= 0 \\ \delta_B(pos(E_1 \sqcup E_2,t), pos(E_1 \sqcup E_2,1)) &= 1+d-t \end{aligned}$$

Since $1-t$ tends to 0 as t approaches 1, the spatial histories of both E_1 and E_2 are HB-continuous at $t = 1$. But $1+d-t$ tends to $d > 0$ as t approaches 1, and so the spatial history of $E_1 \sqcup E_2$ is not HB-continuous then. The reader can easily verify that change in position at $t = 1$ of all three regions (E_1, E_2 and $E_1 \sqcup E_2$) is both A-continuous and H-continuous: it had better be, in view of our earlier results!

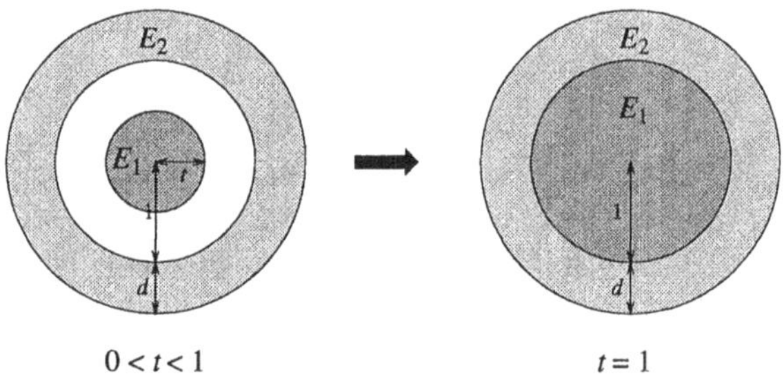

Figure 7.12. Change in boundary-separation of a fusion

The upshot of these results is that in situations in which the atomic theory of spatial change is a reasonable idealization—as for example where the changes in an entity can be described solely in terms of motions of parts of that same entity—the Hausdorff boundary-separation may be an inappropriate measure of closeness on which to base judgements of continuity. The other measures considered here—Hausdorff distance and area-separation—behave much more satisfactorily in this respect.

7.6. Case studies of continuous spatial change

In this section I shall revisit the catalogue of varieties of spatial change in Chapter 6, focusing now on continuity issues.

7.6.1. Change in connectivity

A connected entity becomes disconnected

This is an idealized version of the separation of Britain from the mainland of Europe, illustrated in Figure 7.13*a*. Up to a time t_0, our entity occupies a rectangular region R_0. Between t_0 and t_2, a wedge-shaped indentation appears and grows across the rectangle, reaching the further side at t_2. A typical stage during this process is shown as t_1 in the figure. The erosion continues after t_2, with the entity now having two separated components.

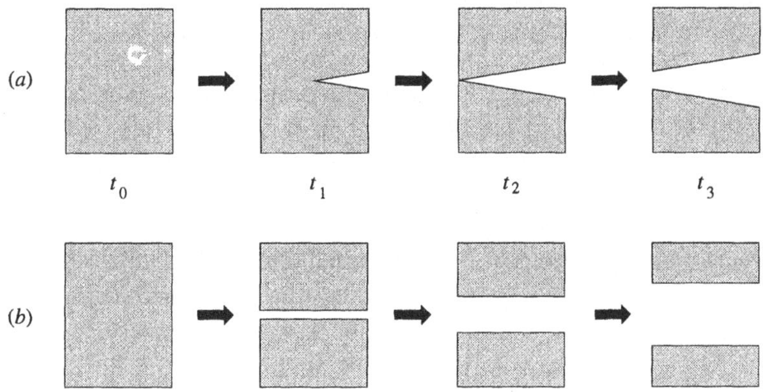

Figure 7.13. A connected entity becomes disconnected

The Hausdorff separation of the figure from R_0 is 0 at t_0 and thereafter is always equal to half the width of the gap in the right-hand edge of the rectangle. It thus increases continuously throughout the history, which is therefore H-continuous.

Between t_0 and t_1, the boundary-separation of the figure from R_0 grows continuously from zero to half the width of the rectangle, the details of its subsequent history depending on the aspect ratio of the rectangle—but in every case it remains continuous. Thus this change is HB-continuous too.

Between t_0 and t_1 the area-separation of the figure from R_0 is the area of the triangular indentation. After that it is the area of the quadrilateral region separating the two components. Since this area grows continuously, the change is A-continuous.

In Figure 7.13*b* we see another way in which disconnection can occur: a narrow channel opens up at all points across the width of the region simultaneously. The width of the channel grows continuously from zero. This history is H-continuous (and therefore also—since the region is of co-dimension zero—A-continuous), but it is not HB-continuous at t_0, since for any $\delta > 0$, the boundary-separation of our figure from R_0 at time $t_0 + \delta$ is at least half the width of the rectangle. This confirms our intuition that this way of disconnecting a region is less continuous than the way shown in Figure 7.13*a*.

A solid entity becomes hollow

A different kind of change in connectivity is the appearance of a cavity within a spherical entity E, illustrated in cross-section in Figure 7.14. The history of E is specified as follows:

> For $t \leq 0$, $pos(E, t)$ is a sphere with centre O and radius 1.
> For $t > 0$, $pos(E, t)$ is a spherical shell $R(t) = R_1(t) - R_2(t)$, where $R_1(t)$ and $R_2(t)$ are spheres with centre O and radii $1 + t$ and t respectively.

The Hausdorff separation between the initial position and the position at time $t > 0$ is simply t, and hence the whole history is H-continuous. Since we are in three dimensions, we must consider continuity with respect to volume-separation rather than area-separation: V-continuity. The volume-separation

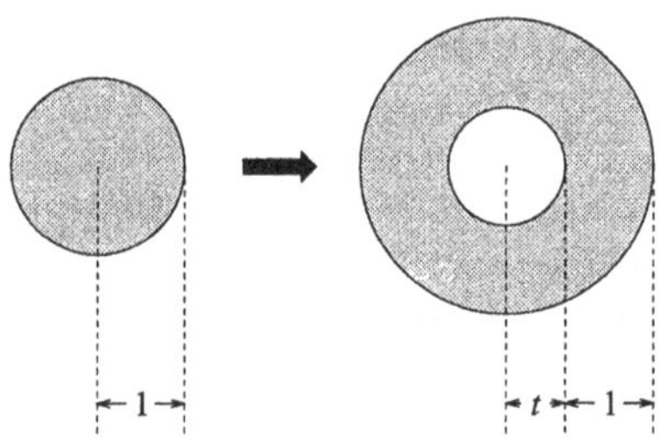

Figure 7.14. A cavity appears and grows in a solid sphere

is, for $0 < t < 1$, the volume of the cavity plus the volume of the spherical shell outside the region occupied by the original sphere; this comes to $\frac{4}{3}\pi t(2t^2 + 3t + 3)$. This is small when t is small and hence the history is V-continuous. The boundary-separation from the initial sphere, on the other hand, has the history

$$\delta_B(R(t), R(0)) = \begin{cases} 0 & (t \leq 0) \\ 1-t & (0 \leq t < \frac{1}{2}) \\ t & (t \geq \frac{1}{2}) \end{cases}$$

which is discontinuous at $t = 0$. Thus the appearance of a cavity is HB-discontinuous; intuitively, this is because a new boundary component appears at distance 1 from the original boundary. Note that the cavity can also be treated as an entity C with a history: $pos(C, t)$ is undefined up to and including $t = 0$, while for $t > 0$ it is the spherical region $R_2(t)$.

Making holes in paper

Here I refer back to the four ways of making a hole in a piece of paper illustrated in Figure 6.3.

In (*a*), the change is somewhat analogous to the two-dimensional version of the example we have just treated: the hole grows from a point and gradually spreads outwards. Some of the paper is pushed out of its original plane, but this happens continuously. This way of making a hole is A- and H-continuous, but HB-discontinuous.

In (*b*), there are two 'magic moments': when the knife first penetrates the paper, and when the cut forms a closed loop so that the enclosed material drops away. The first magic moment is exactly like the previous case, and is therefore A- and H-continuous, but not HB-continuous. The second magic moment, on the other hand, is HB- and H-continuous, but not A-continuous, since as soon as the excised material moves out of the plane of the paper the area-separation increases by the area of the hole (or more, if we take the entity of interest to be the sheet together with the excised material). Taking the two magic moments together, we see that the whole history is H-continuous but neither A- nor HB-continuous.

Case (*c*) differs from (*b*) in that the perforation and the removal of material occur simultaneously, in a single magic moment which combines the discontinuous characters of both the magic moments in (*b*)—that is, it is HB-discontinuous (because a new portion of boundary appears at a distance from the original boundary) and A-discontinuous (because the excised material moves out of the plane). This change is, however, still H-continuous.

Finally, case (*d*), in which the paper is rolled up to make a tube, is both A-continuous and H-continuous, but not HB-continuous, two sections of the original boundary disappearing at the magic moment.

7.6.2. Change in dimension

Consider the case of a linear object spreading out to become an areal object, thus changing its dimension from 1 to 2. This can be interpreted in terms of strict dimension or apparent dimension: these two cases are portrayed as (*a*) and (*b*) respectively in Figure 7.15.

In (*a*) a straight line segment of length l grows into an area containing all points within a certain distance of the original line: say at time $t_0 + \delta$ this distance is δ. Since the line segment has no area, the area-separation between it and its two-dimensional descendent at time $t_0 + \delta$ is equal to the area of the latter. Since this grows continuously from zero, the change from line to area is A-continuous. For the boundary-separation, we have the problem of how to interpret the boundary of the line. If we think of it as consisting of two isolated points (its end-points), then at time δ the boundary-separation of the rectangle from the original line is equal to the distance from a point halfway along one of the long sides of the area to one of the end-points of the line segment. This is never less than $\frac{1}{2}l$, and hence the change in dimension would be HB-discontinuous. If, on the other hand, we determine the boundary of the line segment in the context of the full topology on the plane, then it is its own boundary, and we find that the boundary-separation is the same as the Hausdorff separation. At time δ the Hausdorff separation of the area from the line is δ, which means that the change from line to area is H-continuous, and therefore, according to the second interpretation of boundary, also HB-continuous.

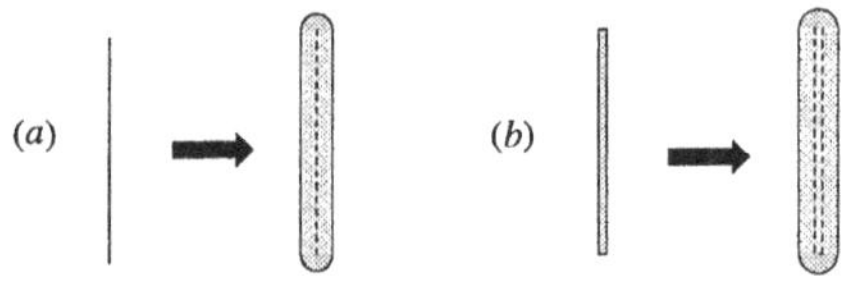

Figure 7.15. Change in dimension: (*a*) strict, (*b*) apparent

In Figure 7.15*b*, the initial figure is not a line but a long thin rectangle. At time δ this has grown into a δ-neighbourhood of the original rectangle. It is easy to see that in this case the change is continuous by all three of our separation measures.

7.6.3. Change in position

Displacement

The purest form of change in location is rigid displacement or *translation*: that is, an object moves in such a way that it undergoes no internal changes in shape or size, and retains the same orientation. Under such displacement, Hausdorff separation and boundary-separation always agree, both being equal to the straight-line distance between the start and end of the motion, which is the same for all points on the interior and boundary of the body, as indicated schematically in Figure 7.16*a*. Hence the motion is H-continuous if and only if it is HB-continuous. As we have seen, the area-separation is sensitive to the co-dimension; if the objects are moving in a space of their own dimensionality (for example, areas moving over a surface, or solids moving through space) then the motion can be A-continuous; but movement out of the dimension of the objects themselves is necessarily A-discontinuous, since even a small displacement leads to an area-separation equal to twice the size of the object itself. (Consider reading Figure 7.16*a* as either representing movement in a plane, in which case the overlap of the two figures represents true overlap of the objects themselves, or as representing movement in depth, in which case the overlap of the figures is merely perspectival, the objects themselves existing in separate parallel planes and hence not overlapping at all in reality.)

(*a*)
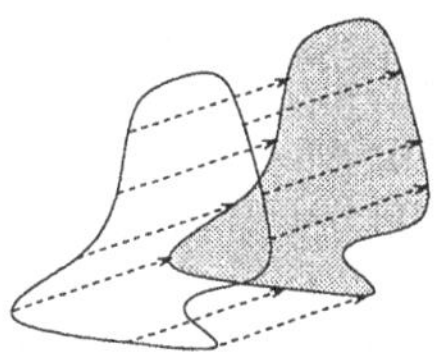

(*b*)
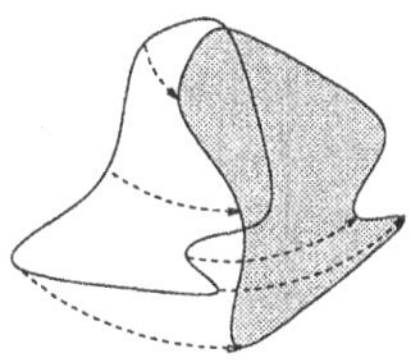

Figure 7.16. Change in position: (*a*) rigid displacement, (*b*) rotation

Rotation

In Figure 7.16*b* we see the same figure as in (*a*), but this time rotated about a point near the top end of the figure. The arrows indicating the trajectory of individual points are now circular arcs, their lengths increasing as we move further away from the centre of rotation. The net displacement of a point under a given amount of rotation is not the length of the arc but of the chord which it subtends; these lengths are, however, approximately in direct proportion, so long as the angle of rotation is small. The maximum value is attained at a point in the figure that is furthest removed from the centre of rotation. This is necessarily on the boundary of the figure, and hence the Hausdorff distance between the two positions is equal to the boundary-separation, and this value increases with the angle through which the figure is rotated, at least initially. If that angle increases continuously from zero, then the rotation is both H-continuous and HB-continuous. Regarding area-separation, so long as the rotation is in the plane of the figure itself (which is assumed planar), then this too will be A-continuous in this case, but as with translations, if the motion takes a planar figure out of its original plane, then it is A-discontinuous. Thus rotation and translation behave alike with respect to our continuity measures.

7.6.4. Change in size

A simple example of change in size is the expansion of a sphere about its centre. At time $t = 0$, object E occupies a closed spherical region of radius 1 and centre O, so $pos(E, 0) = cl(B(O, 1))$. At times $t > 0$, the linear dimensions of the object have grown in proportion to the time elapsed, so $pos(E, t) = cl(B(O, 1 + t))$. It is evident that this change is continuous by all three of our standard measures of separation.

The two kinds of size change illustrated in Figure 6.16 cannot be distinguished without reference to specific quantities of matter. I shall suppose that at $t = 0$ the material of E is a certain quantity of matter q_0, so with the formalism introduced in §3.2 we have $Comp(E, q_0, t)$. For $t > 0$, the material of which E is made is a quantity of matter q_t. For *expansion*, as illustrated in Figure 6.16*b*, we have $q_t = q_0$ for all times t in the period under consideration. For the case of *growth*, as illustrated in Figure 6.16*a*, E increases in size by the accretion of new material. This means that the original quantity of matter q_0 has been supplemented by extra material from elsewhere to give the new quantity q_t; thus q_0 is a proper part of q_t. Specifically, the new matter $\delta q = q_{t+\delta t} - q_t$ acquired in the interval $\langle t, t + \delta t\rangle$ occupies the spherical shell $cl(B(O, t + \delta t)) - B(O, t)$.

If the motion of *matter* is continuous, then discontinuous increase in the size of a material object must happen by growth rather than expansion. The reason for this is that expansion has to occur by the movement of matter already present in the object, and if this is continuous, then the expansion must be continuous as a result. Growth, on the other hand, occurs as a result of new matter being added to the object; there is no reason why a substantial chunk of new matter should not be added in one go, leading to discontinuous growth. As an example, think of attaching a new carriage to a train: it may be difficult to determine precisely at which moment during the coupling process the new carriage counts as being attached, and therefore part of the train, but this is not because the carriage becomes attached incrementally, a little bit at a time—it does not—but because of a familiar uncertainty of criteria. Thus the addition of a carriage does result in discontinuous growth of the train. Similarly, building a wall involves adding a brick at a time, and is thus discontinuous, although it looks continuous when represented at a granularity that is not fine enough to distinguish individual bricks. Organic growth such as we see in living organisms is by contrast more nearly continuous, but the reverse, decrease in size, can be discontinuous, as when a branch is sawn off a tree or a person's limb is amputated. (Of course these are not just changes in size; even restricting ourselves to purely geometrical attributes, there is change in shape as well.)

7.6.5. Change in shape

Under this heading I shall consider a case of simultaneous erosion and deposition, the material eroded from one part of an entity being replaced by matter deposited at another part. The process is illustrated in Figure 7.17.

Throughout the history, our entity is $E_1 \sqcup E_2$. The histories of the two components are specified as follows:

- For $t \leq 0$, $pos(E_1, t)$ is the triangle with vertices $(0, 0)$, $(0, 2)$, and $(2, 0)$, and $pos(E_2, t)$ is undefined.
- For $0 < t \leq 1$, $pos(E_1, t)$ is the quadrilateral with vertices $(0, 0)$, $(2, 0)$, $(0, 2 - t)$, and $(t, 2 - t)$; while $pos(E_2, t)$ is the triangle with vertices $(2, 0)$, $(2, t)$, and $(2 - t, t)$.
- For $t \geq 1$, $pos(E_1, t)$ is the quadrilateral with vertices $(0, 0)$, $(2, 0)$, $(0, 1)$, and $(1, 1)$; while $pos(E_2, t)$ is the triangle with vertices $(2, 0)$, $(2, 1)$, and $(1, 1)$.

Over the interval [0,1], a triangular 'island' has been reconfigured into a rectangular one. Note that at any time $t > 0$ the region occupied by the deposited part, $pos(E_2, t)$, is a triangle congruent to the region formerly occupied by the eroded part, $pos(E_1, 0) - pos(E_1, t)$. It is not assumed that the eroded material subsequently becomes the deposited material, though of course it might.

This change is continuous by all three of our measures, since for $0 \leq t \leq 1$ we have

$$\begin{aligned} \delta_B(pos(E,t), pos(E,0)) &= t \\ \delta_H(pos(E,t), pos(E,0)) &= t \\ \delta_A(pos(E,t), pos(E,0)) &= t^2 \end{aligned}$$

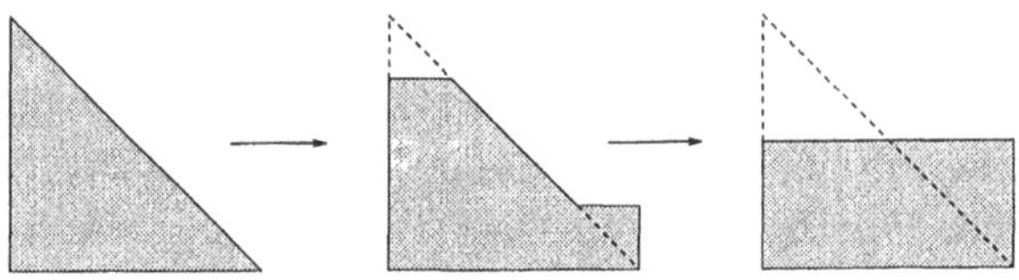

Figure 7.17. Simultaneous erosion and deposition leading to change in outline

7.6.6. Discontinuous change of conventional objects

The creation, modification, and destruction of conventional objects need not involve any material change 'on the ground', but rather only such material change as is necessary to provide a supervenience base for the legislative or other socially defined processes involved in defining the change. Moreover, these changes tend to be associated with specific human acts occurring, or concluding, at precise moments of time. Hence even though all matter might move continuously, spatial change in conventional objects tends to be discontinuous.

This may be illustrated by an example of a kind familiar from recent European history: the redefining of national boundaries. The left-hand illustration in Figure 7.18 shows two neighbouring countries A and B. At midnight on 31 December 1999, a treaty comes into force whereby A cedes to B a certain amount of territory along their common border, as indicated in the central illustration of the figure. Some years later, the east/west tensions in B culminate in the dissolution of B and the establishment of two successor states, C and D, together occupying exactly the territory formerly belonging to B, the new arrangement coming into effect at midnight on 12 July 2004. In our formal framework, we can set up the following history:

- For $t < 31/12/1999$,

$$\begin{aligned} pos(A,t) &= R_1 \sqcup R_2, \\ pos(B,t) &= R_3 \sqcup R_4, \end{aligned}$$

  with $pos(C,t)$ and $pos(D,t)$ undefined.
- For $31/12/1999 \leq t < 12/7/2004$,

$$\begin{aligned} pos(A,t) &= R_1, \\ pos(B,t) &= R_2 \sqcup R_3 \sqcup R_4, \end{aligned}$$

  with $pos(C,t)$ and $pos(D,t)$ still undefined.
- For $t \geq 12/7/2004$,

$$\begin{aligned} pos(A,t) &= R_1, \\ pos(C,t) &= R_2 \sqcup R_3, \\ pos(D,t) &= R_4, \end{aligned}$$

  with $pos(B,t)$ undefined.

These changes are discontinuous by all our measures of separation, and this unequivocal discontinuity is quite characteristic of changes brought about by stipulative fiat rather than as a result of physical movements on the part of the changing entities themselves.

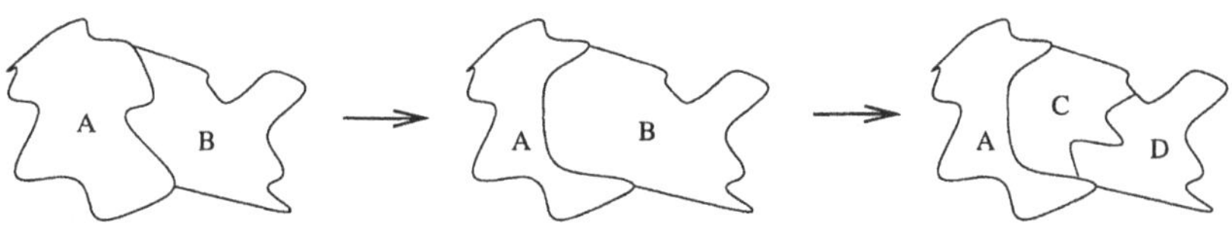

Figure 7.18. Discontinuous change: redefinition of national boundaries

7.7. Summary

In this chapter I have investigated continuity of spatial change from a quantitative point of view, using the standard definition of a continuous function on the real numbers in terms of small increments. In order to apply this to changes in the position of objects in space, one must decide what metric to use to provide a measure of how close the position of an object at one time is to its position at a slightly later time. I discussed a number of different metrics, and noted that in some cases they provide different answers to whether a given motion is continuous.[4]

All the discussion in this chapter could be presented in a more explicitly topological framework if desired. Once a metric has been specified on the space of regions, it can be used to define open 'spherical' neighbourhoods of a given region—for region R, the neighbourhood $B(R, \delta)$ consists of all those regions whose distance from R under the chosen metric is less than δ. This in effect defines a topology on the space of all regions, the open sets of regions being all possible unions of open spherical neighbourhoods. Using the real line as the standard model of time, with the usual topology, the function pos can now be seen as a mapping between two topological spaces; as such, its continuity can be determined using the topological criterion—a function is continuous so long as the inverse image of every open set is itself open (see §8.3). Doing all this will make no material difference to the results obtained; and if we do not want to step outside the limits of this chapter, nothing is gained by recasting it all in topological language. In the next chapter, however, I shall cast aside the quantitative point of view and look at qualitative descriptions of the space of possible positions. From this standpoint the explicit metrics considered in this chapter will no longer be available, and the key to describing changes as continuous or otherwise will be provided by adopting the topological viewpoint in a thoroughgoing way.

[4] Shortly before going to press I received notice of a new, and so-far unpublished, paper by Ernest Davis, entitled 'Continuous Shape Transformation and Metrics on Shapes'. The subject-matter of this paper is very close to that of the present chapter; the reader is advised to look out for publications by Davis on this topic in the near future.

8

Qualitative Continuity

In Chapter 1 I stated that 'the qualitative is supervenient on the quantitative'. There can be no qualitative distinctions without underlying quantitative ones. This does not in itself necessarily imply that qualitative distinctions *arise* from quantitative ones, but that conclusion is all but irresistible in very many cases. In particular, the qualitative spatial attributes I have discussed in the previous chapters can to a very large extent be defined in terms of partitions of an underlying space of quantitative attributes. This was quite explicit in §1.4.1, where I considered the admittedly very simple qualitative division of the range of a real variable into the three qualitative attributes 'negative', 'zero', and 'positive'. Elsewhere in this book we have repeatedly witnessed a pervasive interplay between the qualitative and the quantitative; I have often illustrated qualitative distinctions by means of examples which were, or could have been, fully specified quantitatively. And I have used the language of 'phases' (quantitative) and 'modes' (qualitative) on numerous occasions.

Chapter 7 was rather more quantitative in its approach than might have been expected on the basis of the book's title; this was unavoidable given that the starting-point was the standard mathematical definition of continuity for real-valued functions on the reals. In context, that chapter amounts to a promissory note: to determine what happens to the description of continuous phenomena when we pass from the quantitative level of analysis (in terms of phase spaces and their properties) to the qualitative (in terms of mode spaces). Since mode spaces are discrete, continuity cannot be inherited unaltered in the transition from phase spaces to mode spaces. The question therefore arises in what guise it appears in the latter type of space. That is the subject of this chapter.

8.1. Phase spaces revisited

Here I take up again the phase space for the dynamics of a moving body, as discussed in §1.4.1.

The full system shown in Figure 1.7 is non-deterministic in the sense that given one of the s, v, a-modes it is not in general determined which s, v-mode will occur next. For example, the mode $(-,-,0)$ can be followed by any of $(-,-,+)$, $(-,-,0)$, and $(-,-,-)$. Of course we know that if the system were to be fully described, using as parameters the infinite sequence consisting of derivatives of s of all orders, then the outcome would indeed be deterministic. In the present example, if the third derivative da/dt is non-zero, that suffices to determine the next mode: if $da/dt < 0$ then the next mode must be $(-,-,-)$, whereas if $da/dt > 0$ it must be $(-,-,+)$. If, however, the third derivative is zero then we need to look at the fourth derivative. If this is negative then the next mode will be $(-,-,-)$, whereas if it is positive it will be $(-,-,+)$. If the fourth derivative is zero then we have to look at the fifth derivative, and so on. Only if all the higher derivatives are zero will we remain in the state $(-,-,0)$.

At this point we need to clarify what is meant by the 'next mode'. Let us assume that modes do not change chaotically; by this I mean that any finite period can be divided up into a finite number of subperiods over each of which the mode remains constant. This 'non-intermingling' principle is a standard assumption for outlawing a chaotic intermingling of states (Davis 1992; Galton 1996). Now suppose mode q_1 holds at the instant t_1, but that it does not persist indefinitely. Then by our assumption there is a latest time $t_2 \succ t_1$ such that the mode does not change over $\langle t_1, t_2 \rangle$. Let q_2 be the first mode distinct from q_1 to hold after t_1. Then q_2 might first hold over $\langle t_1, t_2 \rangle$, at t_2, or over some interval $\langle t_2, t_3 \rangle$, giving us the three possibilities:

1. A mode $q_2 \neq q_1$ holds over $\langle t_1, t_2 \rangle$.

2. Mode q_1 holds over $\langle t_1, t_2 \rangle$, but $q_2 \neq q_1$ holds at t_2.

3. Mode q_1 holds over $\langle t_1, t_2 \rangle$ and also at t_2.

This gives us three possible types of transition away from q_1, as illustrated in Figure 8.1. I shall introduce technical terminology in this context as follows: in (1), I shall say that q_1 'pre-dominates' q_2, meaning that the former holds at the beginning of an interval over which the latter holds; in (2), I shall say that

q_1 pre-dominates itself, and that q_2 'post-dominates' q_1, the latter statement meaning that q_2 holds at the end of an interval over which q_1 holds; finally, in (3), I shall say that q_1 both pre- and post-dominates itself, and that q_1 pre-dominates q_2. The key concepts here are pre- and post-dominance, and these will receive specific treatment in the formalization to be given later.

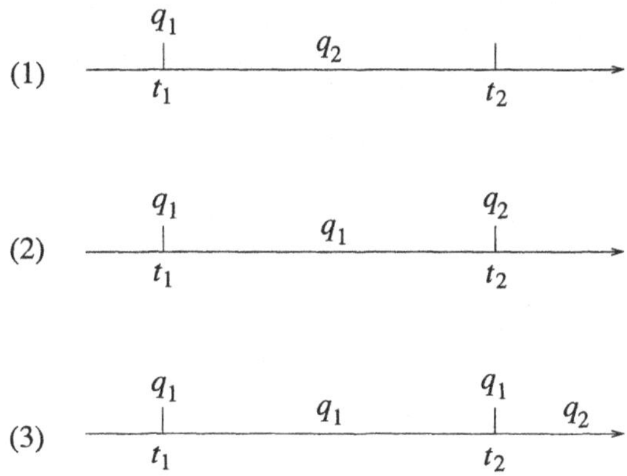

Figure 8.1. Three kinds of transition from mode q_1 to mode q_2

I illustrate these ideas with examples of each of the patterns 1–3:

- Suppose $q_1 = (0, +, +)$. At time t_1, the velocity is positive, so s is increasing, which means that it cannot persist with the value 0. The next mode, $(+, +, +)$, must follow immediately, holding over $\langle t_1, t_2 \rangle$.
- Suppose $q_1 = (-, +, 0)$, and assume the acceleration remains 0. Then v remains positive, and s eventually attains the value 0. It cannot do so immediately, however, since however close to zero s is, it must pass through other non-zero values to get there—this is an instance of the 'continuity rule' of Williams 1990. This will apply to any instant at which q_1 holds, hence the succeeding mode, $(0, +, 0)$, must hold at t_2 itself.
- Suppose $q_1 = (-, +, 0)$ again, only this time assume that before s attains the value 0, the acceleration becomes positive. By the continuity rule there must be a last instant at which $a = 0$: this will be t_2. Then q_1 holds over $\langle t_1, t_2 \rangle$ and at t_2 itself, while the next mode, $(-, +, +)$, holds over some interval $\langle t_2, t_3 \rangle$.

The first of these examples, and others like it, can already be read off from the mode-transition diagram. We know that mode $(0, +, +)$ cannot persist over an interval, since its box has rounded corners, meaning that transition type (1) must apply. But there is nothing in the diagram that enables us to tell when transition types (2) and (3) are possible. This issue will be addressed later on in the chapter.

8.2. Conceptual neighbourhood as a relation on mode space

A comparison of the conceptual neighbourhood diagrams for spatial regions (Figure 2.9) and temporal intervals (Figure 5.2) with the mode spaces for the motion of a particle (Figures 1.6 and 1.7) will reveal both differences and similarities. Both types of diagram consist of a finite set of labelled boxes with arcs running from one box to another. The most obvious difference is that in the mode-space diagrams the arcs are directed—they have a distinguished direction indicated by the presence of an arrowhead at one end—and the boxes are of two kinds, with square or rounded corners. The direction of the arcs indicates temporal sequence: for example, if both the velocity and acceleration of a body are directed towards the left, then the body must move to the left, so its position will pass from positive to zero to negative; hence the leftward-directed arrows on the bottom right-hand edge of Figure 1.7. The different kinds of boxes relate to the possibility of persistence; a body cannot persist in the mode 'position zero, velocity negative, acceleration zero' for more than an instant, since the unchanging negative velocity must immediately carry the object to a negative position; hence the box for the mode $(0, -, 0)$ in Figure 1.7 has rounded corners.

Turning to Figure 2.9, one notes that all the boxes have square corners, and that there are no arrowheads on the arcs. This is readily understood in terms of the mode-space diagrams: there is nothing to prevent two bodies from persisting in any of the RCC relations over an interval, which is why the boxes for these relations are all square-cornered; and any change of RCC relation can be reversed, which is why the arcs are undirected—in fact, reading this diagram as representing possible *changes* in RCC relation, the undirected arcs should rather be understood as bidirectional, as if they had an arrowhead at each end.

With the differences thus accounted for, it becomes obvious that the two kinds of diagram, mode-space transition diagrams and conceptual neighbourhood diagrams, can be regarded as essentially similar. This is even the case with the conceptual neighbourhood diagram for temporal intervals, Figure 5.2. It may seem strange to view the interval relations as states which can change over time, for the temporal intervals of Allen's system do not themselves change, being rather the vehicles for change in other state spaces. However, the mathematical structure of the interval calculus depends only on the fact that intervals are regarded as bounded segments of a one-dimensional space with a favoured orientation, and does not have anything to do with time as such except in so far as time can be viewed as such a space. We can therefore regard the interval calculus as a model for many things other than the temporal intervals for which it was originally intended. In particular, we can consider it as representing *spatial* relations of objects constrained to move under only one degree of freedom (for example, trains along a railway track).

8.3. The topological approach to continuity

The brief introduction to topology in §2.3 omitted to mention what is in fact a central concern of topology: the definition of continuity. The topological notion of an open set can be used to characterize continuity as follows:

> Given two topological spaces S_1 and S_2, a function $f : S_1 \to S_2$ is continuous so long as for every open set $Y \subseteq S_2$, the inverse image $\{x \in S_1 \mid f(x) \in Y\}$ of Y is open in S_1.

To see how this fits in with the definition of continuity given in §7.1, let Y be any open set in $\mathbb{R}$, and let X be the inverse image of Y under f. Suppose $x_0 \in X$. Then $f(x_0) \in Y$, and since Y is open there is some $\epsilon > 0$ such that the neighbourhood $(f(x_0)-\epsilon, f(x_0)+\epsilon) \subseteq Y$. Since f is continuous at x_0, there is a $\delta > 0$ such that if $x \in (x_0-\delta, x_0+\delta)$ then $f(x) \in (f(x_0)-\epsilon, f(x_0)+\epsilon)$, and hence $f(x) \in Y$. This means that $(x_0 - \delta, x_0 + \delta) \subseteq X$. Thus every element of X has a neighbourhood entirely contained in X, which means that X is open. Thus f satisfies the topological criterion for continuity.

Consider for example the simple case of a real variable v with landmark value a. This system has three modes, $v < a$, $v = a$, and $v > a$. Suppose v varies continuously, so there is a continuous function $v : \mathbb{R} \to \mathbb{R}$ giving the value $v(t)$ of the variable v at each time t. For each mode m let $V(m)$ be the set of values falling under m and $T(m)$ the set of times at which m holds.

Then

$$T(m) = \{t \in \mathbb{R} \mid v(t) \in V(m)\}.$$

Since v is continuous, $T(m)$ is open so long as $V(m)$ is. Now in this system there are only three values of m to consider; we see that if m is the mode $v < a$, then $V(m)$, and hence $T(m)$, is open, and likewise if m is $v > a$. For the mode $v = a$, on the other hand, $V(m)$ and $T(m)$ are both closed. Hence we deduce that the modes $v < a$ and $v > a$ must each hold on an open set of instants, hence on the union of a set of open intervals. In particular, if either of these modes holds at a particular instant t, then it must hold over an open interval within which t falls; in particular, it cannot hold at an isolated instant. The mode $v = a$, on the other hand *can* hold at an isolated instant.

These ideas can be related to the phase spaces and mode spaces discussed above by means of the following simple principles:

1. Modes are partitions of phase space; non-landmark modes correspond to open sets in the partition whereas landmark modes correspond to closed sets.
2. Continuous change in the system can be represented by means of a continuous function from the time line (represented by $\mathbb{R}$) to phase space.
3. Since a non-landmark mode corresponds to an open subset of phase space, the set of times on which such a mode holds must be open (that is, a union of open intervals). Thus only landmark modes can hold at isolated instants.

8.3.1. A complication

Thus far, it would appear that we can classify modes according as they hold on open or closed sets of times. This is close to the point of view of Galton 1990, in which qualitative states are classified as 'states of motion' and 'states of position' in almost exactly this way. However, there are complications, which become apparent if we turn now to a topological treatment of the region calculus (RCC). The simple twofold classification is adequate so long as regions cannot change their size or shape. Recall that in §4.4.3 I enumerated six different ways in which the positions of rigid bodies can be related, each characterized in terms of which subset of the RCC relations can hold for them. For such objects, the relations DC, PO, $NTPP$, and $NTPPI$ always hold on open sets of instants—that is, whenever one of them holds at an instant, it must hold over some interval within which that instant falls. The remaining relations, EC, TPP, $TPPI$, and EQ, always hold on closed sets of instants, so that whenever one of them holds over an interval, it must also hold at the end-points of the interval. The former group of relations are 'states of motion' whereas the latter are 'states of position'.

The complication arises when we drop the requirement that body should be rigid. This allows a transition between the states PO, TPP, EQ, as illustrated in Figure 8.2.

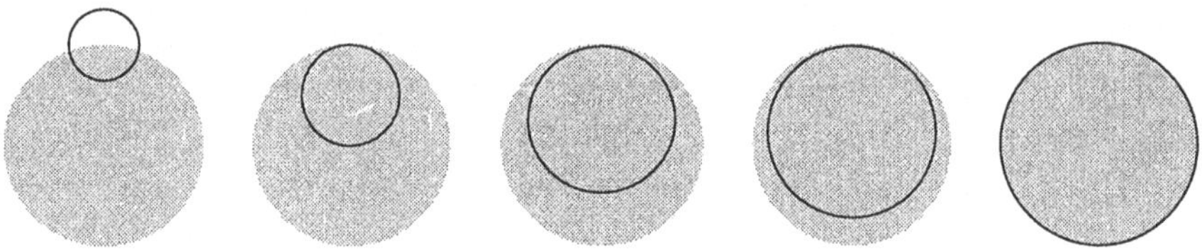

Figure 8.2. Continuous change in region–region relation

Here TPP holds over a period during which the small circle expands till it coincides exactly with the large circle. Since the expansion necessarily takes time, it is not possible for TPP to hold for an isolated instant in the context of a transition from PO to EQ. This contrasts with our earlier observation that TPP *can* hold for an isolated instant in the context of a transition from PO to $NTPP$ (Figure 6.9*a*). Moreover, in the transition considered here, there is a first time at which TPP holds, but no last time. The interval on which TPP holds is open at the end but not at the beginning. This means that the topology of this qualitative state space is more complex than I have considered up to now.

We can obtain a clear view of what is happening if we restrict ourselves to spherical regions: either line segments in space of one dimension, circular discs in two, or spheres in three. Consider the relation which a variable spherical region A bears to a fixed spherical region B. Let us adopt a coordinate system such that B has radius 1 and is centred at the origin. The variable region can be uniquely specified by its radius r and by the position of its centre, giving $n+1$ degrees of freedom, where n is the dimension of the space under consideration. However, as far as concerns the qualitative relation between A and B as expressed in the region calculus, the position of B's centre may be replaced by a single number d giving the distance between the centres of A and B; the direction of the line joining the centres is irrelevant. Thus we are dealing with a system which for our purposes may be regarded as having just two degrees of freedom, represented by the numbers r and d, and it may therefore be represented graphically in the plane; indeed, since $r > 0$ and $d \geq 0$, a quarter-plane suffices. The eight region–region relations can be uniquely characterized in terms of r and d as follows; in this table the relation in question is that borne by A to B:

DC	$d > r + 1$
EC	$d = r + 1$
PO	$\|1 - r\| < d < r + 1$
EQ	$r = 1, d = 0$
TPP	$0 < d = 1 - r$
$NTPP$	$d < 1 - r$
$TPPI$	$0 < d = r - 1$
$NTPPI$	$d < r - 1$

The eight qualitative relations thus correspond to the partitioning of the r, d-plane illustrated in the left-hand diagram of Figure 8.3. Using QP terminology, the qualitative relations are modes, whereas the individual points are phases; the phase space is the quarter-plane $r > 0, d \geq 0$. The right-hand diagram, showing the modes, should be compared with Figure 2.9; the meaning of the arrowheads is explained below in §8.4.2.

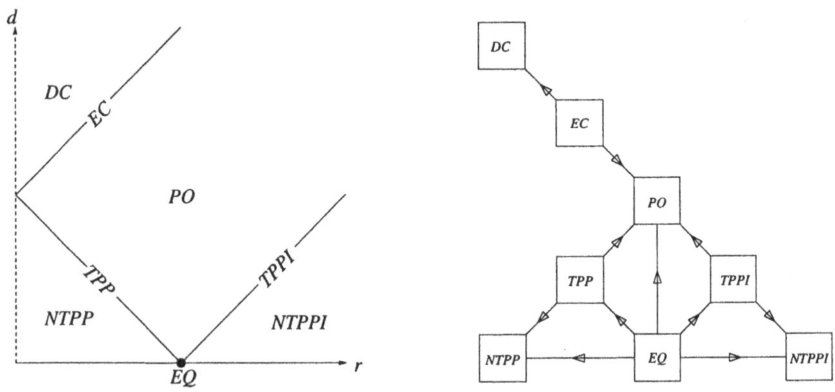

Figure 8.3. The continuous space of region–region relations, partitioned according to the region calculus, and the resulting discrete mode space

Observe that under this representation there are three different kinds of relation, corresponding to areas in the plane (DC, PO, $NTPP$, and $NTPPI$), line segments (EC, TPP, and $TPPI$), and points (EQ). In the topology of the plane, only the areas are open sets; in the context of a particular transition in mode space, however, the relevant topology to consider is the subspace topology on the portion of mode space to which the motion is currently restricted. Consider the subset L of phase space consisting of just the lines in Figure 8.3; this is the union of $EC, TPP, TPPI$, and EQ. In the subspace topology on this set, EC, TPP, and $TPPI$ are all open (note that the point at which EC appears to adjoin TPP is outside the phase space). Only EQ is non-open. To see this in the case of TPP, note that in the full space, the union of PO, TPP and $NTPP$ is open; but TPP is just the intersection of this set with L, so TPP is open in L. It follows that *in the context of a transition from TPP to EQ*, TPP behaves like a non-landmark mode while EQ behaves as a landmark; this explains why in *this* context TPP can only hold over an interval.

The upshot of this is that we require a more detailed taxonomy of qualitative states than just 'landmarks' and 'non-landmarks'. Such a taxonomy will, in fact, provide a powerful tool for investigating the structure of complex qualitative state spaces and the kinds of transitions possible in them.

8.3.2. Mode spaces as topological spaces

In §2.6.2 I defined a class of topological spaces called incidence spaces. They are true topological spaces but are essentially discrete in character, each point having a unique minimal neighbourhood. A space of this kind can be derived from the more familiar type of continuous topological space by quotienting out an appropriate equivalence relation. But this is precisely how the RCC mode space can be derived from the continuous space portrayed on the left of Figure 8.3. Thus the RCC mode space is itself an incidence space, as shown on the right of the same figure, where the arrows indicate bounding, and the positions of the modes are correlated with the positions, in the left-hand diagram, of the phases-sets to which they correspond. This property has been used by Schlieder to *define* the notion of conceptual neighbourhood:

> Let S be the n-dimensional configuration space of a spatial reasoning task and let D be a cell decomposition of S into 0-, 1-, ... n-cells. Two qualitative relations are *conceptual neighbours* iff the corresponding cells are incident.
>
> (Schlieder 1996: 128)

Here incidence is the symmetric closure of bounding: two elements are incident if either bounds the other.

The case of the dynamic position-velocity-acceleration space discussed in §8.1 is a little different, however, because of the temporal asymmetry of some of the transitions. The relation 'post-dominates', for example, cannot be identified with the topological notion of bounding which forms the basis of incidence spaces, although that notion could certainly play a part in how this relation is characterized. For this reason the general theory of mode spaces, and hence of qualitative spatial change, must go beyond the purely topological theory of incidence spaces.

In the remainder of this chapter I develop an appropriate theory that is sufficiently general to encompass the qualitative aspects of all the varieties of continuous spatial change discussed in this book. The abstract spaces I shall use are called *topological mode spaces*; the theory of these spaces can be regarded as an extension of the theory of incidence spaces, but my treatment will not assume any existing results from that theory, preferring rather to develop everything from first principles.

8.4. A theory of topological mode spaces

8.4.1. Formal definitions

By a 'topological mode space' or TM space I shall mean a set Q of qualitative states or 'modes' together with two binary relations $\vdash$ and $\dashv$ on Q. The elements of Q are meant to represent exhaustively the states of some system, and the symbols $\vdash$ and $\dashv$ are to be read as 'pre-dominates' and 'is post-dominated by' (in the sense of §8.1) respectively.

A *model* for a TM space is specified as a triple (P, C, μ), where

- P is a subset of $\mathbb{R}^n$, for some positive integer n, representing the quantitative phase space of some system;
- C is a set of continuous functions from $\mathbb{R}$ to P, each representing a possible evolution, or *history*, of the system over time;[1]

[1] Here I am modelling time by the system of real numbers ordered in the usual way, for which it is natural to use '$<$' to denote the temporal precedence relation. This 'liberates' the symbol $\prec$ for another role, to denote the temporally symmetric dominance relation—for which see below. Correspondingly, from now on I write $t \in (t_1, t_2)$ where previously I should have written $t \sqsubseteq \langle t_1, t_2 \rangle$.

- μ is a surjective mapping from P onto Q giving the (qualitative) mode associated with each (quantitative) phase;

and for each history $f \in C$ the following 'temporal incidence' rules are satisfied:

- *No intermingling.* For each pair of real numbers $t < t'$, there is a finite sequence $t = t_0 < t_1 < \cdots < t_n = t'$ of real numbers such that for $i = 1, \ldots, n$, if $t_{i-1} < u < v < t_i$ then $\mu(f(u)) = \mu(f(v))$;
- *Dominance.* If $\mu(f(t)) = q$ for all $t \in (t_0, t_1)$ then $\mu(f(t_0)) \vdash q$ and $q \dashv \mu(f(t_1))$.

Thus q' can only hold at the beginning of an open interval on which q holds if $q' \vdash q$, and q' can only hold at the end of such an interval if $q \dashv q'$. Note that the symbols '$\vdash$' and '$\dashv$' are therefore somewhat iconic; this may help the reader to remember which is which!

Relations on topological mode spaces

Given a TM space $(Q, \vdash, \dashv)$, further relations may be defined on Q as follows:

- *Perturbation.* The relation $\rightsquigarrow$ is defined as

$$q \rightsquigarrow q' \ \triangleq \ q \vdash q' \vee q \dashv q'.$$

  Thus $q \rightsquigarrow q'$ means that a direct transition is possible from q to q', abstracting away from the topological character of that transition as encoded by $\vdash$ and $\dashv$. We may say that q' is a perturbation of q.

- *Dominance.* The relation $\succ$ is defined as

$$q \succ q' \ \triangleq \ q \vdash q' \vee q' \dashv q.$$

  Thus $\succ$ encodes the topological relation between two modes while abstracting away from the temporal sequence. We may say that q dominates q'.

We have the equivalences

$$\begin{aligned} q \vdash q' &\equiv q \succ q' \wedge q \rightsquigarrow q' \\ q \dashv q' &\equiv q' \succ q \wedge q \rightsquigarrow q' \end{aligned}$$

To say that mode q dominates mode q' is to say that it is possible for q to hold at an instant which limits (at one or the other end) an open interval over which q' holds.

- *Neighbourhood.* The relation $\wr$ is defined as

$$q \wr q' \triangleq q \leadsto q' \vee q' \leadsto q.$$

If $q \wr q'$ then we may say that q and q' are *(conceptual) neighbours.*

Classification of TM spaces

We can classify TM spaces according to the character of the relations on Q, as follows:

- A TM space $(Q, \vdash, \dashv)$ is *symmetric* if $q \vdash q' \leftrightarrow q' \dashv q$ for all modes $q, q' \in Q$. In this case the relations $\vdash$ and $\succ$ coincide. The symmetry here is *time*-symmetry. An alternative characterization is to say that a TM space is symmetric if and only if its $\leadsto$ relation is symmetric. A symmetric TM space $(Q, \vdash, \dashv)$ can be presented as a *dominance space* $(Q, \succ)$, where $\succ = \vdash = \dashv^{-1}$. The dominance rule for a TM space can be simplified, in the case of a dominance space, to:

    If q holds at an instant which limits an interval over which q' holds, then $q \succ q'$.

  A dominance space can be regarded as an incidence space, with $\succ$ as the bounding relation.

- A TM space is *simple* if a mode is always at the same end (dominator or dominated) of every dominance relation it bears to modes distinct from itself:

$$q \succ q' \wedge q' \succ q'' \rightarrow q = q' \vee q' = q''.$$

  In this case, the terminology of Galton 1990 can be used: a dominating mode is a 'state of position', while a dominated mode is a 'state of motion'. This terminology cannot be used in a non-simple TM space. Note that the restriction to distinct modes is necessary, since any mode that is dominated by another mode must dominate itself, as is shown below.

- A TM space is *regular* if $q \vdash q' \wedge q \dashv q'$ implies $q = q'$. Note that a simple TM space is necessarily regular, but a regular TM space need not be simple.

- A TM space is *connected* so long as for any two modes q, q' there is a sequence $q = q_0, q_1, q_2, \ldots, q_n = q'$ such that $q_{i-1} \wr q_i$ for $i = 1, \ldots, n$.

Note that if $q \vdash q'$ then it is possible to have q holding at instant t_0 followed by q' holding over an interval (t_0, t_1). Now let $t \in (t_0, t_1)$. We then have q' holding at t and over both (t_0, t) and (t, t_1). It follows that $q' \vdash q'$ and $q' \dashv q'$. The same conclusion follows if $q' \dashv q$. Hence we have the rule

$$q \succ q' \rightarrow q' \vdash q' \wedge q' \dashv q'.$$

This says that any dominated mode both pre- and post-dominates itself.

Another consequence of the temporal incidence rules is that for every mode q there must be modes q' and q'' such that $q' \leadsto q \leadsto q''$. This is simply because a history is defined as a mapping on the whole set of real numbers, hence whenever q holds on an instant t there must be intervals (t', t) and (t, t'') with some modes q', q'' holding on them (this is because of the non-intermingling principle). There is nothing to prevent one or both of modes q' and q'' being the same as q itself.

8.4.2. Illustrations of the formal definitions

I shall illustrate these ideas by returning to some earlier examples.

TM space for a body moving under constant positive force

My first example is the TM space illustrated in Figure 8.4. There are nine modes,

$$Q = \{q_0, q_1, q_2, q_3, q_4, q_5, q_6, q_7, q_8\},$$

and the dominance and perturbation relations are shown by means of two kinds of arrowhead attached to the arcs. An open arrowhead at the end of an arc indicates perturbation ($\leadsto$) in the direction of the arrow. A triangular arrowhead at the midpoint of the arc indicates dominance ($\succ$). Thus the two kinds of arc in these diagrams are

$$\begin{array}{ll} q \rightarrow\!\!\!\!\!\!\rhd\!\!\rightarrow q' & \text{representing } q \vdash q' \\ q \rightarrow\!\!\!\!\!\!\lhd\!\!\rightarrow q' & \text{representing } q \dashv q' \end{array}$$

The square boxes represent modes q for which $q \leadsto q$ (and, equivalently, $q \succ q$).

It will be observed that the perturbation relation shown here gives us an abstract space identical in structure to the concrete one illustrated in Figure 1.6*a*, which describes the motion of a body acted on by a constant force in

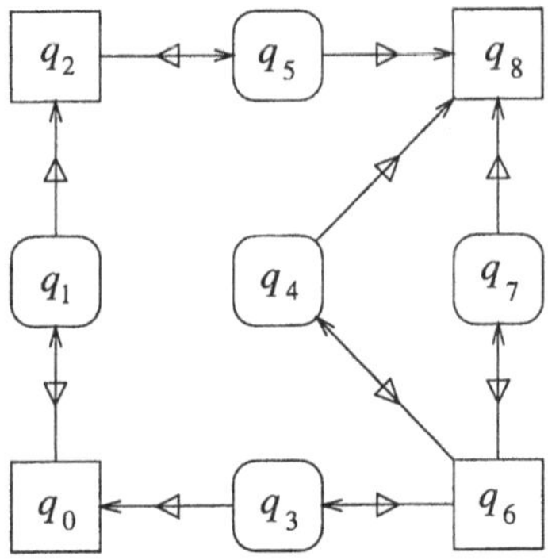

Figure 8.4. A nine-element dominance space

the positive direction. This suggests that the latter should provide a model for the former, and this is indeed the case.

To be precise, the model is (P, C, μ), where

- $P = \mathbb{R}^2$. The elements of P are intended to represent displacement-velocity pairs for our moving body;
- C is the set of functions $f : t \to (s(t), v(t))$, where s is any solution of a differential equation of the form

$$\frac{d^2 s}{dt^2} = a,$$

  where $a > 0$, and $v = ds/dt$. (These are the right-facing parabolas in Figure 1.4.)
- The mapping μ is defined by the following table:

	$s < 0$	$s = 0$	$s > 0$
$v > 0$	q_2	q_5	q_8
$v = 0$	q_1	q_4	q_7
$v < 0$	q_0	q_3	q_6

To check that this is indeed a model, we need to verify the temporal incidence rules.

Since the functions $f \in C$ all describe parabolas in phase space (the s, v-plane) symmetrical about the s-axis, and the nine modes in Q correspond to the origin, the four semi-axes, and the four quadrants in the s, v-plane, it is obvious that no chaotic intermingling of modes can occur in any interval of time.

As regards the dominance rules, we may note that the parabolic paths in phase space cross the axes at isolated points; hence any of the landmark modes involving either $s = 0$ or $v = 0$ can only hold for isolated instants, which must therefore limit intervals over which one of the non-landmark modes holds; and conversely, since the non-landmark modes correspond to open sets in phase space, any instant at which one of these modes holds must fall within an interval throughout which that mode holds. The upshot of all this is that the landmark modes dominate their neighbouring non-landmark modes, which in turn dominate only themselves; this is exactly the pattern of dominance found in Figure 8.4, and we have now established that (P, C, μ) is indeed a model for the abstract TM space $(Q, \vdash, \dashv)$. It will be observed that this TM space is simple and connected, but not symmetric.

The other diagrams in §1.4.1 likewise correspond to TM spaces which are not symmetric. The TM space corresponding to Figure 1.6*c* resembles that for (*a*) in also being connected, whereas those for Figures 1.6*b* and 1.5*b* are not connected; these spaces are all simple. We could also set up a TM space for which the full set of trajectories in s, v, a-space, as shown in Figure 1.7, provides a model. The function-set C for this particular model consists of all functions

$$f : t \to (s(t), v(t), a(t))$$

such that $v = ds/dt$ and $a = dv/dt$—so that f is determined uniquely by s, so long as the latter is continuous and twice differentiable. This TM space is not simple, since, for example, the mode $(0, +, 0)$ dominates the mode $(+, +, 0)$, which in turn dominates $(+, +, +)$. This is illustrated by the function f for which

$$\begin{aligned} s(t) &= \begin{cases} t+1 & (t < 0) \\ t^3 + t + 1 & (t \geq 0) \end{cases} \\ v(t) &= \begin{cases} 1 & (t < 0) \\ 3t^2 + 1 & (t \geq 0) \end{cases} \\ a(t) &= \begin{cases} 0 & (t < 0) \\ 6t & (t \geq 0) \end{cases} \end{aligned}$$

(The graph of the function is shown in Figure 8.5.) The mode-sequence for this function is as follows:

$$\begin{array}{rll} t < -1: & s < 0, v > 0, a = 0 & (-+0) \\ t = -1: & s = 0, v > 0, a = 0 & (0+0) \\ -1 < t \leq 0: & s > 0, v > 0, a = 0 & (++0) \\ t > 0: & s > 0, v > 0, a > 0 & (+++) \end{array}$$

Here we see that $(0+0)$ holds at the beginning of an interval over which $(++0)$ holds, so $(0+0) \vdash (++0)$; and $(++0)$ holds at the beginning of an interval over which $(+++)$ holds, so $(++0) \vdash (+++)$.

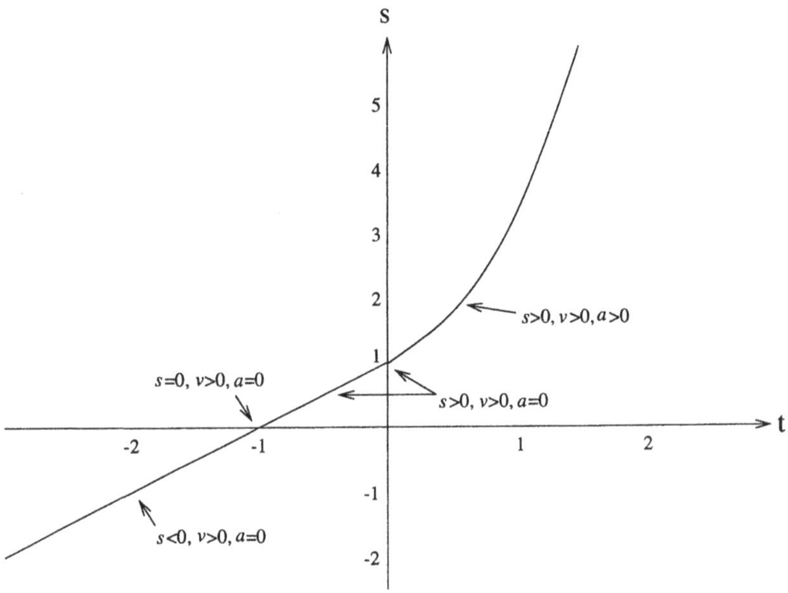

Figure 8.5. $(0+0)$ dominates $(++0)$, which dominates $(+++)$

8.5. Constructing new TM spaces from old

In this section I shall state, prove, and illustrate a series of theorems which have the effect of enabling us to build up complex mode spaces from simple ones. This will throw light on some of the mode spaces encountered in earlier chapters.

8.5.1. The smallest TM spaces

Here I shall enumerate the smallest non-trivial TM spaces, those with just two modes. (TM spaces with only one mode are possible, but they are truly trivial!)

The basis theorem for TM spaces. Up to isomorphism, there are exactly seven two-element TM spaces, as illustrated in Figure 8.6.

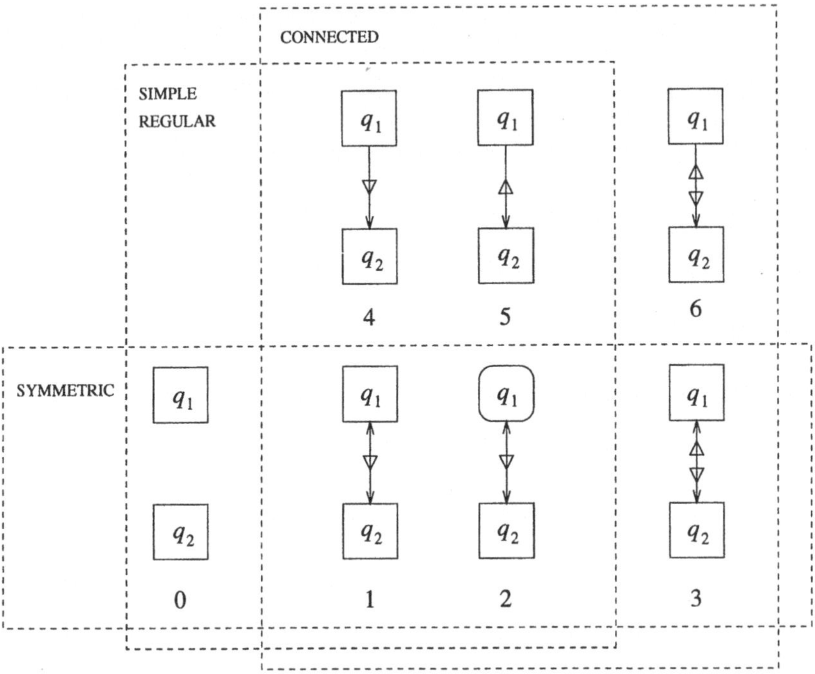

Figure 8.6. The seven distinct two-element topological mode spaces

Proof. Call the two elements of the space q_1 and q_2. The perturbation relation can contain zero, one, or two elements from $\{(q_1, q_2), (q_2, q_1)\}$, giving spaces which are respectively disconnected, non-symmetric connected, and symmetric connected.

Suppose the space is disconnected. Since each mode must have a neighbour, each must be a neighbour of itself, giving us space 0 in Figure 8.6.

Suppose next that the space is symmetric and connected, so $q_1 \rightsquigarrow q_2$

and $q_2 \rightsquigarrow q_1$. Then as regards dominance, since the perturbation $q_1 \rightsquigarrow q_2$ must involve dominance in at least one direction, at least one of $q_1 \succ q_2$ and $q_1 \prec q_2$ must hold. If only one holds then we may assume without loss of generality that it is $q_1 \succ q_2$. Since a dominated mode dominates itself, this implies $q_2 \succ q_2$. If $q_1 \succ q_1$ also, then we have space 1; if not, space 2. If both $q_1 \succ q_2$ and $q_1 \prec q_2$ (making the space non-regular), then we must have both $q_1 \succ q_1$ and $q_2 \succ q_2$ since both modes are dominated, giving space 3.

Suppose finally the space is connected but non-symmetric, so that exactly one of $q_1 \rightsquigarrow q_2$ and $q_2 \rightsquigarrow q_1$ holds. Suppose without loss of generality the former. Since each mode must both have a perturbation and be a perturbation, we must also have $q_1 \rightsquigarrow q_1$ and $q_2 \rightsquigarrow q_2$. As above, at least one of $q_1 \succ q_2$ and $q_1 \prec q_2$ must hold. If only the former, we have space 4; if only the latter, space 5; if both (making the space non-regular), space 6.

Having proved that no more than the seven spaces portrayed in Figure 8.6 are possible, it remains to show that all these seven *are* possible. I shall do this by exhibiting concrete models for the abstract spaces, as follows:

0. A trivial case. Let x be a continuous real variable restricted to two values 0 and 1, and let the two modes be $x = 0$ and $x = 1$. Since the variable is continuous, it cannot jump between 0 and 1 and hence must be constant.

1. Let x be a real variable free to vary in any way, so long as it is continuous, and let the modes be $q_1 : x \leq 0$, $q_2 : x > 0$. That $q_1 \succ q_2$ follows from the continuity rule.

2. As above, but with x restricted to being non-negative, and not constant over any interval, with the modes $q_1 : x = 0$, $q_2 : x > 0$. Since x is nowhere constant, x can only attain the value 0 as an instantaneous minimum, hence q_1 is not a perturbation of itself.

3. Let x, y be two continuous real variables, and let the modes be

$$\begin{aligned} q_1 :&\quad x > 0 \wedge y \geq 0 \\ q_2 :&\quad x \leq 0 \vee y < 0 \end{aligned}$$

   Then each mode must dominate the other, as illustrated by the trajectory given by $x = 2 - t, y = t - 1$, for which the sequence of modes is

$$\begin{aligned} t < 1 :&\quad q_2 \text{ (since } y < 0) \\ 1 \leq t < 2 :&\quad q_1 \\ t \geq 2 :&\quad q_2 \text{ (since } x \leq 0) \end{aligned}$$

Thus we have q_1 at $t = 1$, which limits the interval $(0, 1)$ over which we have q_2; and equally, we have q_2 at $t = 2$, which limits the interval $(1, 2)$, over which we have q_1.

4. Let $x = x_0 - kt$ where $x_0 > 0$ and $k > 0$. Let the modes be $q_1 : x \geq 0, q_2 : x < 0$. For $t \leq x_0/k$, the mode is q_1, whereas for $t > x_0/k$, it is q_2, in accordance with the perturbation and dominance relations for this TM space.

5. The same, but with $q_1 : x > 0$, $q_2 : x \leq 0$.

6. Let x, y be two continuous real variables with y varying in such a way that $dy/dt < 0$ at all times. Let the modes be

$$\begin{aligned} q_1 : &\quad y > 0 \vee (y = 0 \wedge x \geq 0) \\ q_2 : &\quad y < 0 \vee (y = 0 \wedge x < 0) \end{aligned}$$

Then since y can only decrease, transition is possible from q_1 to q_2 but not the other way round. The transition can occur only once, when $y = 0$. At the instant of transition, q_1 holds if $x \geq 0$, but q_2 holds if $x < 0$. Thus each mode dominates the other. □

8.5.2. Products of TM spaces

Given two TM spaces, we can form a larger TM space by taking their Cartesian product. The product theorem shows how this is done.

The product theorem for TM spaces. For $i = 1, \ldots, n$, let (P_i, C_i, μ_i) be a model for the TM space $(Q_i, \vdash_i, \dashv_i)$. Then (P, C, μ) is a model for $(Q, \vdash, \dashv)$, where

- $P = P_1 \times \cdots \times P_n$
- $Q = Q_1 \times \cdots \times Q_n$
- $f \in C$ if and only if there are functions $f_i \in C_i$ $(i = 1, \ldots, n)$ such that for all $t \in \mathbb{R}$, $f(t) = (f_1(t), \ldots, f_n(t))$.
- $\mu : (p_1, \ldots, p_n) \mapsto (\mu_1(p_1), \ldots, \mu_n(p_n))$
- $(q_1, \ldots, q_n) \vdash (q'_1, \ldots, q'_n)$ if and only if $q_i \vdash q'_i$ for $i = 1, \ldots, n$.
- $(q_1, \ldots, q_n) \dashv (q'_1, \ldots, q'_n)$ if and only if $q_i \dashv q'_i$ for $i = 1, \ldots, n$.

Proof. We must check the temporal incidence rules.

No intermingling. Let $t < t'$. Then for each $i = 1, \ldots, n$, since (P_i, C_i, μ_i) is a model for $(Q_i, \vdash_i, \dashv_i)$, there is a finite sequence

$$t = t_{i0} < t_{i1} < \cdots < t_{ik_i} = t'$$

such that for whenever $t_{i(j-1)} < u < v < t_{ij}$ then $\mu_i(f_i(u)) = \mu_i(f_i(v))$. Now let

$$S = \{t_{ij} \mid 1 \leq i \leq n, 0 \leq j \leq k_i\},$$

and label the distinct elements of S as $t_0 < t_1 < \cdots < t_k$. Evidently $t_0 = t_{10} = t_{20} = \cdots = t_{n0} = t$ and $t_k = t_{1k_1} = t_{2k_2} = \cdots = t_{nk_n} = t'$, and the remaining elements are finite in number. Now suppose $t_{i-1} < u < v < t_i$. Since no instant t_{xy} occurs in the interval (t_{i-1}, t_i), for $j = 1, \ldots, n$ there is a number l, $1 \leq l \leq k_j$, such that $t_{j(l-1)} < u < v < t_{jl}$, from which it follows that $\mu_j(f_j(u)) = \mu_j(f_j(v))$. Since this holds for all $j = 1, \ldots, n$, we have

$$\begin{aligned}
\mu(f(u)) &= \mu((f_1(u), \ldots, f_n(u))) \\
&= (\mu_1(f_1(u)), \ldots, \mu_n(f_n(u))) \\
&= (\mu_1(f_1(v)), \ldots, \mu_n(f_n(v))) \\
&= \mu((f_1(v), \ldots, f_n(v))) \\
&= \mu(f(v))
\end{aligned}$$

so the no-intermingling rule is satisfied.

Dominance. Suppose $\mu(f(t)) = (q_1, \ldots, q_n)$ for all $t \in (t_0, t_1)$. Then

$$\mu_1(f_1(t)) = q_1, \ldots, \mu_n(f_n(t)) = q_n$$

for all $t \in (t_0, t_1)$. By the individual dominance rules for the component spaces, this means that

$$\mu_1(f_1(t_0)) \vdash q_1 \dashv \mu_1(f_1(t_1)), \;\ldots,\; \mu_n(f_n(t_0)) \vdash q_n \dashv \mu_n(f_n(t_1)).$$

Hence

$$\begin{aligned}
(\mu_1(f_1(t_0)), \ldots, \mu_n(f_n(t_0))) &\vdash (q_1, \ldots, q_n) \\
&\dashv (\mu_1(f_1(t_1)), \ldots, \mu_n(f_n(t_1))),
\end{aligned}$$

so $\mu(f(t_0)) \vdash (q_1, \ldots, q_n) \dashv \mu(f(t_1))$, as required. □

The product theorem tells us that the Cartesian product of two or more TM spaces is itself a TM space, whose models include the Cartesian products of models for the component TM spaces. To illustrate this theorem, I shall first consider two models of the two-element TM space numbered 1 in Figure 8.6. Since this TM space is symmetric, we can write $\succ$ instead of $\vdash$ and $\dashv$. Let us put:

$$Q_1 = \{-xneg, xneg\}, Q_2 = \{-ypos, ypos\}$$
$$\succ_1 = \{(-xneg, -xneg), (-xneg, xneg), (xneg, xneg)\}$$
$$\succ_2 = \{(-ypos, -ypos), (-ypos, ypos), (ypos, ypos)\}$$

C is the space of all continuous functions from $\mathbb{R}$ to $\mathbb{R}$

$$\mu_1(x) = \begin{cases} -xneg & \text{if } x \geq 0 \\ xneg & \text{if } x < 0 \end{cases}$$
$$\mu_2(y) = \begin{cases} -ypos & \text{if } y \leq 0 \\ ypos & \text{if } y > 0 \end{cases}$$

It is easy to verify that $(\mathbb{R}, C, \mu_1)$ and $(\mathbb{R}, C, \mu_2)$ are models for $(Q_1, \succ_1)$ and $(Q_2, \succ_2)$ respectively. The product theorem then tells us that

$$(\mathbb{R} \times \mathbb{R}, C', \mu),$$

where C' is the set of continuous functions from $\mathbb{R}$ to $\mathbb{R}^2$ and $\mu((x, y)) = (\mu_1(x), \mu_2(y))$, is a model for a four-element TM space $(Q_1 \times Q_2, \succ)$ as illustrated in Figure 8.7*a*. In this figure, since the TM space is symmetric, the perturbation arrows (at the ends of the arcs) have been omitted. I shall follow this convention in all subsequent diagrams of symmetric TM spaces.

It is natural to interpret the mode-pair (q, q') as a logical conjunction, that is, as the mode which holds at exactly those times at which both the modes q and q' hold. This is consistent with interpreting the functions in C as specifying phase histories, each function $f \in C$ representing a particular possible phase history of the system under consideration, such that at time t the system is in phase $f(t)$. Thus the modes of the compound system illustrated in Figure 8.7*a* capture the qualitative behaviour of two independent variables x and y.

The diagram shows a link between $(xneg, ypos)$ and $(-xneg, -ypos)$; this transition is realized, for example, by the phase history f defined by $f(t) = (t, -t)$. For $t < 0$ we have

$$\mu(f(t)) = (\mu_1(t), \mu_2(-t)) = (xneg, ypos),$$

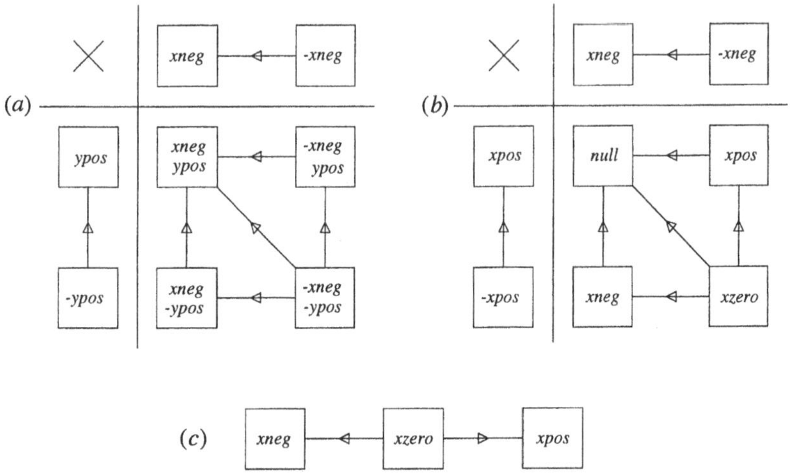

Figure 8.7. Some simple dominance spaces, illustrating the product theorem

while for $t \geq 0$ we have $\mu(f(t)) = (-xneg, -ypos)$. There is a direct transition from the former mode to the latter, and it is the latter which holds at the instant of transition, $t = 0$, as predicted by the dominance arrow in the diagram.

On the other hand, the diagram shows no link between $(xneg, -ypos)$ and $(-xneg, ypos)$. To understand why this is, suppose that in some history $(xneg, -ypos)$ holds over the interval (t_0, t_1) and $(-xeg, ypos)$ holds over the interval (t_1, t_2). On the face of it we have here a direct transition between the former mode and the latter; but this appearance is deceptive, for consider what mode holds at the instant t_1. Since $xneg$ is dominated by $-xneg$, the latter must hold at t_1; but $-ypos$ dominates $ypos$, so the former must hold at that instant. Thus at t_1 we have the mode $(-xneg, -ypos)$: the transition between $(xneg, -ypos)$ and $(-xneg, ypos)$ is thus not direct but is mediated by $(-xneg, -ypos)$, which admittedly holds only for an instant.

8.5.3. Subspaces of topological mode spaces

Suppose we were to replace Q_2 in the example above by $\{-xneg, xneg\}$, everything else remaining the same, *mutatis mutandis*. The four modes of our compound space are now

$$(xneg, xpos),\ (xneg, -xpos),\ (-xneg, -xpos),\ (-xneg, xpos).$$

Interpreting tuples as conjunctions, as always, we note that the first pair, corresponding as it does to the condition $x < 0 \wedge x > 0$, can never hold, and we shall accordingly label it $null$. The remaining pairs can hold, and correspond to the conditions $x < 0$, $x = 0$, and $x > 0$ respectively. In Figure 8.7*b* these modes are labelled $xneg$, $xzero$, and $xpos$ respectively. Since the mode labelled $null$ represents an impossible condition, we can drop it from the diagram, together with all links between it and the other modes, to give us the three-element TM space illustrated in Figure 8.7*c*. In effect, this is obtained from the original space in (*a*) by adding the extra condition $x = y$. This condition collapses the phase space $\mathbb{R} \times \mathbb{R}$ to $\{(x, x) \mid x \in \mathbb{R}\}$, which is isomorphic in an obvious sense to $\mathbb{R}$ itself. Our next theorem provides the full details of 'collapses' of this kind.

The first subspace theorem for TM spaces. Let $(P, \mathcal{C}, \mu)$ be a model for the TM space $(Q, \vdash, \dashv)$. Let $P' \subset P$, and let

$$\begin{aligned} \mathcal{C}' &= \{f \in \mathcal{C} \mid \forall t \in \mathbb{R}, f(t) \in P'\} \\ Q' &= \{\mu(p) \mid p \in P'\} \end{aligned}$$

Let $\vdash', \dashv'$ be the restrictions of $\vdash, \dashv$ to Q', and let μ' be the restriction of μ to P'. Then $(P', \mathcal{C}', \mu')$ is a model for the TM space $(Q', \vdash', \dashv')$.

Proof. As usual, we need to check the temporal incidence rules. The no-intermingling rule follows immediately from the fact that $\mathcal{C}' \subseteq \mathcal{C}$.

For the dominance rule, let $f \in \mathcal{C}'$, and suppose $\mu'(f(t)) = q \in Q'$ for all $t \in (t_0, t_1)$. Then $\mu(f(t)) = q$ for all $t \in (t_0, t_1)$, so $\mu(f(t_0)) \vdash q \dashv \mu(f(t_1))$. Since $f \in \mathcal{C}'$, we have $f(t_0) \in P'$ and $f(t_1) \in P'$, so $\mu(f(t_0)) = \mu'(f(t_0)) \in Q'$ and $\mu(f(t_1)) = \mu'(f(t_1)) \in Q'$. Hence $\mu'(f(t_0)) \vdash' q \dashv' \mu'(f(t_1))$, as required. □

In Figure 8.7, the phase space in (*a*) is $P = \mathbb{R} \times \mathbb{R}$. In (*b*) and (*c*), it is $P' = \{(x, x) \mid x \in \mathbb{R}\}$, which can be replaced by $\mathbb{R}$ without changing the structure of the space.

If we now form the Cartesian product of two copies of the three-element space shown in Figure 8.7*c*, we obtain the nine-element space illustrated in Figure 8.8. This diagram throws some light on a problem discussed by Davis (1990: 171 ff.). Davis considers the case of two independent pendula, labelled A and B. For the purpose of this discussion, the state of each pendulum is described in terms of its angular displacement from the vertical; let this be x for pendulum A and y for pendulum B. Then the modes of pendulum A can

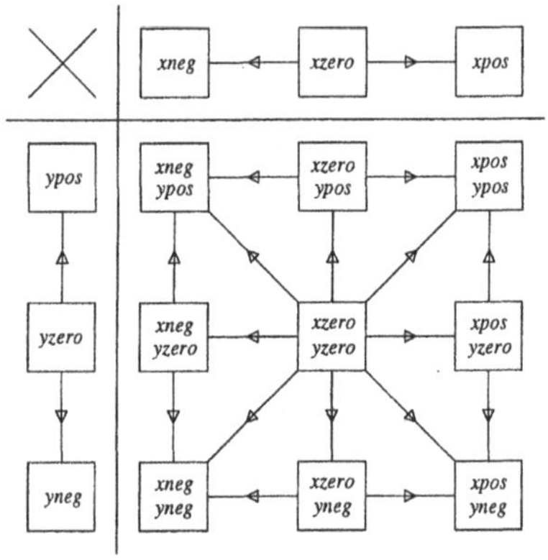

Figure 8.8. A nine-element TM space formed as the product of two independent three-element spaces

be represented as $xneg$ (when the pendulum is inclined to the left), $xzero$ (when it is vertical), and $xpos$ (when it is inclined to the right), and the states of pendulum B are similarly $yneg$, $yzero$, and $ypos$. Davis considers the question: suppose that initially pendulum A is left of centre and pendulum B is at the centre. What are the possibilities for the succeeding state? Davis shows, using a complex set of 'Rules for Mode Transitions', that there are three possibilities: A and B both at the centre, A and B both left of centre, and A left of centre with B right of centre. These possibilities can be read off immediately from Figure 8.8, the transitions in question being:

$$
\begin{aligned}
(xneg, yzero) &\rightarrow (xzero, yzero) \\
(xneg, yzero) &\rightarrow (xneg, yneg) \\
(xneg, yzero) &\rightarrow (xneg, ypos)
\end{aligned}
$$

In particular, no direct transition is possible between the mode $(xneg, yzero)$ and either $(xzero, ypos)$ or $(xzero, yneg)$, which prima facie one might have thought possible. The problem of whether or not there can be a direct transition between two states of the complex system can be *stated*—as it was by Davis in the example just considered—without any reference to the concept of dominance; it is a problem which can exist for someone who is ignorant of, or repudiates, this notion. None the less, it is dominance which most easily allows us to *solve* the problem.

Returning now to the diagrams in Figures 1.5*b* and 1.6*a*–*c*, notice that apart from the labels of the modes each of the four diagrams can be obtained from Figure 8.8 by omitting some of the perturbation arrows. (Recall that since the space in Figure 8.8 is symmetric, each edge is understood as having an arrowhead at each end—hence the edges in the other figures with an arrowhead at one end only can be obtained from these by the omission of an arrowhead.) It is naturally understood that when both arrowheads are removed from an edge, the edge itself, together with any dominance arrows it carries, also disappear. This kind of reduction is handled in the next theorem.

The second subspace theorem for TM spaces. Let $(P, \mathcal{C}, \mu)$ be a model for the TM space $(Q, \vdash, \dashv)$. Let $\mathcal{C}' \subset \mathcal{C}$, and define relations $\vdash', \dashv'$ as follows:

- $q_1 \vdash' q_2$ if and only if there is a function $f \in \mathcal{C}'$ such that for some $t_0 < t_1 \in \mathbb{R}$, $\mu(f(t_0)) = q_1$ and for every $t \in (t_0, t_1)$, $\mu(f(t)) = q_2$.
- $q_1 \dashv' q_2$ if and only if there is a function $f \in \mathcal{C}'$ such that for some $t_0 < t_1 \in \mathbb{R}$, $\mu(f(t_1)) = q_2$ and for every $t \in (t_0, t_1)$, $\mu(f(t)) = q_1$.

Then $(P, \mathcal{C}', \mu)$ is a model for $(Q, \vdash', \dashv')$, and moreover, $\vdash' \subseteq \vdash$ and $\dashv' \subseteq \dashv$.

Proof. The no-intermingling rule follows immediately from the fact that $\mathcal{C}' \subset \mathcal{C}$.

For the dominance rule, let $f \in \mathcal{C}'$, and suppose $\mu(f(t)) = q$ for all $t \in (t_0, t_1)$. It follows immediately from the definitions of $\vdash'$ and $\dashv'$ that $\mu(f(t_0)) \vdash' q \dashv' \mu(f(t_1))$.

Now suppose $q_1 \vdash' q_2$. Then there is a function $f \in \mathcal{C}'$ such that for some $t_0 < t_1 \in \mathbb{R}$, $\mu(f(t_0)) = q_1$ and for every $t \in (t_0, t_1)$, $\mu(f(t)) = q_2$. Since $\mathcal{C}' \subset \mathcal{C}$, we have $f \in \mathcal{C}$, so the above condition also gives us $q_1 \vdash q_2$. Hence $\vdash' \subseteq \vdash$, and a similar argument likewise establishes that $\dashv' \subseteq \dashv$. □

If we take the full nine-element product space shown in Figure 8.8, for which the set of all continuous functions from $\mathbb{R}$ to $\mathbb{R} \times \mathbb{R}$ provides a model (with μ given as above), then models for our earlier nine-element spaces can be obtained, in accordance with the second subspace theorem, by setting $\mathcal{C}'$ as follows:

- Figure 1.5*b*: $\mathcal{C}'$ is the the set of functions of the form

$$f : t \mapsto (A \sin(kt + e), kA \cos(kt + e)).$$

- Figure 1.6*a*: $\mathcal{C}'$ is the the set of functions of the form

$$f : t \mapsto (A + Bt + \tfrac{1}{2}Ct^2, B + Ct),$$

where $C > 0$;

- Figure 1.6b: C' is the the set of functions of the form

$$f : t \mapsto (A + Bt, B);$$

- Figure 1.6c: As for (a) but with $C < 0$;

Note that these choices of C' are not unique; other sets of functions will also provide models for the same TM space; the ones I have given are those suggested by my original motivation for constructing those spaces.

8.5.4. Quotients of topological mode spaces

Returning now to the discussion of pendula, Davis notes, in relation to a single pendulum (which for convenience we shall assume to be the pendulum A of our earlier example), that there are two distinct circumstances in which the pendulum may attain the mode $xzero$: either in passing from $xneg$ to $xpos$, or in passing from $xpos$ to $xneg$. This leads him to refine the partition of the mode space by splitting $xzero$ into two modes, one for 'at centre, moving left', the other for 'at centre, moving right'. Let us denote these $\overleftarrow{xzero}$ and $\overrightarrow{xzero}$. (Here I am ignoring for simplicity the possibility of the pendulum's being stationary at the centre.) The possible transitions are now as shown in Figure 8.9.

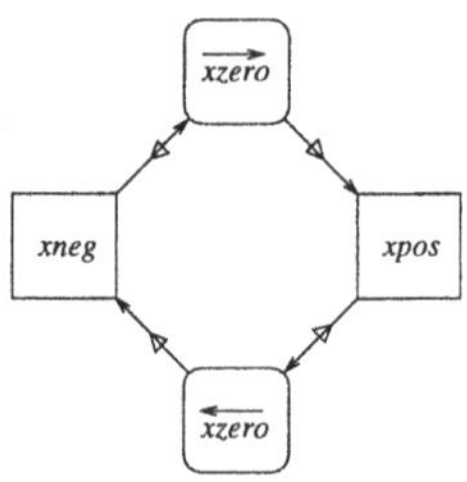

Figure 8.9. A four-element mode space for a single pendulum

The dominance information presented here is obvious enough; but where does it come from? The four-element TM space can be derived from the nine-element TM space shown in Figure 1.5b (itself derived, as we have seen, from the full Cartesian product of two three-element spaces) by the amalgamation of certain modes: $xneg$ combines the three modes $(x < 0, v > 0)$, $(x < 0, v = 0)$, and $(x < 0, v < 0)$ into a single mode, and similarly with $xpos$.

The centre mode, $(x = 0, v = 0)$, has been dropped, as justified by the first subspace theorem.

The amalgamation seen here is justified by the following theorem:

The quotient theorem for TM spaces. Let $(P, \mathcal{C}, \mu)$ be a model for TM space $(Q, \vdash, \dashv)$, and let $\cong$ be any equivalence relation on Q. For each $q \in Q$, define $[\![q]\!] = \{q' \in Q \mid q' \cong q\}$ (this is the equivalence class to which q belongs). Then $(P, \mathcal{C}, \mu')$ is a model for the quotient TM space $(Q/{\cong}, \vdash', \dashv')$, where

- $Q/{\cong} = \{[\![q]\!] \mid q \in Q\}$.
- For each $p \in P$, $\mu'(p) = [\![\mu(p)]\!]$.
- For $q_1, q_2 \in Q$, $[\![q_1]\!] \vdash' [\![q_2]\!]$ if, and only if, there exist $q_1' \in [\![q_1]\!], q_2' \in [\![q_2]\!]$ such that $q_1' \vdash q_2'$.
- For $q_1, q_2 \in Q$, $[\![q_1]\!] \dashv' [\![q_2]\!]$ if, and only if, there exist $q_1' \in [\![q_1]\!], q_2' \in [\![q_2]\!]$ such that $q_1' \dashv q_2'$.

Proof. Clearly any intermingling in the quotient space would show up as intermingling in the original space, so the absence of intermingling in the latter implies its absence in the former.

Now let $f \in \mathcal{C}$, and suppose $\mu'(f(t)) = [\![q]\!] \in Q/{\cong}$ for all $t \in (t', t'')$. By the no-intermingling rule for Q, there are instants

$$t' = t_0 < t_1 < \cdots < t_n = t''$$

such that $\mu(f(t))$ is constant over each interval (t_{i-1}, t_i) $(i = 1, \ldots, n)$, say $\mu(f(t)) = q_i \in Q$ for $t \in (t_{i-1}, t_i)$. Then $q_i \in [\![q]\!]$ for $i = 1, \ldots, n$.

From the fact that $\mu(f(t)) = q_1$ for all $t \in (t', t_1)$, we conclude that $\mu(f(t')) \vdash q_1$, and hence that $\mu'(f(t')) \vdash' [\![q]\!]$. Similarly, we infer $[\![q]\!] \dashv' \mu'(f(t''))$. Hence the dominance rule is satisfied as required. □

The four-element version of the mode space for a single pendulum (Figure 8.9) can be used instead of the three-element version to generate a sixteen-element version of the mode space for two independent pendula, as shown in Figure 8.10. (The nine-element version was Figure 8.8.)

The dominance relations are determined, as usual, by the product theorem. Note how important dominance is in determining the available transitions. If one ignored dominance, and computed transitions on the basis of perturbation alone, one would find eight additional transitions, forming the two orbits

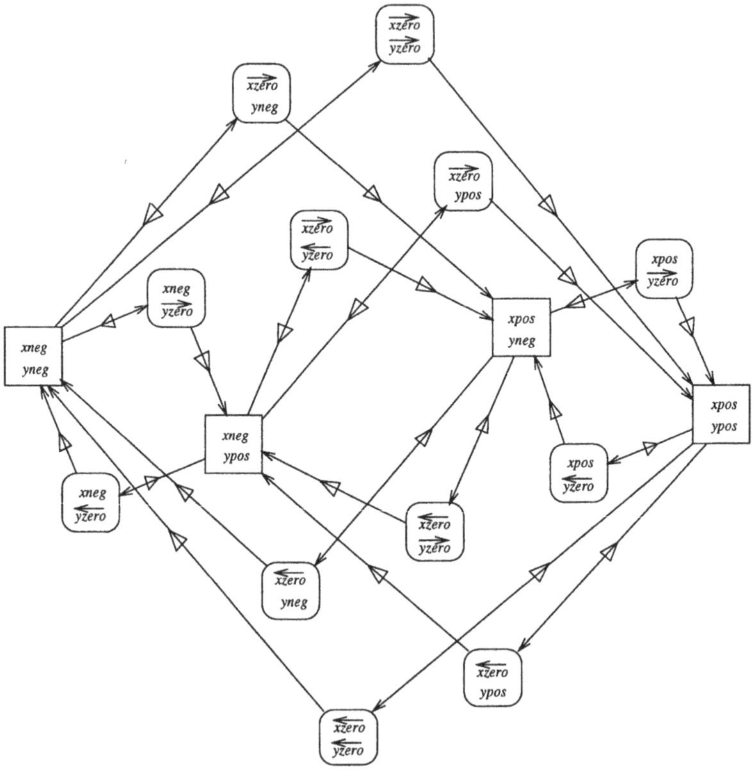

Figure 8.10. TM space for two independent pendula (sixteen-element version)

$$\begin{array}{ccc} (\overrightarrow{xzero}, yneg) & \to & (xpos, \overrightarrow{yzero}) \\ \uparrow & & \downarrow \\ (xneg, \overleftarrow{yzero}) & \leftarrow & (\overleftarrow{xzero}, ypos) \end{array}$$

and

$$\begin{array}{ccc} (\overrightarrow{xzero}, ypos) & \to & (xpos, \overleftarrow{yzero}) \\ \uparrow & & \downarrow \\ (xneg, \overrightarrow{yzero}) & \leftarrow & (\overleftarrow{xzero}, yneg) \end{array} .$$

These transitions are not, in fact, possible, and dominance successfully rules them out.

8.6. Further applications of the TM space theorems

8.6.1. Region–region relations

I return now to the space of RCC relations between spatial regions. In Figure 8.3, the eight relations between two spherical regions A and B are presented as elements of a partition of the quarter-plane determined by the two real variables r (the radius of A) and d (the distance between the centres), it being assumed that B has radius 1 unit.

The associated dominance space can be derived as a product of two simpler spaces. First we transform the variables as follows:

$$\begin{aligned} x &= d - r, \\ y &= d + r. \end{aligned}$$

Next, we note that the eight relations can be expressed in terms of x and y as follows:

DC	$x > 1$	$(y > 1)$
EC	$x = 1$	$(y > 1)$
PO	$\lvert x\rvert < 1$	$y > 1$
EQ	$x = -1$	$y = 1$
TPP	$(\lvert x\rvert < 1)$	$y = 1$
$NTPP$	$(\lvert x\rvert < 1)$	$y < 1$
$TPPI$	$x = -1$	$y > 1$
$NTPPI$	$x < -1$	$(y > 1)$

The conditions in parentheses are superfluous in that they are implied by the other condition in the same row, given the overall constraint that $d \geq 0, r > 0$.

We now set up a dominance space with five elements corresponding to the conditions $x > 1$, $x = 1$, $|x| < 1$, $x = -1$, and $x < -1$, with the second and fourth dominating their immediate neighbours; and a three-element space with $y = 1$ dominating both $y < 1$ and $y > 1$. The Cartesian product of these two spaces has fifteen elements, seven of which are impossible given that $d \geq 0, r > 0$. The product theorem determines the dominance relations amongst these fifteen elements, while the first subspace theorem allows us to drop the impossible ones. In this way we retrieve the correct conceptual neighbourhood diagram for the region–region relations (as shown in Figure 2.9), and have the correct dominance relations in addition. Note that the region calculus is a TM space which is symmetric, regular, and connected, but not simple. Because it is symmetric, we can drop the arrowheads indicating temporal sequence. We observe that EC and EQ dominate all their neighbours, and DC, PO, $NTPP$, and $NTPPI$ are dominated by all their neighbours. This means that TPP dominates PO and $NTPP$, but is dominated by EQ; and likewise $TPPI$ dominates PO and $NTPPI$ and is dominated by EQ: this is what make the space non-simple. The complete picture is as shown in Figure 8.11, in which the 'impossible' modes and their connections are indicated by dotted lines.

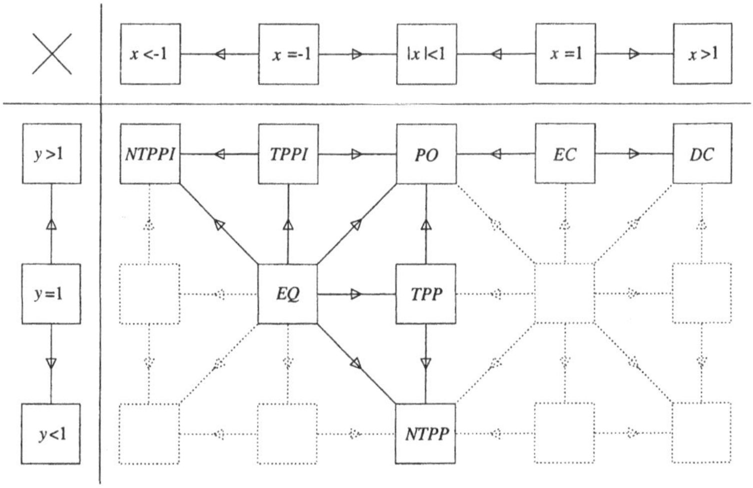

Figure 8.11. The qualitative region–region relations as a dominance space (spherical regions)

Neat though this derivation is, it has the drawback that it depends crucially on the assumption that the regions are spheres. For a more general derivation, we must consider a different set of numerical parameters. Taking our cue from Egenhofer's work, we may note that the nature of the overlap, if any, between regions A and B plays an important part in determining their qualitative relationship. Specifically, we can ask what proportion of A falls within B, and vice versa. This gives us two numerical parameters a and b

which between them suffice to distinguish most of the relations, as follows:

	$a = 0$	$0 < a < 1$	$a = 1$
$b = 0$	DC, EC	–	–
$0 < b < 1$	–	PO	$TPP, NTPP$
$b = 1$	–	$TPPI, NTPPI$	EQ

I am assuming here that the regions are all bounded and of the same dimension. To distinguish those pairs which have the same a, b values, I introduce a third parameter, c, measuring the minimum distance between a boundary point of A and a boundary point of B. We then find that the relations DC, $NTPP$, and $NTPPI$ all have $c > 0$, whereas EC, PO, TPP, $NTPP$, and EQ all have $c = 0$. (An additional assumption here is that each region consists of a single connected component—otherwise the relation between A and B could be PO even though they share no boundary points.)

These parameters, partitioned as above, yield three dominance spaces, having three, three, and two members respectively. The region–region relation space is a subspace of their Cartesian product, as shown in Figure 8.12. This gives rise to exactly the same dominance space as before, showing that the restriction to spherical regions was not so draconian as appeared at first sight.

Having obtained the TM space for the region–region relations, by whatever means, we can then use this to obtain answers to questions concerning more complex relationships. For example, consider the case of a non-rigid body A whose position is to be specified with respect to two given disconnected regions B and C. Here we are dealing with the Cartesian product of two copies of the eight-member space we have been considering. Of the sixty-four prima-facie possibilities, thirty-three are impossible for the reason that since B and C are disconnected from one another, A can only be TPP, $NTPP$, or EQ to one of these regions if it is DC to the other. The full diagram can be seen in Galton 1995: 394. Now suppose that at time t, A bears the relation DC to B and EC to C, so the current mode is represented by the pair (DC, EC). At a later time it is observed that A's position with respect to the two regions is reversed, that is, we have the pair (EC, DC). Can this have been accomplished by a direct transition? Since EC dominates DC, the portion of the dominance space we are concerned with is as shown in Figure 8.13. Here the compound modes are labelled by little pictures illustrating the configurations they represent. It is clear that there can be no direct transition between (DC, EC) and (EC, DC). Either A breaks contact with C before

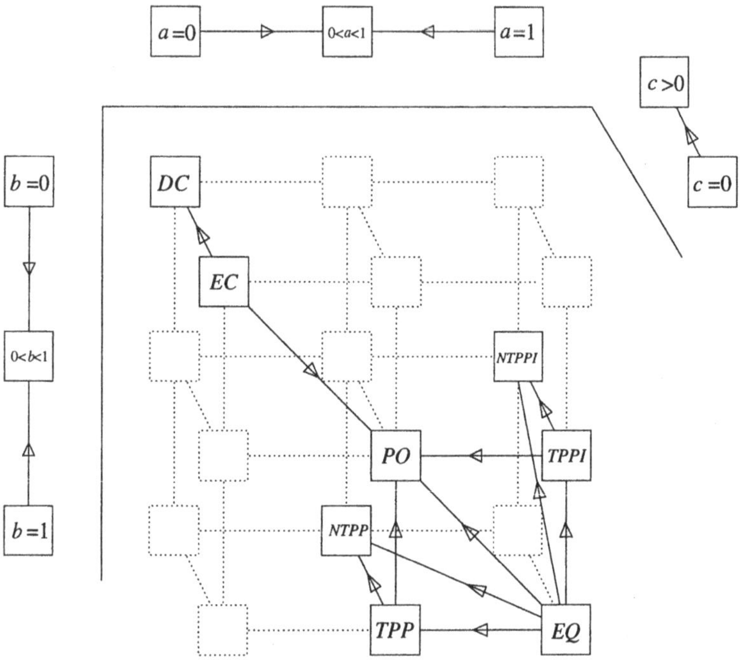

Figure 8.12. The qualitative region–region relations as a dominance space (arbitrary self-connected regions of homogeneous dimension)

it makes contact with B, in which case the mode (DC, DC) holds over the interval between these two events, or it makes contact with B either at the same instant as or before it breaks contact with C, in which case the mode (EC, EC) holds, for a single instant in the former case or over an interval in the latter. All this can be worked out from a careful consideration of continuity, but the benefit of working with the theory of dominance is that it provides us with a uniform and systematic method for obtaining the answer without having to invoke continuity considerations explicitly each time.

If we now compare Figure 8.13 with Figure 6.7, we obtain an insight into the nature of 'connectivity space'. It too is a mode space, and its structure is determined by the same kind of factors as those we have been considering. The two methods of hole-making which give rise to the two distinct maps of connectivity space shown in the latter figure behave differently with respect to dominance: if a hole is created by perforation, then the holeless state dominates the holed state, whereas if the hole is created by enclosure, then it is the other way round. This determines the directions of the dominance arrows

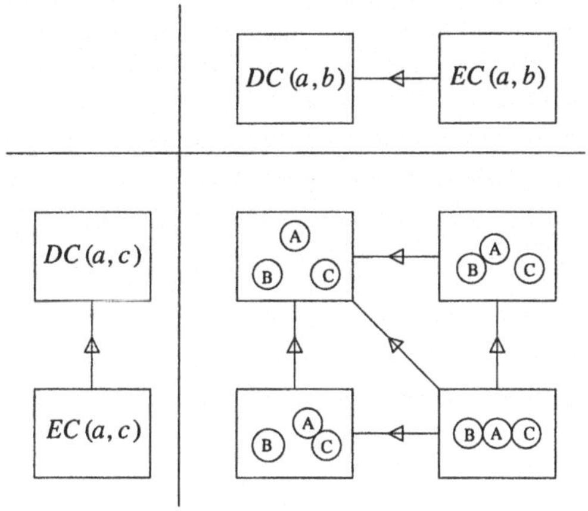

Figure 8.13. A body making and breaking contact with two disconnected regions

to be added to the conceptual neighbourhood diagram (Figure 6.6), which in turn determines which diagonals should be linked. The linkages for the small part of connectivity space shown in Figure 6.7 are as shown in the two parts of Figure 8.14.

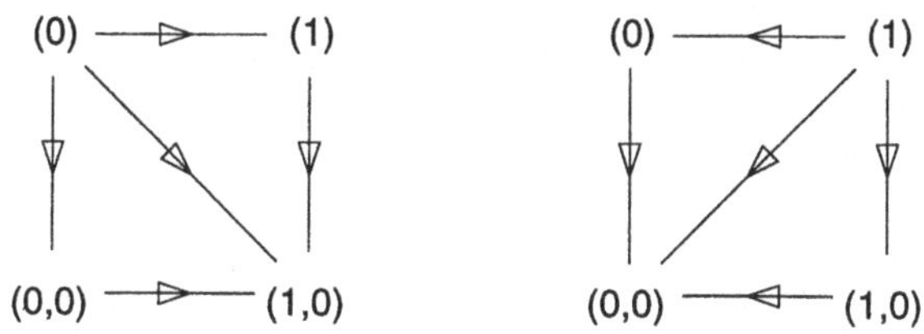

Figure 8.14. Dominance relations in a small part of connectivity space (cf. Figure 6.7)

The pattern of connections forming a square with just one of its diagonals is absolutely characteristic of mode spaces which are constructed by means of the product theorem. Whenever we look at mode spaces of complex qualitative systems, we find this pattern recurring over and over again, eloquent testimony to the far-reaching power of that theorem.

8.6.2. Interval–interval relations

The thirteen qualitatively distinct interval–interval relations listed in Table 5.1 can be specified in terms of the relative sequence of the beginning- and end-points of the two intervals, as indicated in Table 8.1, in which x_1 and x_2 refer to the beginning and ending of the interval x, similarly with y, and the relationships presented are those that x bears to y. Entries enclosed in parentheses are superfluous in the sense that they can be inferred from the unparenthesized entries in the same row.

Table 8.1. *The thirteen interval-interval relations determined by the sequence of end-points*

	x_1, y_1	x_1, y_2	x_2, y_1	x_2, y_2
$<$	$(<)$	$(<)$	$<$	$(<)$
m	$(<)$	$(<)$	$=$	$(<)$
o	$<$	$(<)$	$>$	$<$
s	$=$	$(<)$	$(>)$	$<$
d	$>$	$(<)$	$(>)$	$<$
f	$>$	$(<)$	$(>)$	$=$
$=$	$=$	$(<)$	$(>)$	$=$
fi	$<$	$(<)$	$(>)$	$=$
di	$<$	$(<)$	$(>)$	$>$
si	$=$	$(<)$	$(>)$	$>$
oi	$>$	$<$	$(>)$	$>$
mi	$(>)$	$=$	$(>)$	$(>)$
$>$	$(>)$	$>$	$(>)$	$(>)$

The relations $<, =, >$ on each of the pairs (x_1, y_1), (x_1, y_2), (x_2, y_1), and (x_2, y_2) form four three-element dominance spaces (with $=$ dominating both $<$ and $>$ in each case). If these were independent of one another, we could form an 81-member Cartesian product of the four spaces, with all the dominance relations given by the product theorem. However, as shown in Table 8.1, the spaces are not independent. In particular, because $y_1 < y_2$, we know that if $x_1 \leq y_1$ then $x_1 < y_2$, and if $x_2 \leq y_1$ then $x_2 < y_2$. This means that the pairs (x_1, y_1) and (x_1, y_2) combine to form only a five-member space, as do the pairs (x_2, y_1) and (x_2, y_2). These five-member spaces are shown in Figure 8.15.

The two five-element spaces formed in this way are not wholly independent: if we form the 25-element Cartesian product, we find that twelve

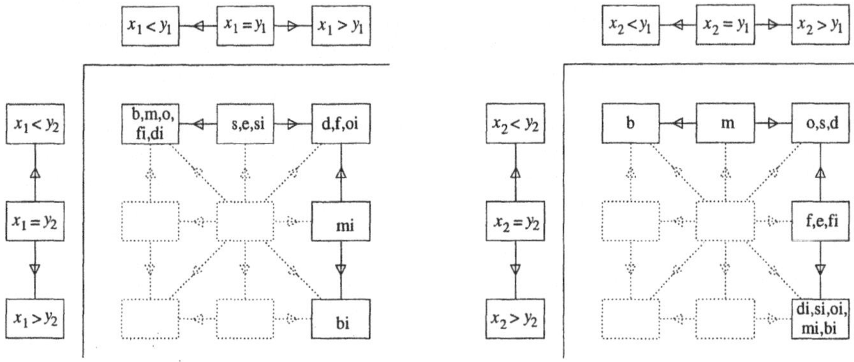

Figure 8.15. Partial dominance diagrams for the interval–interval relations

combinations are impossible; the remaining thirteen correspond precisely to Allen's interval–interval relations, with the dominance relation correctly predicted as in Figure 8.16. The 'impossible' elements are, as usual, drawn with dotted lines. The configuration of the remaining thirteen elements should be compared with Figure 5.2.

There are some interesting connections here with previous work. For example, the dominance relation portrayed in Figure 8.16 coincides exactly with the 'downward granularity conversion' of Euzenat 1995: relation r dominates relation r' if and only if it is possible for a relation which appears as r at a certain level of granularity to appear as r' at a finer level. For example, if interval i appears to meet (m) interval i', it is possible that at a finer resolution it will be seen rather to precede (b) or overlap (o) it—or of course it may still appear to meet it; however far the resolution is increased, however, no other relation can appear. Dominance is also related to the notion of 'topological closure' as expounded by Bessière *et al.* 1996, although the relations are not in fact identical (topological closure is a subset of dominance). It would clearly be worthwhile investigating further the connections between these substantially independently characterized relations.

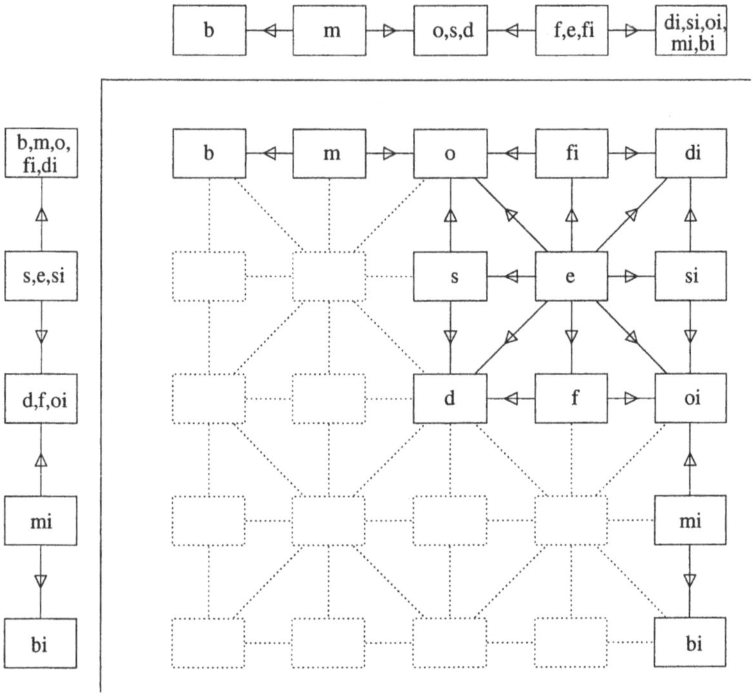

Figure 8.16. Full dominance diagram for the interval–interval relations

8.6.3. Other potential applications

In this section I shall briefly outline a number of other areas of qualitative reasoning to which the theory of topological mode spaces has been or can be applied.

(1) *Relations between regions with 'broad' boundaries.* These have been discussed by Clementini and di Felice (1996), who identify a set of forty-four possible qualitative relations between a pair of such regions, analogous to the eight RCC relations for regions with crisp boundaries. They use the 9-intersection representation of Egenhofer (1991) (see p. 87), adapted for the case in which the boundary of a region is an extended band rather than a line. Cohn and Gotts (1996*a*) present an independently derived set of forty-six relations which include all those of Clementini and di Felice together with two additional relations which the latter authors cannot express using their

chosen formalism. It is possible, in fact, to derive these forty-six relations from a simplified set of relations on regions with crisp boundaries, called RCC-5, by two applications of the product theorem and the first subspace theorem; the derivation reveals that the conceptual neighbourhood diagram given by Clementini and di Felice is missing a great many additional links.[2]

(2) *Reasoning about directions in space.* Frank (Frank 1992; Frank 1996) has investigated calculi of cardinal directions in some ways analogous to the interval calculus of Allen and the region calculi of Cohn and Egenhofer. Related work has been done by Freksa (1992*b*). Since directions can be expressed both in exact quantitative form as continuously variable numerical bearings and in qualitative form using terms such as 'north', 'south', 'east', and 'west' or 'left' and 'right', it is clear that this work comes within the general area treated by the TM space theory.

(3) *Lines of sight.* In Galton 1994, I gave a system for reasoning about the various qualitatively distinct modes of occultation between bodies viewed from a fixed vantage point, such as 'A completely hides B', 'A appears adjacent to B', and so on. Figure 3 of that paper shows the TM space with explicit dominance relations.

(4) *Modes of Overlap.* Regions of arbitrary shape can overlap one another in arbitrarily complex ways. Egenhofer and Franzosa (1995) have developed a system in which all such modes of overlap can be characterized up to topological equivalence by means of a system of indices whose complexity grows in step with the complexity of the overlap modes represented. In Galton 1998, I presented a simplified system in which each mode of overlap is represented by a single 4×4 matrix over the natural numbers, and in which the dominance relations are considered explicitly. The simplified system is naturally unable to discriminate overlap modes up to topological equivalence, but the simplified representation provides useful insights into the structure of 'overlap space', particularly as regards the dynamical aspects of configurations changing over time.

(5) *Triangles in the plane.* Figure 1.2 illustrates both the quantitative space of all triangles (up to similarity) and a qualitative mode space derived from it. This is a clear exemplification of the issues discussed in this chapter.

[2] Detailed support for these claims will appear in a future publication.

(6) *Orientation of three-dimensional shapes.* Here I refer again to the work of Shepard (1988), whose figure 4.3 portrays part of the phase space for the possible orientations of a cube, with its faces labelled R ('right'), L ('left'), T ('top'), U ('underside'), B ('back'), and F ('front'). This can be presented neatly as a compound TM space as follows. We begin with three simple three-element TM spaces: in one, the modes describe which, if either, of the faces L and R are showing, when the cube is viewed from a fixed vantage point; the second treats faces T and U, and the third, B and F. The Cartesian product of these three TM spaces has twenty-seven elements, of which one is impossible (since it cannot be that none of the faces is showing). We thus have the 26-element TM space shown in our Figure 4.15. This diagram captures all the essential qualitative information in the space considered by Shepard, together with the possible transitions as the cube rotates.

(7) *Convexity.* I discussed this briefly as one aspect of change in shape (§6.6). The continuity rule there enunciated amounts to this: that convex dominates concave. The structure of convexity space as a dominance space would seem to be largely determined by this one principle, though as indicated above, the details remain to be worked out.

(8) *Symmetry.* This too was discussed in §6.6. The guiding principle here is that symmetry dominates asymmetry; or, more specifically, any particular type of symmetry dominates the lack of that type.

The topological nature of the dominance relation as a form of *bounding* highlights an essential feature of all this work, which is that the more dominant modes are in a sense more special, more precarious, more unstable, existing as they do in the interstices between their less dominant neighbours. The least dominant modes are *generic* in the sense that an arbitrary mode picked at random is overwhelmingly most likely to be of that kind; any slight perturbation of a dominant mode is highly likely to result in one of the 'weaker' modes it dominates. For the same reason, these less dominant modes are less *interesting*, because they lack the salience that makes landmarks of the other modes and gives us a reason to pick them out.

The motivation for the term 'dominance' is as follows. Suppose modes q_1 and q_2 hold over intervals i_1 and i_2, respectively, where i_1 meets i_2 at the instant t. Suppose further that there is no third mode that can hold at t. Then q_1 and q_2 are as it were in competition as to which of them should hold at t. Whichever of q_1 and q_2 dominates the other wins the competition!

8.7. Summary

The theory of topological mode spaces developed in this chapter provides a theoretical underpinning for a wide range of phenomena of qualitative spatial change, specifically those qualitative transitions which arise as a result of continuous change in some underlying phase space. The theory provides a link between low-level quantitative descriptions in terms of a continuous phase space and higher-level, qualitative descriptions in terms of a discrete mode space derived from the low-level space by abstracting away all distinctions except those considered salient for some particular purpose. I do not pretend that this theory provides a complete account of all aspects of qualitative spatial change; here and in the last chapter I have focused my attention specifically on continuity issues; other aspects of spatial change have been mentioned in some of the earlier chapters, but the treatment was far from comprehensive. There is a vast field of potential research to be mined here, and I have done little more than to take a few aerial snapshots and dig a few exploratory holes. Perhaps some of the readers of this book will be able to take the investigations further.

Bibliography

Ahronovitz, E., Aubert, J.-P., and Fiorio, C. (1995), 'The star-topology: A topology for image analysis', in 'Discrete Geometry for Computer Imagery', Groupe GDR PRC/AMI du CNRS, 107–16.

Allen, J. (1983), 'Maintaining knowledge about temporal intervals', *Communications of the ACM*, **26**, 832–43.

—— (1984), 'Towards a general theory of action and time', *Artificial Intelligence*, **23**, 123–54.

—— and Ferguson, G. (1994), 'Actions and events in interval temporal logic', *Journal of Logic and Computation*, **4**, 531–79.

—— and Hayes, P. J. (1985), 'A common-sense theory of time', in A. Joshi (ed.), 'Proceedings of the Ninth International Joint Conference on Artificial Intelligence (IJCAI'85)', International Joint Conferences on Artificial Intelligence, Inc., 528–31.

—— and —— (1989), 'Moments and points in an interval-based temporal logic', *Computational Intelligence*, **5**, 225–38.

Alt, H., and Godau, M. (1995), 'Computing the Fréchet distance between two polygonal curves', *International Journal of Computational Geometry and Applications*, **5**, 75–91.

—— Behrends, B., and Blömer, J. (1993), 'Approximate matching of polygonal shapes', Technical Report B93-10, Freie Universität Berlin, Institut für Informatik.

—— Fuchs, U., Rote, G., and Weber, G. (1996), 'Matching convex shapes with respect to the symmetric difference', Technical Report B96-03, Institut für Informatik, Freie Universität Berlin.

Asher, N., and Vieu, L. (1995), 'Toward a geometry of common sense: A semantics and a complete axiomatisation of mereotopology', in C. S. Mellish (ed.), 'Proceedings of the Fourteenth International Joint Conference on Artificial Intelligence (IJCAI'95)' (San Mateo, Calif.: Morgan Kaufmann), 846–52.

Attneave, F. (1954), 'Some informational aspects of visual perception', *Psychological Review*, **61**, 183–93.

Bennett, B. (1998), 'Logical Representations for Automated Reasoning about Spatial Relationships', Ph.D. thesis, University of Leeds.

Bennett, D. C. (1975), *Spatial and Temporal Uses of English Prepositions: An Essay in Stratificational Semantics* (London: Longman).

Berlin, B., and Kay, P. (1969), *Basic Color Terms: Their Universality and Evolution* (Berkeley: University of California Press).

Bessière, C., Isli, A., and Ligozat, G. (1996), 'Global consistency in interval algebra networks: tractable subclasses', in W. Wahlster (ed.), 'Proceedings of the Twelfth European Conference on Artificial Intelligence (ECAI'96)' (Chichester: John Wiley), 3–7.

Biederman, I. (1987), 'Recognition-by-components: A theory of human image understanding', *Psychological Review*, **94**(2), 115–47.

Brannan, D. A., Esplen, M. F., and Gray, J. J. (1999), *Geometry* (Cambridge: Cambridge University Press).

Bronowski, J. (1970), 'New concepts in the evolution of complexity', *Synthese*, **21**, 228–46.

Bunt, H. (1985), 'The formal representation of (quasi-)continuous concepts', in J. R. Hobbs and R. C. Moore (eds), *Formal Theories of the Commonsense World* (Norwood, NJ: Ablex Publishing Corporation), 37–70.

Burge, T. (1979), 'Mass terms, count terms, and change', in F. J. Pelletier (ed.), *Mass Terms: Some Philosophical Problems* (Dordrecht: D. Reidel), 199–218.

Cadwell, J. H. (1966), *Topics in Recreational Mathematics* (Cambridge: Cambridge University Press).

Čapek, M. (1987), 'The conflict between the absolutist and relational theory of time before Newton', *Journal of the History of Ideas*, **48**(4), 595–608.

Cartwright, H. (1970), 'Quantities', *Philosophical Review*, **79**, 25–42.

Cartwright, R. (1975), 'Scattered objects', in K. Lehrer (ed.), *Analysis and Metaphysics: Essays in Honor of R. M. Chisholm* (Dordrecht: D. Reidel), 153–71.

Casati, R., and Varzi, A. (1994), *Holes and Other Superficialities* (Cambridge, Mass. and London: MIT Press).

—— and —— (1999), *Parts and Places: The Structures of Spatial Representation* (Cambridge, Mass. and London: MIT Press).

Čech, E. (1966), *Topological Spaces*, revised edition by Zdeněk Frolík and Miroslav Katětov (Chichester: John Wiley).

Chomsky, N. (1965), *Aspects of the Theory of Syntax* (Cambridge, Mass.: MIT Press).

Clarke, B. L. (1981), 'A calculus of individuals based on "connection"', *Notre Dame Journal of Formal Logic*, **26**, 61–75.

Clementini, E., and di Felice, P. (1996), 'An algebraic model for spatial objects with indeterminate boundaries', in P. A. Burrough and A. U. Frank (eds), *Geographic Objects with Indeterminate Boundaries*, vol. 2 of *GISdata* (London: Taylor and Francis), 155–69.

Cohn, A. G. (1995), 'A hierarchical representation of qualitative shape based on connection and convexity', in A. U. Frank and W. Kuhn (eds), 'Spatial Information Theory: A Theoretical Basis for GIS', vol. 988 of *Lecture Notes in Computer Science* (New York: Springer-Verlag), 311–26. Proceedings of International Conference COSIT'95.

—— (1997), 'Qualitative spatial representation and reasoning techniques', in G. Brewka, C. Habel, and B. Nebel (eds), 'KI-97: Advances in Artificial Intelligence', vol. 1303 of *Lecture Notes in Computer Science* (New York: Springer-Verlag), 1–30.

—— and Gotts, N. M. (1996*a*), 'The "egg-yolk" representation of regions with indeterminate boundaries', in P. Burrough and A. U. Frank (eds), *Geographic Objects with Indeterminate Boundaries*, vol. 2 of *GISData* (London: Taylor and Francis), 171–87.

—— and —— (1996*b*), 'Representing spatial vagueness: a mereological approach', in L. C. Aiello, J. Doyle and S. Shapiro (eds), 'Principles of Knowledge Representation and Reasoning: Proceedings of the Fifth International Conference (KR'96)' (San Mateo, Calif.: Morgan Kaufmann), 230–41.

—— and Varzi, A. C. (1998), 'Connection relations in mereotopology', in H. Prade (ed.), 'Proceedings of the Thirteenth European Conference on Artificial Intelligence (ECAI'98)' (Chichester: John Wiley), 150–54.

—— Bennett, B., Gooday, J., and Gotts, N. M. (1997), 'Representing and reasoning with qualitative spatial relations about regions', in O. Stock (ed.), *Spatial and Temporal Reasoning* (Dordrecht: Kluwer Academic Publishers), 97–134.

—— Gotts, N. M., Cui, Z., Randell, D. A., Bennett, B., and Gooday, J. M. (1998), 'Exploiting temporal continuity in qualitative spatial calculi', in M. J. Egenhofer and R. G. Golledge (eds), *Spatial and Temporal Reasoning in Geographic Information Systems* (New York and Oxford: Oxford University Press), 5–24.

—— Randell, D. A., and Cui, Z. (1995), 'Taxonomies of logically defined qualitative spatial relations', *International Journal of Human-Computer Studies*, **43**, 831–46.

Couclelis, H. (1992), 'People manipulate objects (but cultivate fields): Beyond the raster-vector debate in GIS', in A. U. Frank, I. Campari, and U. Formentini (eds), 'Theories and Methods of Spatio-Temporal Reasoning in Geographic Space', vol. 639 of *Lecture Notes in Computer Science* (Berlin: Springer-Verlag), 65–77.

Coventry, K. R. (1998), 'Spatial prepositions, functional relations, and lexical specification', in P. Olivier and K.-P. Gapp (eds), *Representation and Processing of Spatial Expressions* (Mahwah, NJ: Lawrence Erlbaum Associates), 247–62.

Cui, Z., Cohn, A. G., and Randell, D. A. (1992), 'Qualitative simulation based on a logical formalism of space and time', in 'Proceedings of the Tenth National Conference on Artificial Intelligence (AAAI-92)' (Menlo Park, Calif.: AAAI Press/MIT Press), 679–84.

Dahl, Ö. (1981), 'On the definition of the telic-atelic (bounded-nonbounded) distinction', in P. Tedeschi and A. Zaenen (eds), *Tense and Aspect* (New York: Academic Press), 79–90.

Davidson, D. (1967), 'The logical form of action sentences', in N. Rescher (ed.), *The Logic of Decision and Action* (Pittsburgh, Pa.: University of Pittsburgh Press). Reprinted in D. Davidson, *Essays on Actions and Events* (Oxford: Clarendon Press, 1980).

Davis, E. (1990), *Representations of Commonsense Knowledge* (San Mateo, Calif.: Morgan Kaufmann).

—— (1992), 'Infinite loops in finite time: some observations', in B. Nebel, C. Rich, and W. Swartout (eds), 'Principles of Knowledge Representation and Reasoning: Proceedings of the Third International Conference (KR'92)' (San Mateo, Calif.: Morgan Kaufmann), 47–58.

—— (1995), 'Approximations of shape and configuration space', Technical Report TR1995-703, Computer Science Department, New York University.

Davis, L. (1974), 'Disembodied brains', *Australasian Journal of Philosophy*, **52**, 121–32.

Dehaene, S. (1997), *The Number Sense: How the Mind Creates Mathematics* (Harmondsworth: Allen Lane).

de Laguna, T. (1922), 'Point, line, and surface, as sets of solids', *Journal of Philosophy*, **19**, 449–61.

Dennett, D. C. (1989), *The Intentional Stance* (Cambridge, Mass.: MIT Press).

Dornheim, C. (1998), 'Undecidability of plane polygonal mereotopology', in A. Cohn, L. Schubert, and S. Shapiro (eds), 'Principles of Knowledge Representation and Reasoning: Proceedings of the Sixth International Conference (KR'98)' (San Mateo, Calif.: Morgan Kaufmann).

Dowty, D. (1979), *Word Meaning and Montague Grammar* (Dordrecht: D. Reidel).

Dryden, I. L., and Mardia, K. V. (1998), *Statistical Shape Analysis* (Chichester: John Wiley).

Dummett, M. (1973), *Frege: Philosophy of Language* (London: Duckworth).

Egenhofer, M. J. (1989), 'A formal definition of binary topological relationships', in W. Litwin and H. Schek (eds), 'Third International Conference on Foundations of Data Organisation and Algorithms (FODO), Paris, France', vol. 367 of *Lecture Notes in Computer Science* (New York: Springer-Verlag), 457–72.

—— (1991), 'Reasoning about binary topological relations', in O. Günther and H.-J. Schek (eds), *Advances in Spatial Databases* (New York: Springer-Verlag), 143–60.

—— and Franzosa, R. D. (1995), 'On the equivalence of topological relations', *International Journal of Geographical Information Science*, **9**, 133–52.

—— and Mark, D. M. (1995), 'Naive geography', in A. U. Frank and W. Kuhn (eds), 'Spatial Information Theory: A Theoretical Basis for GIS', vol. 988 of *Lecture Notes in Computer Science* (New York: Springer-Verlag), 1–15. Proceedings of International Conference COSIT'95.

—— and Sharma, J. (1993), 'Topological relations between regions in R^2 and Z^2', in D. Abel and B. C. Ooi (eds), 'Advances in Spatial Databases — Third International Symposium on Large Spatial Databases SSD'93', vol. 692 of *Lecture Notes in Computer Science* (New York: Springer-Verlag), 316–36.

Euzenat, J. (1995), 'An algebraic approach to granularity in qualitative time and space representation', in C. S. Mellish (ed.), 'Proceedings of the Fourteenth International Joint Conference on Artificial Intelligence (IJCAI'95)' (San Francisco, CA: Morgan Kaufmann), 894–900.

Feldman, J. (1993), 'Causal models of spatial categories', *Behavioral and Brain Sciences*, **16**, 244–245. Commentary on Landau and Jackendoff 1993.

Forbus, K. D. (1984), 'Qualitative process theory', *Artificial Intelligence*, **24**, 85–168.

Forrest, P. (1996), 'From ontology to topology in the theory of regions', *The Monist*, **79**(1), 34–50.

Frank, A. U. (1992), 'Qualitative spatial reasoning about distances and directions in geographic space', *Journal of Visual Languages and Computing*, **3**, 343–71.

—— (1996), 'Qualitative spatial reasoning: cardinal directions as an example', *International Journal of Geographical Information Science*, **10**, 269–90.

Freksa, C. (1992*a*), 'Temporal reasoning based on semi-intervals', *Artificial Intelligence*, **54**, 199–227.

—— (1992*b*), 'Using orientation information for qualitative spatial reasoning', in A. U. Frank, I. Campari, and U. Formentini (eds), 'Theories and Methods of Spatio-Temporal Reasoning in Geographic Space', vol. 639 of *Lecture Notes in Computer Science* (Berlin: Springer-Verlag), 162–78.

Friedrich, P. (1970), 'Shape in grammar', *Language*, **46**, 379–407.

Galton, A. P. (1984), *The Logic of Aspect: an Axiomatic Approach* (Oxford: Clarendon Press).

—— (1990), 'A critical examination of Allen's theory of action and time', *Artificial Intelligence*, **42**, 159–88.

—— (1991), 'Reified temporal theories and how to unreify them', in J. Mylopoulos and R. Reiter (eds), 'Proceedings of the Twelfth International Joint Conference on Artificial Intelligence (IJCAI'91)', International Joint Conferences on Artificial Intelligence, Inc., 1177–82.

—— (1993), 'Towards an integrated logic of space, time, and motion', in R. Bajcsy (ed.), 'Proceedings of the Thirteenth International Joint Conference on Artificial Intelligence (IJCAI'93)' (San Mateo, Calif: Morgan Kaufmann), 1550–5.

—— (1994), 'Lines of sight', in M. Keane, P. Cunningham, M. Brady, and R. Byrne (eds), 'AI and Cognitive Science '94', Dublin University Press, 103–13. Proceedings of the Seventh Annual Conference, September 8-9, 1994, Trinity College Dublin.

—— (1995), 'Towards a qualitative theory of movement', in A. U. Frank and W. Kuhn (eds), 'Spatial Information Theory: A Theoretical Basis for GIS' (New York: Springer-Verlag), 377–96. Proceedings of the International Conference COSIT'95.

—— (1996), 'An investigation of "non-intermingling" principles in temporal logic', *Journal of Logic and Computation*, **6**, **2**, 271–94.

—— (1997), 'Continuous change in spatial regions', in S. C. Hirtle and A. U. Frank (eds), 'Spatial Information Theory: A Theoretical Basis for GIS', vol. 1329 of *Lecture Notes in Computer Science* (New York: Springer-Verlag), 1–13. Proceedings of International Conference COSIT'97.

—— (1998), 'Modes of overlap', *Journal of Visual Languages and Computing*, **9**, 61–79.

—— (1999), 'The mereotopology of discrete space', in C. Freksa and D. M. Mark (eds), 'Spatial Information Theory: Cognitive and Computational Foundations of Geographic Science' (New York: Springer-Verlag), 251–66. Proceedings of International Conference COSIT'99.

—— (2000*a*), 'Continuous motion in discrete space', in A. G. Cohn, F. Giunchiglia, and B. Selman (eds), 'Proceedings of the Seventh International Conference on Principles of Knowledge Representation and Reasoning (KR2000)' (San Francisco, Calif.: Morgan Kaufmann), 26–37.

Galton, A. P. (2000*b*), 'Transitions in continuous time, with an application to qualitative change in spatial relations', in H. Barringer, M. Fisher, D. Gabbay, and G. Gough (eds), *Advances in Temporal Logic* (Dordrecht: Kluwer Academic Publishers), 279–97.

—— and Meathrel, R. C. (1999), 'Qualitative outline theory', in T. Dean (ed.), 'Proceedings of Sixteenth International Joint Conference on Artificial Intelligence (IJCAI'99)', vol. 2 (San Francisco, Calif.: Morgan Kaufmann), 1061–6.

Gooday, J. M. (1994), 'A Transition-Based Approach to Reasoning about Action and Change', Ph.D. thesis, University of Exeter.

—— and Galton, A. P. (1997), 'The transition calculus: a high-level formalism for reasoning about action and change', *Journal of Experimental and Theoretical Artificial Intelligence*, **9**, 51–66.

Goodman, N. (1951), *The Structure of Appearance* (Cambridge, Mass.: Harvard University Press. 3rd edn., Dordrecht: D. Reidel, 1977).

Gotts, N. M. (1996), 'An axiomatic approach to topology for spatial information systems', Technical Report 96-25, University of Leeds, School of Computer Studies.

Hajnicz, E. (1995), 'Some considerations on non-linear time intervals', *Journal of Logic, Language, and Information*, **4**, 335–57.

Hamblin, C. L. (1971), 'Instants and intervals', *Studium Generale*, **24**, 127–34.

Hammer, P. C. (1955), 'General topology, symmetry, and convexity', *Transactions of the Wisconsin Academy of Sciences, Arts and Letters*, **44**, 221–55.

—— (1960), 'Kuratowski's closure theorem', *Nieuw Archief voor Wiskunde*, **8**, 74–80.

—— (1964), 'Extended topology: Continuity I', *Portugaliae Mathematica*, **25**, 77–93.

—— (1969), 'Continuity', in P. C. Hammer (ed.), *Advances in Mathematical Systems Theory* (University Park, Pa., and London: Pennsylvania State University Press), 93–105.

Hawkes, J. (1946), *Early Britain* (London: Collins).

Hayes, P. J. (1985), 'Naive physics I: Ontology for liquids', in J. R. Hobbs and R. C. Moore (eds), *Formal Theories of the Commonsense World* (Norwood, New Jersey: Ablex Publishing Corporation), 71–107.

Heidorn, P. B. (1998), 'Shapes from natural language in VerbalImage', in P. Olivier and K.-P. Gapp (eds), *Representation and Processing of Spatial Expressions* (Mahwah, NJ: Lawrence Erlbaum Associates), 119–32.

Herskovits, A. (1986), *Language and Spatial Cognition* (Cambridge: Cambridge University Press).

Hilbert, D. (1987), *Color and Color Perception: A Study in Anthropocentric Realism*, number 9 in 'CLSI Lecture Notes' (Stanford University: Center for the Study of Language and Information).

Hobbs, J. R. (1985), 'Granularity', in A. Joshi (ed.), 'Proceedings of the Ninth International Joint Conference on Artificial Intelligence (IJCAI'85)', International Joint Conferences on Artificial Intelligence, Inc., 432–5.

Hoffman, D. D., and Richards, W. A. (1982), 'Representing smooth plane curves for recognition', in 'Proceedings of the National Conference on Artificial Intelligence (AAAI-82)', American Association for Artificial Intelligence, 5–8.

Hurvich, L. M. (1981), *Color Vision* (Sunderland, Mass.: Sinauer Associates Inc.).

Jackendoff, R. (1996), 'The architecture of the linguistic-spatial interface', in P. Bloom, M. Peterson, L. Nadel, and M. Garrett (eds), *Language and Space* (Cambridge, Mass.: MIT Press).

Johnston, V. S. (1999), *Why We Feel: The Science of Human Emotions* (Reading, Mass.: Perseus Books).

Johnstone, P. T. (1982), *Stone Spaces*, number 3 in 'Cambridge Studies in Advanced Mathematics' (Cambridge: Cambridge University Press).

Jordan, D. W., and Smith, P. (1987), *Nonlinear Ordinary Differential Equations*, 2nd edn. (Oxford: Clarendon Press).

Khalimsky, E., Kopperman, R., and Meyer, P. R. (1990), 'Computer graphics and connected topologies on finite ordered sets', *Topology and its Applications*, **36**, 1–17.

Kong, T. Y., and Rosenfeld, A. (1989), 'Digital topology: Introduction and survey', *Computer Vision, Graphics and Image Processing*, **48**, 357–93.

—— Kopperman, R., and Meyer, P. R. (1991), 'A topological approach to digital topology', *The American Mathematical Monthly*, **98**(10), 901–17.

Koppelberg, S. (1989), 'Elementary arithmetic', in J. D. Monk and R. Bonnet (eds), *Handbook of Boolean Algebras* (Amsterdam: Elsevier Science Publishers), 5–48.

Kovalevsky, V. A. (1989), 'Finite topology as applied to image analysis', *Computer Vision, Graphics, and Image Processing*, **46**, 141–61.

Kowalski, R. A., and Sergot, M. J. (1986), 'A logic-based calculus of events', *New Generation Computing*, **4**, 67–95.

Kraak, M.-J., and Ormeling, F. (1996), *Cartography: Visualization of Spatial Data* (Harlow: Longman).

Ladkin, P. (1986), 'Time representation: a taxonomy of interval relations', in 'Proceedings of the Fifth National Conference on Artificial Intelligence (AAAI-86)', American Association for Artificial Intelligence, 360–6.

Lakoff, G. (1987), *Women, Fire, and Dangerous Things: What Categories Reveal about the Mind* (Chicago: Chicago University Press).

Landau, B., and Jackendoff, R. (1993), '"What" and "where" in spatial language and spatial cognition', *Behavioral and Brain Sciences*, **16**, 217–65.

Lang, E. (1989), 'The semantics of dimensional designation of spatial objects', in M. Bierwisch and E. Lang (eds), *Dimensional Adjectives: Grammatical Structure and Conceptual Interpretation* (Berlin: Springer-Verlag), 263–417.

—— Carstensen, K.-U., and Simmons, G. (1991), *Modelling Spatial Knowledge on a Linguistic Basis: Theory — Prototype — Integration* (Berlin: Springer-Verlag).

Langacker, R. W. (1991), *Foundations of Cognitive Grammar* (Stanford, Calif.: Stanford University Press), 2 volumes.

Leonard, H. S., and Goodman, N. (1940), 'The calculus of individuals and its uses', *Journal of Symbolic Logic*, **5**, 45–55.

Lewin, K. (1936), *Principles of Topological Psychology* (New York: McGraw-Hill).

Lewis, D. K., and Lewis, S. R. (1970), 'Holes', *Australasian Journal of Philosophy*, **48**, 206–12.

Leyton, M. (1988), 'A process-grammar for shape', *Artificial Intelligence*, **34**, 213–47.

Link, G. (1983), 'The logical analysis of plurals and mass terms: A lattice-theoretical approach', in R. Bäuerle, C. Schwarze and A. von Stechow (eds), *Meaning, Use, and Interpretation of Language* (Berlin and New York: Walter de Gruyter), 302–23.

Lockwood, E. H. (1963), *A Book of Curves* (Cambridge: Cambridge University Press).

Lyons, J. (1977), *Semantics* (Cambridge: Cambridge University Press), 2 volumes.

McCarthy, J., and Hayes, P. J. (1969), 'Some philosophical problems from the standpoint of artificial intelligence', in B. Melzer and D. Michie (eds), *Machine Intelligence 4* (Edinburgh: Edinburgh University Press).

McDermott, D. (1982), 'A temporal logic for reasoning about processes and plans', *Cognitive Science*, **6**, 101–55.

Manley, G. (1952), *Climate and the British Scene* (London: Collins).

Margulis, L. (1970), *The Origin of Eukaryotic Cells* (New Haven: Yale University Press).

Mark, D. M. (1993), 'Toward a theoretical framework for geographic entity types', in A. U. Frank and I. Campari (eds), 'Spatial Information Theory: A Theoretical Basis for GIS' (New York: Springer-Verlag), 270–83. Proceedings of European Conference COSIT'93.

Marr, D., and Nishihara, H. K. (1978), 'Representation and recognition of three dimensional shapes', *Proceedings of the Royal Society of London, Series B*, **200**, 269–94.

Meathrel, R. C., and Galton, A. P. (2000), 'Qualitative representation of planar outlines', in W. Horn (ed.), 'Proceedings of the Fourteenth European Conference on Artificial Intelligence (ECAI2000)'.

Miller, G. A., and Johnson-Laird, P. (1976), *Language and Perception* (Cambridge, Mass.: Harvard University Press).

Moens, M., and Steedman, M. (1988), 'Temporal ontology and temporal reference', *Computational Linguistics*, **14**, 15–28.

Montello, D. R. (1993), 'Scale and multiple psychologies of space', in A. U. Frank and I. Campari (eds), 'Spatial Information Theory: A Theoretical Basis for GIS' (New York: Springer-Verlag), 312–21. Proceedings of European Conference COSIT'93.

—— (1995), 'How significant are cultural differences in spatial cognition?', in A. U. Frank and W. Kuhn (eds), 'Spatial Information Theory: A Theoretical Basis for GIS', vol. 988 of *Lecture Notes in Computer Science* (New York: Springer-Verlag), 485–500. Proceedings of International Conference COSIT'95.

Moore, T., and Carling, C. (1982), *Understanding Language: Towards a Post-Chomskyan Linguistics* (Basingstoke: Macmillan).

Muller, P. (1998*a*), 'Eléments d'une théorie du mouvement pour la formalisation du raisonnement spatio-temporel de sens commun', Ph.D. thesis, Université Paul Sabatier, Toulouse III.

—— (1998*b*), 'A qualitative theory of motion based on spatio-temporal primitives', in A. Cohn, L. Schubert and S. Shapiro (eds), 'Principles of Knowledge Representation and Reasoning: Proceedings of the Sixth International Conference (KR'98)' (San Mateo, Calif.: Morgan Kaufmann).

Newton-Smith, W. H. (1980), *The Structure of Time* (London: Routledge and Kegan Paul).

Nicod, J. (1930), *Geometry in the Sensible World* (London: Kegan Paul, Trench, Trubner, & Co.).

Pedoe, D. (1957), *Circles*, vol. 2 of *International Series of Monographs on Pure and Applied Mathematics* (London: Pergamon).

Peuquet, D. J. (1984), 'A conceptual framework and comparison of spatial data models', *Cartographica*, **21**(4), 66–113.

Pratt, I., and Francez, N. (2000), 'A decidable temporal logic for temporal prepositions', in H. Barringer, M. Fisher, D. Gabbay, and G. Gough (eds), *Advances in Temporal Logic* (Dordrecht: Kluwer Academic Publishers), 255–78.

—— and Lemon, O. (1997), 'Ontologies for plane, polygonal mereotopology', *Notre Dame Journal of Formal Logic*, **38**(2), 225–45.

—— and Schoop, D. (1997), 'A complete axiom system for polygonal mereotopology of the real plane', Technical Report UMCS-97-2-2, University of Manchester.

Putnam, H. (1981), *Reason, Truth and History* (Cambridge: Cambridge University Press).

Quine, W. V. O. (1960), *Word and Object* (Cambridge, Mass.: MIT Press).

—— (1974), *The Roots of Reference* (La Salle, Ill.: Open Court).

Randell, D. A., and Cohn, A. G. (1992), 'Naive topology: modelling the force pump', in P. Struss and B. Faltings (eds), *Advances in Qualitative Physics* (Cambridge, Mass.: MIT Press), 177–92.

—— Cui, Z., and Cohn, A. G. (1992), 'A spatial logic based on regions and connection', in B. Nebel, C. Rich, and W. Swartout (eds), 'Principles of Knowledge Representation and Reasoning: Proceedings of the Third International Conference (KR'92)' (San Mateo, Calif.: Morgan Kaufmann), 165–76.

Ray, C. (1991), *Time, Space and Philosophy* (London and New York: Routledge).

Reichenbach, H. (1947), *Elements of Symbolic Logic* (New York: Macmillan).

Reid, C. (1996), *Hilbert* (New York: Springer-Verlag).

Robinson, I. (1975), *The New Grammarians' Funeral: A critique of Noam Chomsky's linguistics* (Cambridge: Cambridge University Press).

Rosenfeld, A. (1979*a*), 'Digital topology', *American Mathematical Monthly*, **86**(8), 621–30.

—— (1979*b*), *Picture Languages: Formal Models for Picture Recognition* (New York: Academic Press).

Sampson, G. (1980), *Making Sense* (Oxford: Oxford University Press).

Schlieder, C. (1996), 'Qualitative shape representations', in P. A. Burrough and A. U. Frank (eds), *Geographic Objects with Indeterminate Boundaries*, vol. 2 of *GIS-data* (London: Taylor and Francis), 123–40.

Schoenheimer, R. (1942), *The Dynamic State of Body Constituents* (Cambridge, Mass.: Harvard University Press).

Shanahan, M. (1990), 'Representing continuous change in the Event Calculus', in L. C. Aiello (ed.), 'Proceedings of the Ninth European Conference on Artificial Intelligence (ECAI'90)' (London: Pitman), 598–603.

Sharvy, R. (1971), 'Things', in E. Freeman and W. Sellars (eds), *Basic Issues in the Philosophy of Time* (Lasalle, Ill.: Open Court), 164–80.

Shepard, R. N. (1988), 'The role of transformations in spatial cognition', in J. Stiles-Davis, M. Kritchevsky, and U. Bellugi (eds), *Spatial Cognition: Brain Bases and Development* (Hillsdale, NJ: Lawrence Erlbaum Associates), 81–110.

—— and Metzler, J. (1971), 'Mental rotation of three-dimensional objects', *Science*, **191**, 701–3.

Shoemaker, S. (1969), 'Time without change', *Journal of Philosophy*, **66**, 363–81. Reprinted as chapter 3 of S. Shoemaker, *Identity, Cause, and Mind* (Cambridge: Cambridge University Press, 1984).

Simons, P. (1987), *Parts: a Study in Ontology* (Oxford: Clarendon Press).

Smith, B. (1993), 'Ontology and the logistic analysis of reality', in N. Guarino and R. Poli (eds), 'Proceedings of the International Workshop on Formal Ontology in Conceptual Analysis and Knowledge Representation' (Padova, Italy: Institute for Systems Theory and Biomedical Engineering of the Italian National Research Council), 51–68.

—— (1994), 'Topological foundations of cognitive science', in C. Eschenbach, C. Habel, and B. Smith (eds), 'Topological Foundations of Cognitive Science', Kognitionswissenschaft, Hamburg, 3–22. Papers from the FISI-CS Workshop, Buffalo NY, 6-10 July, 1994.

—— (1995), 'On drawing lines on a map', in A. U. Frank and W. Kuhn (eds), 'Spatial Information Theory: A Theoretical Basis for GIS', vol. 988 of *Lecture Notes in Computer Science* (New York: Springer-Verlag), 475–84. Proceedings of International Conference COSIT'95.

—— (1999), 'Agglomerations', in C. Freksa and D. M. Mark (eds), 'Spatial Information Theory: Cognitive and Computational Foundations of Geographic Science' (New York: Springer-Verlag), 267–82. Proceedings of International Conference COSIT'99.

Smyth, M. B. (1995), 'Semi-metrics, closure spaces and digital topology', *Theoretical Computer Science*, **151**, 257–76.

Stell, J. G., and Worboys, M. F. (1997), 'The algebraic structure of sets of regions', in S. C. Hirtle and A. U. Frank (eds), 'Spatial Information Theory: A Theoretical Basis for GIS', vol. 1329 of *Lecture Notes in Computer Science* (New York: Springer-Verlag), 163–74. Proceedings of International Conference COSIT'97.

Stewart, I., and Golubitsky, M. (1993), *Fearful Symmetry: Is God a Geometer?* (Harmondsworth: Penguin).

Stroll, A. (1988), *Surfaces* (Minneapolis: University of Minnesota Press).

Tarski, A. (1956), *Foundations of the geometry of solids* (Oxford: Clarendon Press), 24–9.

Taylor, B. (1977), 'Tense and continuity', *Linguistics and Philosophy*, **1**, 199–220.

—— (1985), *Modes of Occurrence*, vol. 2 of *Aristotelian Society Series* (Oxford: Basil Blackwell).

Taylor, J. R. (1989), *Linguistic Categorization: Prototypes in Linguistic Theory* (New York: Oxford University Press).

Van Benthem, J. F. A. K. (1991), *The Logic of Time: A Model-Theoretic Investigation into the Varieties of Temporal Ontology and Temporal Discourse*, 2nd. edn. (Dordrecht: Kluwer Academic Publishers).

Vandeloise, C. (1986), *L'Espace en français* (Paris: Editions du Seuil). English translation, *Spatial Prepositions* (Chicago: Chicago University Press, 1991).

Varela, F. J., Thompson, E., and Rosch, E. (1991), *The Embodied Mind: Cognitive Science and Human Experience* (Cambridge, Mass.: MIT Press).

Vogel, S. (1999), *Cats' Paws and Catapults* (Harmondsworth: Penguin).

Wells, H. G. (1927), *The Complete Short Stories of H. G. Wells* (London: Ernest Benn).

White, M. J. (1992), *The Continuous and the Discrete: Ancient Physical Theories from a Contemporary Perspective* (Oxford: Clarendon Press).

Williams, B. C. (1990), 'Temporal qualitative analysis: Explaining how physical systems work', in D. S. Weld and J. de Kleer (eds), *Qualitative Reasoning about Physical Systems* (San Mateo, Calif.: Morgan Kaufmann), 133–77.

Williamson, T. (1994), *Vagueness* (London: Routledge).

Wittgenstein, L. (1966), *Lectures and Conversations on Aesthetics, Psychology and Religious Belief* (Oxford: Basil Blackwell). Compiled from Notes taken by Yorick Smythies, Rush Rhees, and James Taylor. Edited by Cyril Barrett.

—— (1975), *Philosophical Remarks* (Oxford: Basil Blackwell). Edited from his posthumous writings by Rush Rhees and translated into English by Raymond Hargreaves and Roger White.

Worboys, M. F. (1994), 'Unifying the spatial and temporal components of geographic information', in T. C. Waugh and R. G. Healey (eds), 'Advances in Geographic Information Systems: Proceedings of the Sixth International Symposium on Spatial Data Handling, Edinburgh' (London: Taylor and Francis), 505–17.

Yaglom, I. M., and Boltyanskii, V. G. (1961), *Convex Figures* (New York: Holt, Rinehart and Winston). Translated from the Russian by Paul J. Kelly and Lewis F. Walton.

Yonge, C. M. (1949), *The Sea Shore* (London: Collins).

Yuan, M. (1994), 'Wildfire conceptual modelling for building GIS space-time models', in 'GIS/LIS in Phoenix, USA', 860–9.

Index of Special Symbols

Index

The manufacturer's authorised representative in the EU for product safety is Oxford University Press España S.A. of El Parque Empresarial San Fernando de Henares, Avenida de Castilla, 2 - 28830 Madrid (www.oup.es/en or product.safety@oup.com). OUP España S.A. also acts as importer into Spain of products made by the manufacturer.
Printed and bound by CPI Group (UK) Ltd, Croydon, CR0 4YY
14/09/2026
02216286-0010